AF775880

Challenges in Risk Analysis for Science and Engineering

Development of a common language

Online at: https://doi.org/10.1088/978-0-7503-3643-7

Challenges in Risk Analysis for Science and Engineering

Development of a common language

Edited by
Tracey J Temple and Melissa K Ladyman
Cranfield Defence Chemistry, Cranfield University at Shrivenham, Shrivenham, United Kingdom

IOP Publishing, Bristol, UK

© IOP Publishing Ltd 2022

All rights reserved. No part of this publication may be reproduced, stored in a retrieval system or transmitted in any form or by any means, electronic, mechanical, photocopying, recording or otherwise, without the prior permission of the publisher, or as expressly permitted by law or under terms agreed with the appropriate rights organization. Multiple copying is permitted in accordance with the terms of licences issued by the Copyright Licensing Agency, the Copyright Clearance Centre and other reproduction rights organizations.

Permission to make use of IOP Publishing content other than as set out above may be sought at permissions@ioppublishing.org.

Tracey J Temple and Melissa K Ladyman have asserted their right to be identified as the editors of this work in accordance with sections 77 and 78 of the Copyright, Designs and Patents Act 1988.

ISBN 978-0-7503-3643-7 (ebook)
ISBN 978-0-7503-3641-3 (print)
ISBN 978-0-7503-3644-4 (myPrint)
ISBN 978-0-7503-3642-0 (mobi)

DOI 10.1088/978-0-7503-3643-7

Version: 20221001

IOP ebooks

British Library Cataloguing-in-Publication Data: A catalogue record for this book is available from the British Library.

Published by IOP Publishing, wholly owned by The Institute of Physics, London

IOP Publishing, No.2 The Distillery, Glassfields, Avon Street, Bristol, BS2 0GR, UK

US Office: IOP Publishing, Inc., 190 North Independence Mall West, Suite 601, Philadelphia, PA 19106, USA

Contents

Introduction to challenges in Risk Analysis for science and engineering: development of a common language

M Ladyman and T Temple

0.1 Introduction

Risk Analysis is required in many areas of science and engineering, for example pharmaceutical processing, the application of pesticides, the use of nano-particles, saturation of toxicokinetic processes, working with machinery, manufacturing goods, managing staff in the workplace etc [1–4]. All of these activities have potential hazards that can lead to wider impacts on the organisation, i.e. harm to the environment, flora, fauna, individuals, populations of human beings as well as economic and reputational threats. However, risk assessing prior to any activity being conducted and identifying potential hazards early can help to avoid impacts, such as excessive clean-up and remediation costs, breaching of legislation and damage to reputations. Also, early assessments can prevent occupational risks, can provide information to staff and workers, create training opportunities and enables organisations to implement suitable management and mitigation measures. The generic steps of risk assessment usually include the following [5]:

Step 1 – identifying the hazards;
Step 2 – assessing the risks from the hazards;
Step 3 – controlling the risks;
Step 4 – recording the findings;
Step 5 – ensuring that the entire process is reviewed continuously.

0.1.1 Risk Management and the challenges

The steps involved to undertake Risk Analysis appear to be straightforward although the actual process often poses challenges. These challenges may be related to different cultures, up-front cost implications, communication or holistic commitment to the continual review process. To explore some of these issues in more detail, three years ago, Cranfield University launched the first international Risk Assessment symposium for decision making. This was an opportunity to bring together those professionals involved in Risk Assessment from a variety of disciplines. This event attracted a number of delegates that had a wide range of experiences and were from different disciplines involve in Risk

Assessment. There were many perspectives on Risk Assessment discussed, including the methods used, the processes involved and the interesting outcomes from this vibrant and knowledge sharing gathering. The key messages that came from the event were concise and clear, 'terminology for Risk Assessments' are not standardised, meaning that most of the words used in individual organisations are different and this leads to confusion and poor communication. It was agreed that some way of bringing this all together should be undertaken and the result was this book.

The chapters in this book are organised thematically beginning with broad concepts associated with Risk Analysis presented in chapter 1 and chapter 2 on the defintion of Risk Analysis (chapter 1) and tolerability (chapter 2). The focus then moves to Risk Assessment in the safety domain, with case studies from industries with high-risk activities such as those involving explosives (chapters 3 and 4) and the nuclear industry (chapter 5). A discussion of the importance of representative data in environmental Risk Analysis and management (chapter 6) is followed by two examples in the explosives industry, one from a defence perspective (chapter 7) and another from mine action (chapter 8). Chapter 9 merges the need for public safety with environmental protection demonstrating the need for communication between internal and external parties, this concept is extended in chapter 10, which argues for a sustainable approach to Risk Assessment. Similarly, chapter 11 explores the idea that isolating one type of Risk Assessment, i.e. safety, is not always the most effective way to respond to risk. Chapter 12 evaluates the future of science and engineering education in terms of the potential risk posed to students from online higher education. Towards the end of the book, the focus is on holistic approaches to business risks (chapter 13) and how Risk Assessment should be a continuous and dymanic process (chapter 14) with lessons learned between disciplines (chapter 15). The concluding chapter reviews the challenges in Risk Assessment identified by all contributors specifcally focussing on variation in the terminology used by the different discplines. Abstracts for each chapter are outlined below.

Chapter 1: An introduction to Risk Analysis for scientists and engineers

Miss E Gutierrez-Carazo and Miss F Persico

In this chapter, a literature review is presented showing differences in processes and terminology related to the concept of 'Risk Analysis'. Research shows that there is no clear definition for Risk Analysis as it can be interpreted differently as the study field changes. Different case studies showed what entails a failure in a proper Risk Analysis, they highlight as a guiding thread most malfunctions due to a miscarried Risk Communication. Consequently, missing a proper terminology or process in Risk Analysis ensures that the failure continues leading into misunderstandings of the risks within and outside organisations. Therefore, literature emphasizes an prompt need to ensure that a standardisation in Risk Analysis is carried out due to numerous benefits that can keep businesses and companies safe and also compliant with legal, social, economic and environmental standards.

Chapter 2: On the tolerability of risk, public and private

Dr Mike Williams

The word 'Risk' commonly associates a hazard with the likelihood of exposure to that hazard and of that hazard causing harm to those exposed. Different individuals will accept voluntary exposure to hazards in line with their perceptions of danger and the benefits to themselves. For a Risk Assessment to be more than just a hazard some concept of 'tolerability' has to be developed by those preparing and approving the Risk Assessment, especially where the risk is not entirely voluntary.

Various nations have developed concepts of tolerable risk, for example the UK, USA and Australia but the boundaries of likelihood adopted in these may be so unlikely as to have little meaning to either the general public or even to the 'experts' preparing and approving the Risk Assessment. Other societies adopt an almost 'fatalistic' approach to risk.

This chapter will explore different approaches to tolerability and how these link to perceptions of likelihood. It will consider likelihoods of harmful and beneficial events and how these are perceived by the public. Often words are used to encompass likelihoods especially in qualitative Risk Assessments. This chapter will consider if these words have any commonly held numerical meaning but will do so from the viewpoint of a practitioner rather than a statistician.

Chapter 3: Storage of military ammunition and explosives- risk evaluation and hazard management

WO2 D Skelly

Whether conducted by the HSE, MOD, or NATO, licencing of an ammunition/explosive's storage area is essentially a Risk Assessment process based on the hazard presented by the ammunition/explosives which are to be stored. The prescribed safety distances are founded upon scaled distance (for blast) or the expected fragmentation/thermal effects of accidental initiation of ammunition/explosives. However, the trials data and assumptions which underpin the safety distances are decades-old (and as it stands these distances are under review).

This chapter will critically analyse both the current (and the proposed future) HSE, MOD and NATO regulations relating to ammunition/explosives storage. In particular, the data and assumptions underpinning the regulations will be examined in detail to determine their validity and usefulness.

Additionally, these regulations will be applied to a number of case studies of ammunition/explosives storage accidents around the world to consider holistically the consequences of both appropriate and inappropriate application of ammunition/explosives storage risk assessment.

Chapter 4: The application of Systems Thinking to Risk Assessment: left shifting safety

Dr D Holley, Dr P Gill, Dr N Mai and Dr R Vrcelj

Failure Modes and Effects Analysis (FMEA) has previously been used successfully as part of Quality Control for production processes. The development of the FMEA systems thinking model for identification of quality issues automatically lends itself to viewing a production process from a safety perspective.

In this chapter, the concept of using the FMEA approach to analysing the safety and thus assessing the risks of a process are described. Two examples are given, the first for a seemingly simplistic home-based hobby, incorporating easily accessible concepts and processes. This is then extended to an apparently more complex example, where processing of energetic materials is considered. Both approaches show that with good practice and a well-defined FMEA approach, not only is Quality Control enhanced, but also a more robust and reliable consideration of the Safety Features can be made, leading to a better assessment of the risk for each process.

Chapter 5: Influence of automation on human factor integration in Risk Assessment systems

Mr B Thawani

Risk Assessment processes have played an important role in the safe functioning of risk critical industries such as nuclear power plants. Over the years, the growth in industry created the need for development of Risk Assessment systems to account for more than just the technical factors. The inclusion of human factors in qualitative and quantitative Risk Assessment process allowed stakeholders in the industry to gauge the influence of human error on workplace safety. Subsequently, the development of automation and online process control systems added a level of complexity to industrial undertakings, resulting in further growth in Risk Assessment approaches.

This chapter will discuss three case studies: Three Mile Island; Chernobyl; and Fukushima incidents to provide context for the influence of human error, automation, and other non-technical factors in Risk Assessment at the time of the incidents.

Chapter 6: Minimising the risks to decision-making by selecting representative experimental methods for environmental science research

Miss E Gutierrez-Carazo, Miss F Persico, Dr F Coulon, Dr M Ladyman and Dr T Temple

Representative experimental methods for environmental science are essential to ensure that environmentally relevant decisions such as scale of remediation efforts, mitigation plans and management strategies are supported by accurate and reliable data. Risks associated with these decisions can determine the success or failure of the project, and thus, experimental work needs to be representative of the real environment under study.

The aim of this chapter is to highlight the importance of incorporating representative methods into environmental science research in order to minimise the risks associated with poor decision-making.

Poor decisions can lead to a variety of impacts varying in significance from mostly insignificant to severe economic, legislative, environmental and social consequences.

Chapter 7: The Importance of the conceptual site model in the assessment of land contamination

Mr P Burden, Mrs A-M Deloughry and Mrs J Morgan

The conceptual site model (CSM) lies at the heart of the assessment of land contamination in the UK. It was central to the earliest UK land contamination guidance and remains the primary method to both represent and understand the site relationships between potential sources of contamination and the receptors who may be affected by them. There are a wide range of potential sources of chemical contamination to be considered; receptors tend to fall into either human or water environments (surface water or groundwater).

This chapter will set out the legislative background to UK guidance, how this guidance works in practice when assessing potential land contamination and finishes with a case study which demonstrates the approach on a real-life site.

Chapter 8: Land release: A Risk Management approach for mine action

Mr R Evans

Explosive ordnance left behind after conflict presents significant human security, economic and environmental risks. Removing this contamination draws on finite resources. How those resources are allocated to reduce the ongoing risk to civilians in the most efficient and effective way practicable remains an ongoing challenge for the humanitarian mine action sector.

Removing explosive ordnance is not a straightforward process since it is invariably not always clear exactly where the contamination is located. This is especially so if that contamination was concealed when it was emplaced, as is often the case with minefields. Time spent clearing land that isn't contaminated is time not spent clearing land which is, and implicitly carries the risk of more civilian casualties that may not have happened if that land had been cleared earlier. In short, there is an opportunity cost in not accurately identifying the contaminated land prior to clearance.

This chapter discusses how then to approach the problem of efficiently allocating resources to clear the contamination when the location of the explosive ordnance itself is often hard to confirm until after clearance has taken place? Or put another way, how best to manage the risk that, with limited initial information, the wrong priority for clearance is chosen, even with the best of intentions? The Mine Action sector's attempt to answer this ongoing dilemma is a Risk Management process called Land Release.

Chapter 9: Risk Assessment and management at a heritage site

Dr G Hooper

The Ministry of Defence is steward of the largest historic estate in Government ownership, protected for its respective historic, architectural, archaeological and artistic value. Much of this estate is no longer required for defence purposes and has been, or will be, released for alternative uses. Several have unique historical interest, including the former Royal Gunpowder Mills at Waltham Abbey. Trustees of this heritage site have two principal tasks, firstly to preserve and hopefully enhance all the various historic buildings, flora and fauna on site and secondly to make sure that the staff, volunteers, contractors and of course visitors are kept safe and healthy.

This chapter will discuss how Risk Assessment is carried out on all aspects of the management of the site, and the role of a variety of external regulators monitoring the company's activities. These include Historic and Natural England, the Health and Safety Executive and the Office of Rail and Road.

Chapter 10: Amazonian non-timber forest products and ways forward to ensure their sustainability at low risk: Acai case-study

K Yamaguchi, J Campos Zeballos, S Janzen, E F G Galante and V F Veiga-Junior

The Acai fruit consumption from the Brazilian Amazon region has been rapidly expanding in the national and international context. This Non-Timber Forest Product (NTFP) plays an important role for sustainability of the region, as its fast market expansion is reducing pressures from land use change. However, research on the Acai industry as well as on the risks faced by the industry today and in the future, is lacking.

This chapter presents factors to consider in an integrated Risk Assessment, as a first step towards such analysis. The two-fold approach of global review and local case study reveals that there is an important gap in assessing risks to emerging NTFP industries. The global literature review found no paper on risk assessments and NTFPs, and only one paper specifically on Acai and risk. This lack of research may result in inadequate preparedness of the industry to future challenges. The case study of the Middle Solimões municipality (Coari, Codajás, Tefé and Alvarães) in the state of the Amazonas (Brazil) shows the dependence of local livelihoods on the Acai market as well as the strong price fluctuations linked to seasonality and productivity. By contributing towards integrated risk assessments of the NTFP industry and especially the Acai industry, this paper makes an important contribution to filling the research gap and informing developments of the industry.

Chapter 11: Development of a risk matrix for low-cost engineering solutions: a systematic approach

L Roberts, R Critchley, D Salberg, S Bloodworth-Race and R Hazael

The development of a selection matrix that seeks to mitigate a risk is often difficult as it is a subjective assessment of available facts and individual experiences. It is acknowledged that while these are important sources from which to draw information, ideally a scientific underpinning is necessary to provide credibility and reliability of applying that Risk Assessment.

This chapter considers a systematic development of a methodology to assess suitability of Personal Protection Equipment to the specified threat and the Risk Assessment process which therefore could be applied. A case study approach, with examples drawn from relevant literature will be used to apply this proposed methodology.

Chapter 12: Assessing the academic and experiential risk to students from the transition to hybrid and online delivery

Dr M Ladyman, Dr T Temple and Mr A Temple

The Covid-19 pandemic forced higher education establishments to convert their teaching and deliver by the use of online platforms, such as online teleconferencing platforms. The initial transition was rapid, and largely successful thanks to available technology, but with teaching returning to normal in the UK many providers are faced with the decision of whether to continue to provide remote online teaching.

This chapter assesses the risks to students from the transition to hybrid and online delivery, a Risk Assessment was developed using the mandatory higher education standard and quality descriptors and Postgraduate Taught Experience Survey. The Risk Assessment focussed on the risk to student failure and poor student experience and demonstrated that while it is feasible to deliver level 7 qualifications in a hybrid or online format, significant mitigation would be required for both hybrid and fully online cohorts. From the Risk Assessment, it was found that while mitigations for hybrid and online delivery are similar, it may be more challenging to achieve equality of opportunity and experience for hybrid cohorts where some students are face-to-face and some are remote online.

Chapter 13: Practical application of a decision framework to mitigate environmental risks posed by the treatment of explosive contaminated wastewater

W Gilroy-Hirst, J Padfield, M Ladyman, F Coulon and T Temple

The increased demand for Insensitive High Explosives (IHE), i.e. those less sensitive to accidental impact, in militaries around the world has required munitions manufacturers to adapt their processes to accommodate changes to the explosive formulations. IHE formulations may combine up to three compounds, often a mixture of traditional legacy explosives and compounds previously unused in explosive formulation. Therefore, when explosive manufacturers start to process with IHE, they must often review the entire processing lifecycle, including the environmentally responsible disposal of side products and excess explosive material to ensure current procedures are suitable for the new formulation. Failure to assess the risk of changing standard procedures on the environment may have significant impacts to business if the existing procedures cause environmental impacts such as pollution, which may lead to fines, or 'cease' notices from environmental regulators.

Therefore, in this chapter a four-stage decision framework was applied to an explosive manufacturing facility to determine whether the facility would be able to process IHE using the existing disposal procedures without negatively impacting the environment. The decision framework was based on the manufacturer's existing environmental management system and extended the assessment to evaluate the safety, compliance, reputational and financial risks that a change to disposing of IHE using existing procures would impose on the facility. The aim of the decision framework was to identify significant impacts on the environment that imposed a significant risk on the business and use an evidence-based approach to propose short-term and long-term disposal solutions in line with environmental best practice. The decision framework facilitates an evidence-based approach for making decisions relating to environmental risk as well as providing a transparent and defensible record of how that decision was reached that can be applied to any activity that has the potential for an environmental impact regardless of industry.

Chapter 14: Risk recognition and continuous risk management

Mr Z Rustom

Having a mature safety culture is a good thing, but only to a point. Regular safety conversations and posters serve a purpose until they become so familiar we fail to recognise the risk and hazard they are alerting us to. The system does not alter risk tolerance or aversion, it tends to make an organisational statement, whereas the protection of individuals is contingent on the individual(s) recognising the hazard and acknowledging the consequence.

This chapter builds an assessment and continuous review model that recognises the experiential needs of andragogy and combats the safety fatigue that can impact an over-delivered safety culture. The model builds on the experience of a Commercial Response Organisation (CRO) established to respond to incidents and events in hostile working environments. It accepts event similarity can lead to hazard complacency and a workforce's tendency to see the hazard and not the consequence. The model recognises common failings in workgroup Risk Management and avoids miscommunication by recognising stakeholders and concurrent events/ risks. The model recognises that Risk Assessment significantly loses value if it is without Risk Management.

Stakeholder recognition and engagement allows the model to provide an understanding of the risk consequences and the mechanics that might allow occurrence; in this manner, the model enables proactive Risk Management.

Conclusion: Challenges in Risk Analysis

M Ladyman and T Temple

A concluding chapter summarising the challenges in Risk Analysis as identified by the contributors.

References

[1] Singh S, Kumar Naik T S S, Anil A G, Dhiman J, Kumar V and Dhanjal D S *et al* 2022 Micro (nano) plastics in wastewater: A critical review on toxicity risk assessment, behaviour, environmental impact and challenges *Chemosphere* **290** 133169

[2] Sundararaman S, Aravind Kumar J, Deivasigamani P and Devarajan Y 2022 Emerging pharma residue contaminants: Occurrence, monitoring, risk and fate assessment—A challenge to water resource management *Sci. Total Environ.* **2022** 153897

[3] Tan Y M, Barton H A, Boobis A, Brunner R, Clewell H and Cope R *et al* 2021 *Opportunities and challenges related to saturation of toxicokinetic processes: Implications for risk assessment* Reg. Toxicol. Pharmacol **127** 105070

[4] Robertson L J, Jore S, Lund V and Grahek-Ogden D 2021 Risk assessment of parasites in Norwegian drinking water: opportunities and challenges *Food Waterborne Parasitol* **22** e00112

[5] HSE 2021 *Managing risks and risk assessment at work* www.hse.gov.uk/pubns/indg163.htm

Acknowledgement

This book would not have been possible without the support and knowledge sharing from attendees at the first International Symposium on Risk Assessment Approaches for Decision Making, where the need for this book was first recognised. The Editors would first like to thank all contributors to this book, without whom it would not have been possible to collate such a variety and diversity of case studies and experience on Risk Assessment.

The Editors would also like to thank Cranfield University for supporting this initiative and the numerous staff members and partners who have given their time and expertise to the development of the symposium and consequently this book.

Finally, the Editors thank IOP for providing the opportunity to write and publish this volume.

Editor biographies

Tracey Temple

Dr Tracey Temple is a senior lecturer in environmental science at Cranfield University and has held this position since 2006. Prior to Cranfield, Tracey was an environmental consultant primarily for UK Defence, working in the UK and Cyprus. Before returning to academia Tracey served in the RAF for 10 years. Tracey gained her PhD in the fate and transport of explosives from Cranfield University, and her MSc in Environmental Science and BA in Geography from Queen's University, Belfast. Tracey is currently the Course Director for the *Explosives Ordnance and Engineering* (EOE) MSc at Cranfield University. Tracey frequently chairs and attends NATO panel meetings, specifically related to the environmental impact of explosives, which include sampling for contamination, toxicity from military training ranges and environmental munition regulations.

Melissa Ladyman

Dr Melissa Ladyman is a lecturer in Environmental Science at Cranfield University where she is course lead for a suite of Environmental Management and Sustainability courses and several EOE modules. Her main research is in the fate and transport of explosives in the environment and environmental risk. She graduated from Edinburgh University with an MChem degree in 'Chemistry with a Year in Europe' in 2010. Her undergraduate degree included a year placement at the Ecole Nationale Supérieure de Chimie de Lille where she undertook research into potential medications for tuberculosis and depression. She gained a PhD in 2014 in organic and medicinal chemistry focussing on the development of fluorescent assays for biological analysis, also from The University of Edinburgh.

Together Tracey and Melissa lead the teaching and research of Environmental Science at Cranfield University at Shrivenham. They teach a diverse mix of students and actively engage with the explosives and defence industries. From their research and teaching into the behaviour of explosives in the environment, the importance of Risk Assessment became increasingly evident from an environmental and safety perspective. Recognising the challenges of Risk Assessment in their area, prompted the idea to develop a cross-disciplinary network of risk practitioners starting with the International Symposium on Risk Assessment Approaches for Decision Making, followed by the creation of this book and several open access special editions on Risk in academic journals. The future aspiration is to develop a holistic and effective approach to Risk Assessment and Risk Management to improve cross-disciplinary communication.

List of contributors

S Bloodworth-race
Cranfield University, Forensic Institute, Defence Academy of the United Kingdom, Shrivenham, SN6 7LA, UK

P Burden
Defence Infrastructure Organisation, Room 106, Building 420, MOD, Westwells Road, Corsham, SN13 9NR, UK

J Campos Zeballos
United Nations University—Environmental and Human Security, UN Campus Platz der Vereinten, National 1, 53113, Bonn, Germany

F Coulon
Cranfield School of Water, Energy and Environment, Cranfield University, MK43 0AL, UK

R Critchley
Cranfield University, Forensic Institute, Defence Academy of the United Kingdom, Shrivenham, SN6 7LA, UK

A-M Deloughry
Defence Infrastructure Organisation Head Office, St. George's House, DMS Whittington, Lichfield, Staffordshire, WS14 9PY, UK

R Evans
Cranfield University, Centre for Defence Chemistry, Defence Academy of the United Kingdom, Shrivenham, SN6 7LA, UK

E F Galante
Chemical Engineering Department, Instituto Militar de Engenharia, Praça General Tibúrcio, 80, Praia Vermelha, Urca, 22290-270 Rio de Janeiro, Brazil

P P Gill
Roxel Group Engineering Laboratory, Kidderminster, DY11 7RZ, UK

W Gilroy-Hirst
Cranfield University, Centre for Defence Chemistry, Defence Academy of the United Kingdom, Shrivenham, SN6 7LA, UK

E Gutierrez-Carazo
Cranfield University, Centre for Defence Chemistry, Defence Academy of the United Kingdom, Shrivenham, SN6 8LA, UK

Dave Holley
University of Wales, Trinity St. David Kings Road, Swansea, SA1 8PH, UK

R Hazael
Cranfield University, Forensic Institute, Defence Academy of the United Kingdom, Shrivenham, SN6 7LA, UK

G Hooper
Royal Gunpowder Mills, Beaulieu Drive, Waltham Abbey, Essex, EN9 1JY, UK

S Janzen
United Nations University—Environmental and Human Security, UN Campus Platz der Vereinten, National 1, 53113, Bonn, Germany

M K Ladyman
Cranfield University, Centre for Defence Chemistry, Defence Academy of the United Kingdom, Shrivenham, SN6 7LA, UK

M Mai
Cranfield University, Centre for Defence Chemistry, Defence Academy of the United Kingdom, Shrivenham, SN6 7LA, UK

J Morgan
Defence Infrastructure Organisation Head Office, St. George's House, DMS Whittington, Lichfield, Staffordshire, WS14 9PY, UK

J Padfield
RWM Italia SpA, Località Matt-è Conti 09015, Domusnovas (SU), Italy

F Persico
Cranfield University, Centre for Defence Chemistry, Defence Academy of the United Kingdom, Shrivenham, SN6 7LA, UK

L Roberts
Cranfield University, Forensic Institute, Defence Academy of the United Kingdom, Shrivenham, SN6 7LA, UK

Z Rustom
Komak Global, 1 Keats Close, Horsham, West Sussex, RH12 5PL, UK

D Salberg
Cranfield University, Forensic Institute, Defence Academy of the United Kingdom, Shrivenham, SN6 7LA, UK

D R Skelly
Stevington, Milton Keynes, MK43 7QT, UK

A Temple
Shrivenham, SN6 8BG, UK

T J Temple
Cranfield University, Centre for Defence Chemistry, Defence Academy of the United Kingdom, Shrivenham, SN6 7LA, UK

B Thawani
Cranfield University, Forensic Institute, Defence Academy of the United Kingdom, Shrivenham, SN6 7LA, UK

V F Velga-junior
Chemical Engineering Department, Instituto Militar de Engenharia, Praça General Tibúrcio, 80, Praia Vermelha, Urca, 22290-270 Rio de Janeiro, Brazil

R Vrcelj
Cranfield University, Centre for Defence Chemistry, Defence Academy of the United Kingdom, Shrivenham, SN6 7LA, UK

M Williams
Cranfield University, Centre for Defence Chemistry, Defence Academy of the United Kingdom, Shrivenham, SN6 7LA, UK

K K L Yamaguchi
Institute of Health and Biotechnology, Amazonas, Federal University, Estrada, Coari/Mamiá, 305, Espírito Santo, 69460-000, Coari, AM, Brazil

Message from the editors

The idea to compile this book came following discussions during the International Symposium on Risk Assessment Approaches for Decision Making in 2019. The delegates consisted of risk practitioners from a wide range of disciplines who concluded that many of the challenges in Risk Assessment were the result of poor cross-disciplinary communication often due to differences in understanding of terminology. This meant that good practices in one discipline were not able to be translated for the benefit of other disciplines. Therefore, in discussion with several of the delegates from the symposium and other colleagues the idea for this book was generated: to collate a series of case studies and evaluations of Risk Assessment in different disciplines and for different risk to determine how terminology really differed. We hope this book will provide insight into different types of Risk Assessment and the challenges faces by different disciplines to open the conversation on how to universally improve communication of risk.

The editors have thoroughly enjoyed working with the contributors to produce this book, and hope the network will continue to grow in the future.

Abbreviations

3D	Three dimensional
AC	Air conditioning
AE	Ammunition and explosives
AN	Ammonium nitrate
AP	Anti-personnel
AXO	Abandoned ordnance
BAC	Battle area clearance
BAM	Bundesanstalt fir Materialprìfung
BBN	Bayesian belief networks
BFI's	Bulk fuel installations
BPEO	Best practicable environmental option
CA	Critical analysis
CG	Compatibility Groups
CHA	Confirmed hazardous area
COVID	Coronavirus disease
CRO	Commercial Response Organisation
CSM	Conceptual site model
D-GPS	Differential GPS
DNAN	2,4-Dinitroansile
DNT	Dinitrotoluene
DQRA	Detailed quantitative Risk Assessment
DSC	Differential scanning calorimetry
DWEL	Drinking water exposure level
EED's	Electro-explosive devices
EM	Elcctromagnctic
EOD	Explosive ordnance disposal
EPA	Environmental Protection Agency
ERA	Environmental Risk Assessment
ERW	Explosive remnants of war
ESD	Electrostatic discharge
FMEA	Failure modes and effects analysis
FMECA	Failure modes and effects critical analysis
FSC	Forest Stewardship Council
GAC	Generic assessment criteria
GQRA	Generic quantitative Risk Assessment
HAZOP	Hazard and operability analysis
HD	Hazard division
HFI	Human factor integration
HRA	Human reliability assessment
HRC	Human robot collaboration
HSE	Health and safety executive
HWT	Hampshire and Isle of Wight Wildlife Trust
ICRC	International Committee of the Red Cross
IED	Improvised explosive devices
IHE	Insensitive high explosives
IM	Insensitive munition
IMSMA	Information Management System for Mine Action

LCRM	Land contamination: Risk Management
LQA	Land quality assessment
MCE	Maximum credible event
MoD	Ministry of defence
NEM	Net explosive mass
NMAA	National Mine Action Authority
NPPF	National Planning Policy Framework
NTFP	Non-timber forest product
NTO	3-Nitro-1,2,4-triazol-5-one
ONS	Office for National Statistics
ORR	Office of Railroad Regulator
PBX	Polymer bonded explosive
PC	Polycarbonate
PFMEA	Process-failure modes and effects analysis
PPE	Personal protective equipment
PRA	Preliminary (qualitative) Risk Assessment
PSA	Probabilistic safety assessment
PTES	Postgraduate taught experience survey
QAA	Quality assessment and assurance
QD	Quantity distance
QFD	Quality function deployment
QRA	Quantitative Risk Assessment
RA	Risk Assessment
RDX	1,3,5-Trinitroperhydro-1,3,5-triazine
RF	Radiofrequency
SACE	Alert system to critical events
SAH	Hydrologic alert systems
SAM	Surface-to-air missiles
SAM	Scheduled ancient monument
SCJ	Shaped charge jet
SHA	Suspected hazardous area
SOP	Safe operating procedure
SsD	Storage sub-divisions
SSSI	Site of special scientific interest
SWIFT	Structured what if technique
SWOT	Strength, weakness, opportunity and threat
TEPCO	Tokyo Electric Power Company
TGA	Thermogravimetric analysis
TNMA	Technical note for mine action
TNT	2,4,6-Trinitrotoluene
UEBT	Union for Ethical Biotrade
UK	United Kingdom
UN	United Nations
US/USA	United States of America
USSR	United States of Soviet Russia
UV	Ultraviolet light
UXO	Unexploded ordnance
WP	White phosphorous

IOP Publishing

Challenges in Risk Analysis for Science and Engineering
Development of a common language
Tracey J Temple and Melissa K Ladyman

Chapter 1

An introduction to risk analysis for scientists and engineers

F Persico and E Gutierrez Carazo

A literature review is presented showing differences in processes and terminology related to the concept of 'Risk Analysis'. Research shows that there is no clear definition for Risk Analysis as it can be interpreted differently as the study field changes. Different case studies showed what entails a failure in a proper Risk Analysis, they highlight as a guiding thread most malfunctions due to a miscarried Risk Communication. Consequently, missing a proper terminology or process in Risk Analysis ensures that the failure continues leading into misunderstandings of the risks within and outside organisations. Therefore, literature emphasises an in-prompt need to ensure that a standardisation in Risk Analysis is carried out due to numerous benefits that can keep businesses and companies safe and also compliant with legal, social, economic and environmental standards.

1.1 Risk Analysis: definition and uncertainties

'Risk Analysis' has many different meanings depending on the subject, the specific field, the year it was defined, the person who gives the definition or the purpose of the investigation. Therefore, it has been used as an interchangeable term embedded in Risk Assessment or other standard procedures (Gratt 1987, Fletcher 2015). Risk Analysis has been defined as the process of estimating the parameters of the '*probability density function*' used to represent risk (Gratt 1987). The risk is, therefore, considered a random variable described by a probability distribution and it can be represented by continuous, discrete or combined functions. Alternatively, some organisations define Risk Analysis as the identification and management of any uncertainties or risks that could impact business or financial goals. This definition considers the possibility of adverse processes (earthquakes) or events caused by malicious or inadvertent human activities (e.g. arson) and, in some cases, identifies the harm of the event and likelihood that it will occur.

doi:10.1088/978-0-7503-3643-7ch1

© IOP Publishing Ltd 2022

Fletcher (2015), defines Risk Analysis as part of, or the Risk Assessment procedure, embedded within the Risk Management process. Aven (2015), describes Risk Analysis as the central part of Risk Management i.e. the Risk Assessment (evaluation of severity/likelihood) the Risk Treatment (assessment and identification of mitigation/control measures). The Society of Risk Analysis has defined this term as a science that covers or studies '*Risk Assessment, perception, communication, management, governance and policy in the contest of risks to individuals, public and private sectors, organisations and society at a local, regional, national or global level*' (https://www.sra.org/risk-analysis-introduction/). This last definition is broader and more general, to encompass actions that can be used for every field. Depending on the source, Risk Analysis can be defined as (a) the first step of the risk process; (b) the evaluation or assessment step of the risk process or (c) in some cases, the whole risk process (figure 1.1), however the identification of potential risks is a recurrent theme.

In the first case, where Risk Analysis is assumed to be the first step of the process, the analysis is just focussed on the identification of the hazards. For example, a security company may identify that a security breach might happen due to poor firewalls or because an employee is providing data to a third party. In the second case, Risk Analysis is assumed to be part of Risk Assessment, so once a potential security breach is spotted, its likelihood and severity are evaluated within the Risk Assessment procedure. Last, in the third case, the Risk Analysis is defined as the entire process i.e. the identification of hazards (e.g. a breach of data), the likelihood and severity of the event (e.g. very high), how it would be managed (e.g. training on how to protect personal data) and how it should be communicated within the organisation (e.g. effective antivirus) are all incorporated in the Risk Analysis definition.

Conversely, published research frequently states that Risk Analysis is often mistaken for Risk Assessment and Risk Management and argues that the three concepts are different and must be used according to the action that is being carried out. The word 'analysis' comprises the separation of a whole into its component

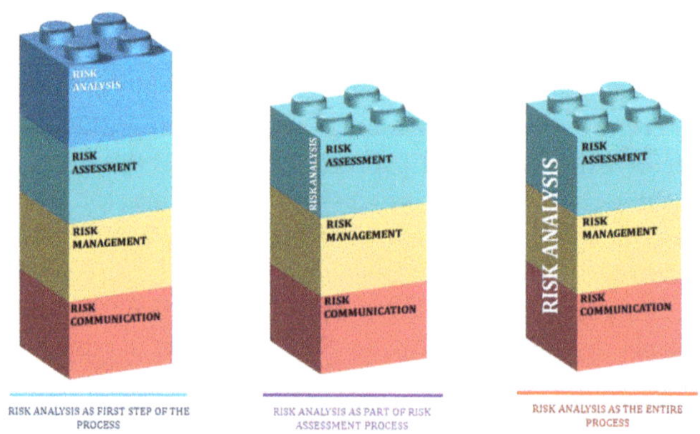

Figure 1.1. Different Risk Analysis definitions from left to right: Risk Analysis as first step of the process; as part of Risk Assessment; as a description of the entire process.

parts, an examination of the complex, its elements and their relationship. By contrast, the word 'assessment' is the act of determining the importance, size or value of something. Last, the word 'management' refers to the action of directing or controlling something to achieve a purpose. Hence, the main problem is inconsistent terminology for use by risk analysts becoming a greater issue as more practitioners with different areas of expertise contribute their results to this expanding field.

Over the years, different researchers and institutions have attempted to provide widely accepted definitions of key terms related to the Risk Analysis field (Gratt 1989, Aven 2015, Fletcher 2015, Jensen and Aven 2018). Yet, experience has proven that agreement on a unified set of definitions is not realistic as none of the attempts has achieved success (Yoe 2019). This raises the challenge of how to use the correct terminology when performing a Risk Analysis in a way that is clear for all the users and readers and leaving no ambiguity that can lead to misunderstandings and fatal mistakes.

Therefore, the aim of this review is to introduce Risk Analysis as it is interpreted by scientists and engineers focussing on the diversity of terminology by discussing several case studies.

1.2 Types of Risk Analysis for science and engineering

Ultimately, as highlighted above, for some, Risk Analysis provides information concerning adverse events; the Health and Safety Executive (HSE) defines different steps that should be taken when doing Risk Analysis, which requires a comprehensive examination of the probability of events and expected consequences for the identified risks. This process is strictly related to standard procedures, which incorporates Risk Assessment and aims to identify risks to avoid further consequences within and outside the organisation.

A mismeasurement of the risks can lead to serious incidents (disasters), such as the financial crisis in 2008. Because the risks were both miscalculated and underestimated, the consequences of the financial crush were seen for over 10 years around the globe. Missing data on the probability of crisis led industries and organisation to believe that the market was continuously growing, guiding them to invest in riskier assets. Although, in reality, less stringent financial checks prior to mortgage loans and ambiguously structured securitised mortgages became the main cause of the financial crash that has seen as one of the consequences an increase in poverty and job losses (Acharya and Richardson 2009). Therefore, a well-prepared Risk Analysis can ensure that all the bases are covered to train the organisation for all the different risks that it might face.

Possible threats that can be identified during a Risk Analysis can involve a wide range of possibilities that differ as the activity changes (table 1.1).

The importance of a Risk Analysis is ultimately highlighted by incidents that may have been less significant if a thorough plan had been in place to understand risks and consequences of all the different activities. Although, major disasters have resulted in changes to law and policy, for example following the Great Smog of London the Clean Air Act was established (Brimblecombe 2006); following the

Table 1.1. Several examples of a brief risk analyses and identification of unforeseen disaster that could have been avoided.

Case study	Hazards	Risk identified	References
Great Smog of London 1952	• Coal burning increase from households and industries • Decrease in coal quality • Unforeseen weather conditions	Natural	Brimblecombe (2012)
Chernobyl 1986	• Missing communication between the working and safety teams • Project not informed by previous incidents • Radiation Exposure	Procedural	Saenko *et al* (2011)
Great Recession 2008	• Indebtedness through uncontrolled mortgages • Missing communication on the related risks of different assets • Public perception of the market	Financial	Aziz and Hadia (2009)
Deepwater Horizon 2010	• Blowout Preventer (BOP) failure • Missing prompt communication with the public • Underestimation of the problem	Reputational	Skogdalen and Vinnem (2012)
Beirut Explosion 2020	• Unsafe storage large quantities of ammonium nitrate • Miscommunication between the authorities • City centre vicinity	Technical	Yu *et al* (2021)
The Capitol Assault 2021	• Unclear message • Violation of democratic boundaries • Public incitation	Political	Scacco and Coe (2021)

Great Recession the Consumer Protection Act was passed (Kim and Muldoon 2015); after Chernobyl new legislations and awareness were raised internationally (Schwartz 2006).

For different group of scientists, as highlighted before, Risk Analysis comprises the entire process from Risk Assessment to Risk Communication. In that case, the evaluation of the risks follows into the Risk Assessment process which also assess the likelihood of an event to occur. There are multiple different methods of Risk Assessment.

Qualitative methods evaluate the risk based on the knowledge of the person that is carrying out the Risk Assessment. Qualitative Risk Assessments are commonly performed to assist a risk manager in priority setting or policy decision-making (Coleman and Marks 1999). Some of these methods include activities such as mind-mapping, interviews or questionnaires and judgement of specialists or experts. Some methods define the risk using descriptive categories (e.g. low likelihood, high severity) (figure 1.2). As they are described verbally and are highly subjective, careful attention must be paid to the scale used to define the risk in order to avoid misunderstandings or misinterpretations of the results. This method is frequently used to make decisions in business when the risk is low and does not warrant the time and resources necessary for undertaking a full analysis. In addition, a qualitative assessment is performed when the numerical data is poorly understood or non-existent. Although qualitative assessments might not provide a clear answer to evaluate risks, they provide a general evaluation of the importance and probability of the occurrence of an event and can be used as a basis for subsequent quantitative analysis.

In contrast, using quantitative methods allows the assessment of a risk by assigning values or numbers to the severity and the likelihood of the event. It enables calculation of expected values so different risks can be directly compared and generates numerical

		Consequences/Impacts				
		Negligible	Minor	Moderate	Major	Catastrophic
Likelihood	Almost Certain	Medium	High	High	Very High	Very High
	Likely	Medium	Medium	High	High	Very High
	Possible	Low	Medium	High	High	Very High
	Unlikely	Low	Medium	Medium	High	High
	Rare	Low	Low	Medium	Medium	High

Figure 1.2. Example of the structure of a Risk Assessment matrix. The terminology used to describe likelihood and consequences/impacts is subjective and may mean something different for different people.

data that can mathematically support or assist decision-making. In this case, uncertainties are evaluated numerically in both the probability and severity of the risk making the confidence more explicit (Aven 2011).

Risk exposure = Probability × Severity

Several examples of how to quantitatively assess risks can be found in literature (Creemers *et al* 2014, Rycroft *et al* 2019, Guindon and Wtight 2020) and most of them include computer simulations and probabilistic approaches such as the Uses Monte Carlo Method. This method allows the user to explore reality through a mathematical risk model in such a way that by assigning values randomly to the variables (risks) of the model, different scenarios and results are obtained. Quantitative Risk Assessment methods allow the performance of statistical studies by evaluating the results from each iteration and the extraction of conclusions from statistical parameters. Therefore, parameters such as the mean, maximum and minimum values, standard deviations, variances and likelihood of occurrence of the different variables determined on which to measure the risk are included (Firestone *et al* 1997). Obviously, the quantitative Risk Analysis is more detailed as it describes a full range of possible risks, enabling a more thorough assessment of the variability and uncertainty associated to a risk.

Risk Analysis, as intended to be the entire process, includes the subprocesses of Risk Assessment, Risk Management and Risk Communication (Yoe 2019), as a methodology to follow to ensure that uncertainty related to different activities are evaluated. For instance, Erena and Worku (2018) explained the entire process in terms of a flood Risk Analysis research, from analysing causes, risks and consequences to discussing mitigation strategies to avoid possible unforeseen consequences.

Ultimately, for Risk Analysis to be effective it must be communicated in such a way that the majority of people understand, however this highly dependent on the media used, cultural background, simplicity of the report, awareness and so on. Miscommunication can result in unexpected events due to underestimation or system failure within and outside the organisation, therefore, is essential that communication is provided in an efficient manner.

1.3 Communication in Risk Analysis: a standardised terminology

Communication plays a critical role in providing balanced information to ensure preparedness of the targeted audience within and outside an organisation. The literature has highlighted that, for years, several problems related to not properly communicating the risks associated with a particular activity had unexpected negative consequences (Keeney and Winterfeldt 1986, MacDiarmid and Pharo 2003). An example of these events is the accident following a test carried out in the reactor of the Chernobyl nuclear complex in Ukraine. The accident followed different fires that released radioactive material to the air for several days, which resulted in human deaths following radiation poisoning (Saenko *et al* 2011). The events could have been avoided if proper communication of the risks were in place. For instance, in 1975 another explosion accident occurred in Leningrad, in a nuclear plant in the Soviet Union territory. Had processes and a positive safety culture been in place to enable engineers to inform colleagues of weaknesses in the reactors' designs, further incidents of a similar nature may have been avoided. Moreover, in the case of Chernobyl, the crew performing the test were unable to coordinate with the group responsible for safety at the reactor, therefore, no precautions were in place due to lack of effective communication.

Often it is only after an accident occurs that there is the realisation that communication should have the same importance as determination and acceptance of the risks, because, ultimately within the organisation, the people that are facing those risks can be in danger if they do not fully understand the problem (MacDiarmid and Pharo 2003). This is essential also during crisis situations where effective communication can help avoid a chaotic response (De Wolf 2013). An example is the Deepwater Horizon oil spill in which the owner (BP) did not immediately communicate the severity of the accident to the public. Moreover, public communication from the company was not sensitive to the loss of lives and misrepresented the environmental damage following the explosion, which further resulted in a great economic loss for the industry itself. Different studies have assessed failures in communication and shown how public perception of the company changed due to poor communication with the public during and after the accident occurred (De Wolf 2013, Ye and Ki 2017).

The effectiveness of communicating risks depends on the source, the message content and the perspective of the target audience involved which can be also influenced by the type of organisation or cultural background (Abunyewah *et al* 2018) (figure 1.3). Social media has become the main channel for spreading biased information, which leads to poor communication and ineffective Risk Communication because of misinterpretation of information (Jagiello and Hills 2018). This is further complicated because the communication often needs to express complex risks associated with different activities in simple and effective language, especially when communicating between disciplines. Communication can be eased if bureaucratic and technical language is minimised making it understandable across a broader group of

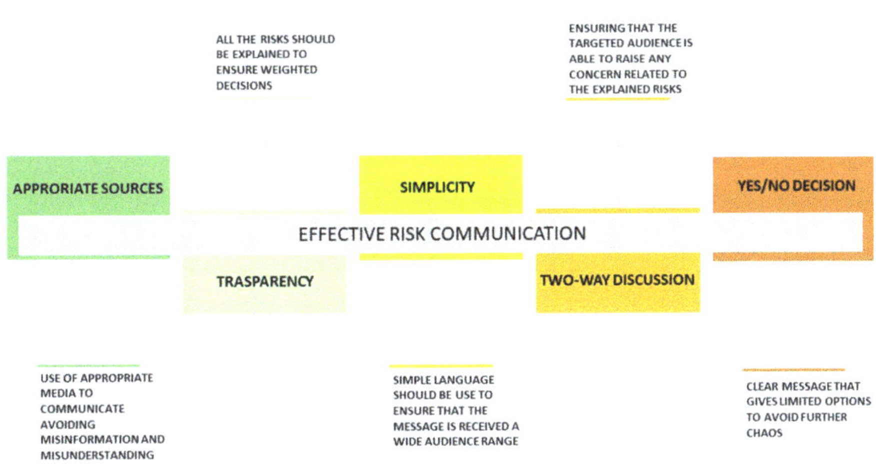

Figure 1.3. How to render Risk Communication effective during various situations.

people when communicating risk (Keeney and Winterfeldt 1986). For example, when regulating a discharge plan within an organisation, multiple issues are included in the Risk Analysis: health issues, environmental impact, remediation cost, political implication and so on. All these risks, as a whole, are difficult to communicate within the organisation and the communities involved due to the complexity of the matter. Informing workers within an organisation that some of the activities may lead to safety issues involving potential risks to their health or the environment, might trigger a series of emotional responses (Fisher 1991); in fact, the situation can worsen if the risks are not correctly communicated, resulting in people being unprepared for the risk. Communication to the public is often reduced to a simple yes/no risk associated decision as the complexity of communicating a broader body of alternatives can interfere with the meaning of the message within and outside an organisation (Keeney and Winterfeldt 1986). Poor communication can result in lack of preparedness due to miscommunication or misunderstanding of the risks; although, personal motivation is the main driver for the audience to prepare for facing the hazards (Abunyewah *et al* 2018).

Rejection of the risks is largely related to misunderstanding the consequences related to bad communication and the use of complicated and technical language that varies as the field changes (Fisher 1991). Usually, the public responds to the risk of an activity differently due to various inherent bias. For instance, smoking has high risks of cancer-related health consequences, although different people will perceive the risks differently as they interpret the word ‘high’ in different ways (Fisher 1991).

Effective communication of risk can reduce public anxiety, which can be influenced by how the information is processed (Jagiello and Hills 2018). Ideally, Risk Communication should involve a group discussion with the public because an efficient conversation involves full transparency from the organisation regarding potential risks by using a science-based approach. Therefore, some of the hazards may not be fully understandable by the entire targeted audience, causing chaos among people, rendering necessary the inclusion of the audience when risks are reported to ensure that questions that may arise are fully answered (Palenchar and Heath 2007). This process is essential to enable effective management of crisis events.

An example of this is the COVID-19 crisis. With the pandemic, the role of Risk Communication became essential to promote protective and risk-mitigating actions to be taken to ensure the safety of the others (Heydari *et al* 2021). However, the source of information was shifted to being mainly online, making social media the main source that people accessed to obtain information and to evaluate risks associated with contracting COVID-19 (Adanlawo 2020). Whilst a useful source of information, it also enabled uncontrolled misinformation from unreliable sources to be circulated, which led many people to believe that COVID-19 was not a real threat and changing social perception (Balog-Way *et al* 2020, Heydari *et al* 2021). Government’s faced steep challenges to ensure that the public was fully aware of the risks and how it was possible to mitigate them. Therefore, simple and concise messages, such as ‘wear a mask’—‘save lives’—‘wash your hands’ were essential to ensure that the population was following actions considered to be efficient and reliable (Heydari *et al* 2021).

In summary, a simple and concise message can ensure that risks are efficiently communicated, and that mitigations are placed within and outside the organisation.

1.4 Benefits of performing a Risk Analysis

The performance of a Risk Analysis for science and engineering purposes may seem time consuming at a first glance, but it comes with short and long term benefits for employees, companies and customers (Simister 1994). Firstly, a Risk Analysis allows the identification and control of potential hazards in the workplace and creates awareness among employees. In addition, it can be used as a method to provide training to employees, making them more qualified and responsible to prevent an accident or respond accordingly in case a risk is realised (Sabanov *et al* 2020). A properly executed Risk Analysis reduces the risks of incidents in the workplace as for instance, damaged or misused equipment can be detected and retired, mitigating the number of accidents, injuries and minimising potential law suits for the company. Therefore, despite the resource requirement Risk Analysis can been seen as a cost effective for most companies.

In addition, Risk Analysis can prevent secondary problems that might affect customers or third parties. Companies and business are more socially reliable when they follow all safety protocols. The absence or scarcity of safety protocols has further impact on the industry, because, if risks are not properly evaluated consequences can result in serious fines to the company and loss of confidence in the organisation. In even worse cases, it can result in the loss of life. Moreover, ensuring the proper analysis of the risks can be beneficial during a crisis situation (e.g. COVID-19) because by standardising procedures and terminology, it is ensured that risks are evaluated, analysed, managed and communicated in an efficient manner.

Therefore, a Risk Analysis is not only highly recommended to keep businesses and companies safe but also to comply with legal, social, economic and environmental standards.

1.5 Conclusion

The lack of a single definition for 'Risk Analysis' has resulted in a divergence of research across different disciplines. Therefore, literature shows that Risk Analysis can play different roles as interpreted in different subject fields: as a first step of a longer process; as embedded in the Risk Assessment evaluation or as a process that comprises Risk Assessment, Management and Communication. Through different case studies, research has shown that Risk Analysis is interpreted differently in different disciplines, consequently, there is an inconsistency in terminology in the literature that makes it difficult to effectively communicate the impact of different risks within an organisation, and makes it challenging to effectively communicate risks to those outside an organisation.

Several examples have shown that poor Risk Analysis drives the development of unforeseen events (e.g. Beirut explosion 2020, The Great Recession 2008) leading to global disasters due to the lack of suitable and standardised Risk Analysis processes.

Moreover, numerous cases in the literature showed that a consistent failure was due to how the risks were communicated within and outside the organisations. In fact, communication within a company is essential when performing Risk Analysis in order to identify, evaluate and manage the risks.

Lack of communication can lead to severe consequences (e.g. Chernobyl explosion). The failures in communications can be related to the use of overly complicated and technical language, source or field, the message content or the public perception. Therefore, it is essential to convey the message simply and concisely, although this cannot be achieved if there is no standardised methodology and terminology while carrying out a Risk Analysis.

Aligning the Risk Analysis processes and terminologies will be beneficial for different organisations as it will involve several advantages: acceptable safe practices and legal requirements, proper qualifications and awareness within the organisation, concise communication within and outside the organisation, avoiding unforeseen consequences and potential unreversible consequences due to misevaluation. Therefore, organisations should work together to define terminologies to ensure that a proper Risk Analysis in carried out with the purpose of avoiding future consequences related to misunderstanding and miscalculations of the risks.

References

Abunyewah M, Gajendran T and Maund K 2018 Conceptual framework for motivating actions towards disaster preparedness through risk communication *Procedia Eng.* **212** 246–53

Adanlawo M M R 2020 Crisis risk communication of Covid-19: the role of the media in behavioural change *PalArch's J. Archaeol. Egypt/Egyptol.* **17** 15567–75

Acharya V V and Richardson M 2009 Causes of the financial crisis *Crit. Rev.* **21** 195–210

Aven T 2015 *Risk Analysis.* (New York: Wiley)

Aven T 2011 *Quantitative Risk Analysis.* (Cambridge: Cambridge University Press)

Aziz A and Hadia H 2009 Introducing successful financial innovations: rewriting the rules in light of the global financial crisis *World Acad. Sci. Eng. Technol.* **54** 1205–11

Balog-Way D, McComas K and Besley J 2020 The evolving field of risk communication *Risk Anal.* **40** 2240–62

Brimblecombe P 2012 *The Big Smoke (Routledge Revivals): A History of Air Pollution in London Since Medieval Times.* (Milton Park: Routledge)

Brimblecombe P 2006 The clean air act after 50 years *Weather* **61** 311–14

Coleman M E and Marks H 1999 Qualitative and quantitative risk assessment *Food Control* **10** 289–97

Creemers S, Demeulemeester E, Van and de Vonder S 2014 A new approach fror quantitative analysis *Ann. Oper. Res.* **213** 27–65

De Wolf, D 2013 Crisis communication failures: the BP case study *Int. J. Adv. Manag. Econ.* **2** 48–56

Erena S H and Worku H 2018 Flood risk analysis: causes and landscape based mitigation strategies in Dire Dawa city, Ethiopia *Geoenviron. Disasters* **5** 1–19

Firestone M, Fenner-Crisp P, Barry T, Bennett D, Chang S and Callahan M 1997 Guiding principles for Monte Carlo analysis. Report number EPA/630/R-97/001. USEPA

Fisher A 1991 Risk communication challenges *Risk Anal.* **11** 173–9

Fletcher W R J 2015 Review and refinement of an existing qualitative risk assessment method for application within an ecosystem-based management framework *ICES J. Mar. Sci.* **72** 1043–56

Gratt L B 1987 Risk analysis or risk assessment; a proposal for consistent definitions *Uncertainty in Risk Assessment, Risk Management and Decision Making* (New York: Plenum Press) pp 241–9

Gratt L B 1989 The definition of risk associated terminology for risk analysis *Risk Assessment in Setting National Priorities* (New York: Plenum Press) pp 675–80

Guindon A A and Wtight D J 2020 Analytical approach to quantitative risk assessment for solar power projects *Renew. Sustain. Energy Rev.* **133** 110262

Heydari S T, Zarei L, Sadati A K, Moradi N, Akbari M, Mehralian G and Lankarani K B 2021 The effect of risk communication on preventive and protective Behaviours during the COVID-19 outbreak: mediating role of risk perception *BMC Public Health* **21** 1–11

HSE n.d. *Managing risks and risk assessment at work* https://hse.gov.uk/simple-health-safety/risk/more-detail-on-managing-risk.htm

Jagiello R D and Hills T T 2018 Bad news has wings: dread risk mediates social amplification in risk communication *Risk Anal.* **38** 2193–207

Jensen A and Aven T 2018 A new definition of complexity in a risk analysis setting *Reliab. Eng. Syst. Saf.* **171** 169–73

Keeney R L and Winterfeldt D V 1986 Improving risk communication *Risk Anal.* **6** 417–24

Kim S H and Muldoon C 2015 The Dodd-Frank Wall Street Reform and Consumer Protection Act: accomplishments and shortcomings *J. Appl. Bus. Econ.* **17** 3

MacDiarmid S C and Pharo H J 2003 Risk analysis: assessment, management and communication *Rev. Sci. Tech.* **22** 397–408

Palenchar M J and Heath R L 2007 Strategic risk communication: adding value to society *Public Relat. Rev.* **33** 120–29

Rycroft T, Hamilton K, Haas C N and Linkov I 2019 A quantitative risk assessment method for synthetic biology prodcuts in the environment *Sci. Total Environ.* **696** 133940

Sabanov S, Madani N, Mukhamedyarova Z and Tussupbekov Y 2020 A risk analysis method for estimation of financial benefits of the existing mine ventilation system *Mining, Metall. Explor.* **37** 1137–49

Saenko V, Ivanov V, Tsyb A, Bogdanova T, Tronko M, Demidchik Y and Yamashita S 2011 The Chernobyl accident and its consequences *Clin. Oncol.* **23** 234–43

Scacco J M and Coe K 2021 Securing the guardrails of democracy? Accountability and presidential communication in the 2020 election *Quart. J. Speech* **107** 423–29

Schwartz J A 2006 International nuclear third party liability law: the response to Chernobyl *Int. Nuclear Law in the Post-Chernobyl Period* 41–4

Simister S J 1994 Usage and benefits of project risk analysis and management *Int. J. Project Manage.* **12** 5–8

Skogdalen J E and Vinnem J E 2012 Quantitative risk analysis of oil and gas drilling, using Deepwater Horizon as case study *Reliab. Eng. Syst. Saf.* **100** 58–66

Ye L and Ki E J 2017 Organizational crisis communication on Facebook: a study of BP's Deepwater Horizon oil spill *Corp. Commun.* **22** 80–92

Yoe C 2019 *Principles of Risk Analysis: Decision Making under Uncertainty.* (Boca Raton, FL: CRC Press)

Yu G, Wang Y, Zheng L, Huang J, Li J, Gong L, Chen R, Li W, Huang J and Duh Y S 2021 Comprehensive study on the catastrophic explosion of ammonium nitrate stored in the warehouse of Beirut port *Process Saf. Environ. Prot.* **152** 201–19

IOP Publishing

Challenges in Risk Analysis for Science and Engineering
Development of a common language
Tracey J Temple and Melissa K Ladyman

Chapter 2

On the tolerability of risk, public and private

M R Williams

'*For the stronger that we our houses do build—the less chance we have of being killed.*' William McGonagall, at the conclusion of his *The Tay Bridge Disaster*.

The author of this chapter does not claim to be an expert on probabilistic Risk Assessment but he has had to consider the factors that go towards such matters while running Risk Assessment courses. Some of these factors are covered here. There are published guidelines for 'public tolerability' of risk but these are really for those managing risk. What the real public will tolerate is not so easy to quantify and may only become clear after the event.

2.1 Introduction

Years ago, I was told by a colleague, whose word I would not doubt, of a project he was involved in some time in the 1970s or 1980s. It was a hazard and Risk Assessment of a major infrastructure project in the UK. When they took a draft of their study to the government minister concerned, they asked the minister how they would know if the risks that they had identified would be tolerable. The minister replied to the effect that it would be intolerable if the magnitude of the incident were such that he, or his successor, should have to resign during the lifetime of a parliament or, heaven forbid, that it should bring down the government.

I have this as a private communication and had it confirmed recently but the details of the communicant, the study or the minister are not important to my case. The story indicates that what is called public tolerability of risk is often political tolerability of risk. It also suggests that if a Government has to worry about public opinion for re-election then the government's perception of public tolerability is as important as the actual level of public tolerance. I imagine that the converse is true.

When I promised to write this chapter, this is what I said I would do. I said that the chapter would explore different approaches to tolerability and how these link to perceptions of likelihood. The work will consider likelihoods of harmful and

doi:10.1088/978-0-7503-3643-7ch2

© IOP Publishing Ltd 2022

beneficial events and how these are perceived by the public. Often words are used to encompass likelihoods especially in qualitative Risk Assessments. This chapter will consider if these words have any commonly held numerical meaning but will do so from the viewpoint of a practitioner rather than a statistician.

2.2 Risk, hazard and peril

The word ‘risk’ is often co-defined with ‘hazard’. Both have a long etymology and seem to have swapped meanings at some stage, probably in the recent past.

The word ‘hazard’ has a long history, originally coming from the Arabic al-zahr, meaning a dice (or die). This developed into the name for a game of chance.

‘Risk’ seems to derive from a Latin word meaning a cliff or a reef and may even derive from Greek. A much better word for what we now take as hazard would be peril (from Latin periculum—danger or jeopardy). These days, in English, peril is always bad, chance can be good or bad. In the tales of the fictional court of the fictional King Arthur there was a ‘Siege Perilous’. This siege (or seat) was reserved for the knight who would find the Holy Grail and was fatal to anybody else who sat in it. Peril is bad.

I shall treat ‘hazard’ as a source of potential harm and ‘risk’ as the chance that that particular harm can occur. This is in line with the definitions of the Health and Safety Executive. Peril may creep in. The main question is who does the peril affect and how do they view their vulnerability to it.

We live in a world of perilous hazards and risks associated with those hazards. The word ‘risk’ commonly associates a hazard with the likelihood of exposure to that hazard and of that hazard causing harm to those exposed. Different individuals to different extents will accept voluntary exposure to hazards in line with their perceptions of danger and the benefits to themselves. They consider the risk ‘tolerable’. For a Risk Assessment to be more than just a hazard assessment some concept of ‘tolerability’ must be developed by those preparing and approving the Risk Assessment, especially where the risk is not entirely voluntary.

For instance, when the initial data about the risk of death during the COVID-19 pandemic was made public, the peril to those of my age and sex was such that the risk of death from COVID-19 was between 1 and 2 in 100 should we catch the disease. I did the computation that for this to be tolerable I had to reduce my risk of *catching* the disease to a similar level or lower so that my individual risk of death was nearer 1 in 10000. So, I started hand washing and social distancing earlier than advocated by the UK Government and adopted a face covering very much earlier than advocated by the UK Government (but respiratory protection is one of my specialist subjects).

But I was taking a highly focussed approach. One might use the word ‘blinkered’ instead of focussed. The fact that the risk of death for a man of my age in the UK anyway is between 1 and 2 in 100 did not enter my head. In my career at a university, I taught a statistical approach to risk but risk is not likely to be a statistical computation for the majority of individuals. I had done a statistical calculation (had I had an envelope I would have used it) but I did not take a holistic approach. The same is true of most Risk Assessments in that only one source of peril is considered.

2.3 Political tolerability of risk

Various nations have developed concepts of tolerable risk, for example the UK, USA and Australia but the boundaries of likelihood adopted in these may be so unlikely as to have little meaning to either the general public or even to the 'experts' preparing and approving the Risk Assessment[1].

The UK 'standards' were set out in 'Reducing Risks, Protecting People' but were in existence before then (HSE 2001). They were aimed at individual occupational risk. They also included guidelines for societal risk, which is more of a group risk concept. I do not intend to look at societal risk in this piece.

The concept of three ranges of risk, unacceptable, tolerable and broadly acceptable are used in Reducing Risks, Protecting People with the boundary between the first and second being 1 in 1000 per annum for death of all workers and 1 in 10 000 for members of the public who have a risk of death imposed on them 'in the wider interests of society'.

The HSE note that these levels 'rarely bite', that is to say that the historical data available at that time suggested that no industry or group of members of the public would be in the unacceptable region. Deep sea fishing had an individual risk of about 1 in 850 per annum in 1982 when the annual risk of death in all industries was put at about 1 in 2000 (Williams 1998). More recent data put the annual risk of death in all occupations at around 1 in 160 000 for all occupations in 2018 (HSE 2021).

The HSE placed the boundary between tolerable risk and broadly acceptable risk at 1 in a million deaths per annum for both workers and members of the public, that is below 1 in a million is broadly acceptable. As the HSE is a public body funded by government it might be assumed that their concepts of risk can be taken as reflecting political tolerability of risk.

2.3.1 Road and rail traffic as exemplars of this approach

Risks of travel are often cited 'per journey' or per vehicle km. The criteria in Reducing Risks, Protecting People are per year, no matter how many days worked or how many days the individual member of the public experiences risk 'in the wider interests of society'. With most members of the population of the UK travelling by road each year and a sizable proportion travelling by rail any level of deaths above around 50 per annum for road and 25 per annum for rail would place the risk of travel above 'broadly acceptable'. Anything above 5000

[1] Other societies are regarded as adopting an almost 'fatalistic' approach to risk (Bowering (1997) or Ringgren (1967) on Islam for example) though their philosophical arguments are at odds with the experience of the author, who has delivered courses on Risk Assessment to the armed forces of Gulf States who take it very seriously.

for road and 2500 for rail would be intolerable. Current numbers are about 1600 deaths per annum for road (DoT 2021) and around 12 for rail (ORR)[2].

Travel is often regarded as a voluntary risk, a risk that the road user chooses to take, yet many of those travelling by road or rail are doing so because (say) their parents decide that road or rail is the means of travel that they shall use. That is to say that many road deaths are of people who do not travel in a voluntary fashion but have been coerced or at least encouraged to do so by others[3].

2.4 Public mortality

The number of people dying per annum in the UK has remained at about half a million per year since 1870 (ONS). This reflects the improving life expectancy due to improved healthcare and as death in infancy used to be a large contributor to the statistics must indicate improvements that took place back in the 1930s and 1940s.

The annual risk of death per annum in the UK from all causes for all ages has been declining since before 1993. It stands at about 9 per thousand before COVID and was about 14 per thousand in 1994 (ONS). These numbers should be set against the levels of risk for occupations set out in Reducing Risks Protecting People, of an upper level of tolerable of 1 in 1000 per annum. What the HSE levels of tolerability do not reflect, I suspect, are chronic conditions developed as a peril from a particular occupation. Again, the ONS publishes extensive data on causes of death. While causes which may be a combination of lifestyle and workplace exposure have declined since 1994 (lung disease) as have others that have far better treatments available (prostate) others have increased (dementia and Alzheimer's), probably due to these being late onset diseases (no fatality can be prevented forever) and an increased recognition of them as causes of death.

2.5 Mortality and fiscal cost

Figures exist for consideration in Risk Assessments that use a 'value per prevented fatality' (VPF). At a VPF of £1 million (probably an underestimate) then prevention of 300 000 deaths by actions in the UK against COVID-19 (as some modelling suggested) is worth 300 billion pounds. The government spent £400 billion on their response to COVID (as reported by the BBC in October 2021). This was for everything including furlough. Although, at that point, the cost seemed vast, the cost per predicted fatality was not excessive.

---

[2] The numbers of journey-ists is an estimate. I judge that most of the population of the UK would travel somewhere by road in the course of the year. Rail travel is recorded by journeys made rather than journey-ists. Deaths on the road have fallen significantly since 2009, having been in the 3000s per annum before then. This seems to be due to increased protection—air bags, braking systems, etc. In 2019 there were no deaths in train accidents in the UK. The most dangerous form of road transport in the UK is the motor bike (or equivalent). This seems to be due to the lack of protection not recklessness on the part of the rider as many motor bike accidents are due to motorists not the rider.

[3] The risk of death on the road in the UK is about 1 in 37 500 of the population per annum, nearly all of whom travel. For comparison—Francois Caron 'Histoire des Chemins de Fer en France' page 67, notes the risk of death for travellers on French Fast Mail coaches around 1850 was 1 in 35 378 travellers (which I suspect means journeys). Things are getting better.

2.6 Public perception of risk

Ortwin Renn (2009) declares that 'The invention of risk as a mental construct is contingent on the belief that human action can prevent harm in advance'. This builds on a quote from OECD (2003, p 67) that Risk is a mental construction, not real but originating in the human mind. Later the author notes that human behaviour is primarily driven by perception and not by facts (or what are perceived as facts by risk analysts and scientists) (Renn). The concept of 'dread' often comes into play to prevent the individual accepting a situation where there is known to be peril and far outweighs any probabilistic re-assurance. Others may take a misguided view of statistical risk that allows them to pursue actions that they wish to do[4].

The perception of 'The public' in its wider sense encompasses many individual views. The information (or lack of it) that forms these views may come from personal experience, the experiences of a trusted person or the opinions of somebody completely unknown to the individual. They may be formed by the last piece of information that the individual read or heard or the first. Often, in a developing situation, the former is based on little information and the latter is far more nuanced. In the UK, for example, the Government's initial resistance to face coverings during the early phases of COVID has never been replaced in some minds by later evidence showing their benefit.

Returning to the information on COVID-19, it may prove to have been a mistake to release, in the early stages of the pandemic, figures for the relative risk of death for various age groups and the two sexes (data that I relied on for my personal Risk Assessment). These will have been based on the original cases in China. While this estimation of risk informed the responses of a range of the population who saw themselves to be in a high risk group (into lowering their exposure to the hazard long before the authorities had insisted on it) it may also have informed the responses of those who saw themselves to be in a low risk to carry on as normal and to and not respond to the more detailed information that comes with studies of collective experience. Once an individual has formed an opinion it is difficult for them to change it, especially if the evidence to counter that opinion is complex and hard to explain.

Having said all that and speaking as somebody who has worked with risk as a probabilistic concept, Wachinger, Renn and others, in a meta study of published data (The Risk Perception Paradox) note that neither the likelihood of a disaster nor the perceived magnitude of a disaster 'play a very important role in the risk perception of natural hazards' (Wachinger *et al* 2013). Other factors play a far more important part in the individual perception of risk. They also note an apparent paradox, that individuals with experience of a hazard and thus, presumably, a high risk perception 'seldom take appropriate preparedness actions'.

The European Commission (2015) conducted a survey on the public perception of environmental risk. They used a group of experts and lay focus groups. In general,

[4] There is a story, undoubtedly untrue, of an explosives expert who tried to take his own bomb onto a plane because 'the chances of there being two bombs on the same plane are infinitesimally small'.

the two groups agreed on the risk of eleven of the fourteen categories presented for consideration, with air pollution highest risk for both groups. The lay groups were more concerned about chemicals in products than the experts (this may be an example of 'dread') but far less concerned about individual consumption (a voluntary risk). The study notes that lay people tend to look to an institution or individual responsible for an issue and its management rather than to themselves. They also note that the lay respondents expressed low trust in institutions and experts and that this went hand in hand with a loss of trust in mass media. The study suggests that participants would look for information from other (trusted) sources. In spite of this study being published well into the Facebook and Twitter era, on-line sources of information are not mentioned specifically.

2.7 Peril and the information super highway

At an earlier stage in my life I had a temporary position as a nursing assistant in a psychiatric hospital. In the long-term male ward was a gentleman who I shall call Mr T. He was physically harmless but delusional and very imaginative. He would come from his room for afternoon tea and announce that he'd just got back from Africa. One day he said that he had a 40 mile long submarine in the English Channel off Dover.' 'If it's 40 miles long', we said, 'How do you turn it round'? 'It's telescopic' he replied, without pausing for a moment. These days Mr T would have an on-line presence and his imagination would be producing fantasies that would affect and be followed by thousands of others.

The trouble with the information super highway is that not everything on it is road worthy. There is no MOT system for information. During the COVID-19 pandemic a large number of people seemed more inclined to believe self-declared experts (with self-declared expertise and qualifications) rather than known academics with known expertise and qualifications. On the other hand, I have known a number of professors in my academic career and I would be lying if I said that every single one of them was trustworthy and devoid of deviousness.

Although the internet is full of diverse views it is hard to find a balanced selection. Links offered by Facebook, for example, are selected by algorithms and are chosen to fit with the user's perceived interests[5] and the user never knows that contrary information exists. So it would be wrong to assume that the most popular websites and influencers reflect public opinion, especially on peril and risk, as those viewing that website may be sent there not of their own volition but by an algorithm.

On the other hand, even when mainstream news media try to give a balanced piece on a particular subject they invite two proponents of completely opposite opinions to confront each other[6] rather than producing an objective summary by a non-involved reporter.

[5] I like alcohol, cream, chocolate and orange but did I like the orange truffle cream liqueur we were given for Christmas? Not in the slightest. A Facebook algorithm would suggest it was perfect for me.

[6] Those of one opinion think that the proponent of the opposite opinion is bonkers and those of us in the middle think they're both bonkers.

A very small group of vaccine doubters, in a BBC Panorama programme in February 2021, were presented with an on-line collage of opinions of self-declared experts against the vaccine and a chance to talk to a professor of Immunology about their doubts. All declared themselves more convinced by the live professor than by the recorded opinions. Face-to-face contact had more of an influence than remote viewing probably because the participants felt that their individual concerns were being addressed individually.

2.8 Peril versus lack of benefit as an arbiter of tolerability

One of the aspects of the COVID-19 pandemic that troubled me was the rejection of health interventions, public or private, on the basis that there was no evidence that they do any good. This was true of early vaccinations using one of the major vaccines in France and Germany. It was true of the efficacy of vitamin D. It was true of the use of face coverings in the UK.

The resistance in Government and some academia to wearing face coverings, as opposed to respirators (which were in short supply), was inexplicable to me. I am a board member of the International Society for Respiratory Protection and it was well known, in our community, that face coverings were never meant to protect the wearer but were to protect others from the wearer. Surgical masks have been used for over 100 years to achieve this purpose. This is, I think, another instance of the Risk Assessment being based on a perceived lack of benefit.

Some of this has persisted. The early word often catches the worm of perception. Younger people were told that their risk of serious disease from the virus was low and this belief has persisted even when variants have arrived which may show a different epidemiology. This leads to vaccine resistance because 'I don't need it' in spite of the various arguments in favour, one of which is that it would protect others from you.

This 'perceived lack of benefit' may well run into the public tolerability of environmental risks. The peril is not acute but chronic. Chronic peril, especially chronic peril that won't affect those who are being asked to limit their activities today, is much more difficult to 'sell'. Some politicians' attitudes to climate change and international treaties is essentially one of 'jam today' rather than 'no bread tomorrow'.

2.9 How rare is rare: how to describe unlikely events

'More or Less', a BBC radio programme about statistics quoted that a 'rare disease' is one that affects less than 0.05% of the European population by definition. This is less than 1 in 2000. Yet between 6% and 8% of the European population suffer from a 'rare disease' so although each disease may be rare the chance of knowing somebody with a rare disease is quite high. We notice unlikely things happening because there are a lot of unlikely things that could happen.

Professor Neil Stewart (and his colleagues) of Warwick University psychology department, studied what people understood by the meaning of words such as 'rare' and 'unlikely'. They compiled a table, using volunteers, linking one of 71 phrases to the likelihood of drawing a red balls from an urn of 100 balls (red and another colour) where there were different numbers of red balls in the urn. This enabled them to link risk phrases to probabilities but their lowest probability (p) was 0.0376 or 1 in 26 ('extremely doubtful'). Other examples were 'improbable', $p = 0.133$ and 'more likely', $p = 0.64$.

Of the 71 phrases 32 were less than $p = 0.33$ and 24 were for p greater than 0.66. All were greater than $p = 0.03$ except for words such as 'never' where p was zero. These sorts of probabilities are far too frequent for Risk Assessment, although they are for single events not for a sequence of 'AND' events that all have to happen before any particular peril is realised.

Stewart *et al* (2006), conducted a properly constructed peer reviewed study. In a rather less formal study, I asked a series of attendees at meetings and courses to relate a series of guide words to the chance of predicting a series of outcomes for the original UK national lottery, per annum, playing just over 100 times per year (chance of getting six correct numbers out of 6 once in a year becomes 1 in 140 000). Here we found that 'Frequent' meant it was nearly certain in any year, 'Probable' was twice a year, 'Occasional'' as once in 10 years, 'Remote was once in 500 years, 'Improbable' was once in 23 000 years and 'Incredible' was once in 140 000 years. I chose these words as they occur in an MOD qualitative Risk Assessment scheme. This was not a properly constructed study, hasn't been peer reviewed and suffered because it only including respondents who had the disadvantage for this task of education, including by the author, experience or (usually) both.

2.10 Conclusion

Having considered these things, especially in the light of COVID-19, I am aware that what I always took to be public tolerability of risk is really political tolerability. Where politicians will place this boundary is a matter for them. There are no rules. I doubt if any would be so unwise as to let the rest of us know where there boundaries are but it will depend on far more than just an assessment of the peril to the public. The peril to the politicians is a large part of the equation.

I do not believe that what the public will tolerate can be predicted until they actually experience the particular peril (at which point it ceases to be a prediction). For a new peril some will be cautious, some foolhardy. The same will even apply to those who have experienced the peril before.

Some experts have indicated that, in the current pandemic, early (or excess) deaths should be the arbiter of the success or failure of national response. For the individual, the death of one close to them will be far more telling.

Information is important but it must be clear, consistent[7], authoritative and not confusing. This is my view, but I have heard similar requirements from representatives of the World Health Organisation. Those giving information must remember that the internet never forgets. A mistake in a live news conference can be corrected

[7] When the recommended level of alcohol intake fell suddenly from 21 to 14 units per week those of us who liked a glass of wine or two thought 'let us know when you've made up your minds.'

but both mistake and correction will exist forever and can be used to support opposite views.

I believe that the public will react more positively to measures designed to lower their risk of peril if given positive physical benefits to consider even if these have to be stated simply and in broad terms. Statistical benefits, especially if they are in the areas of 'improbable' or 'incredible' noted above, confuse. An example of this was when the instance the rare risk of a blood clot from a particular vaccine was stated to be in the order of the risk of blood clot from taking the contraceptive pill. Knowing this did not re-assure some potential vaccines but only made them question using the contraceptive pill. When I was doing Risk Assessments for explosive storage sites I might have predicted that the chance of a resident, who had risks imposed on them 'in the wider interests of society', being killed following an incident was less than 1 in 100 000 per annum using sound statistical probability data and documented trials evidence yet that resident would be quite right to ask 'why me?'.

References

Bowering G 1997 The concept of time in Islam *Proc Am. Philos. Soc.* **141** 55–66

DoT 2021 https://www.gov.uk/government/statistics/reported-road-casualties-great-britain-provisional-results-2021

European Commission 2015 Survey on public perceptions of environmental risks ENV.F.3/ETU/2014/0037, EU

HSE 2001 *Reducing Risks, Protecting People* https://www.hse.gov.uk/risk/theory/r2p2.pdf

HSE 2021 *Workplace fatal injuries in Great Britain* https://hse.gov.uk/statistics/pdf/fatalinjuries.pdf

OECD 2003 *Emerging Systemic Risks. Final Report to OECD Futures Project OECD, Paris*

ORR 2021 https://dataportal.orr.gov.uk/media/1999/rail-safety-2020-2021.pdf

Renn O in Bouder F, Slavin D and Ragnar E 2009 *The Tolerability of Risk* (London: Earthscan)

Ringgren H and Fatalism I 1967 Islamic fatlism *Scr. Inst. Donneriani Aboensis* **2** 52

Stewart N, Chater N and Brown G D A 2006 Decision by sampling *Cogn. Psychol.* **53** 1–26

Wachinger G, Renn O, Begg C and Kuhlicke C 2013 'The risk perception paradox—implications for governance and communication of natural hazards' *Risk Anal.* **33** 1049–1065

Williams D R 1998 *What Is Safe* (London: The Royal Society of Chemistry)

IOP Publishing

Challenges in Risk Analysis for Science and Engineering
Development of a common language
Tracey J Temple and Melissa K Ladyman

Chapter 3

Storage of military ammunition and explosives—risk evaluation and hazard management

D Skelly

The need to store military ammunition and explosives for any period of time brings with it a number of related risks and hazards. Not only does the presence of explosives pose inherent dangers, those explosives are also vulnerable to external influences unless correctly protected. Such risks must be recognised and reduced as far as practicable. Even where risks are minimised there always remains the possibility of an accident occurring; in large-scale storage the consequences of an explosives accident can be catastrophic unless appropriate hazard mitigation measures are implemented. This chapter presents an overview of risks associated with ammunition and explosives storage, potential hazards from an unintended explosion, and means of both alleviating those risks and reducing the impact of an accident. A series of case studies are included, examining recent explosives storage incidents. The assumptions underpinning current risk evaluation and hazard management practices are explored, and opportunities for harmonisation and standardisation are highlighted.

3.1 Introduction

The manufacture, storage, handling, and use of explosives are inherently dangerous undertakings. Explosives are metastable at best and can be sensitive to accidental initiation from a variety of stimuli, qualities which present unique challenges in terms of managing the risks they pose.

Explosives and their characteristics are thoroughly defined in existing literature (Akhavan 2011, Klapötke 2019) and so will not be repeated here. In terms of storage safety the most relevant properties of explosives are their capacity to produce blast, heat, and fragmentation (be that primary fragmentation from a munition case, or

doi:10.1088/978-0-7503-3643-7ch3

© IOP Publishing Ltd 2022

secondary fragmentation from the surrounding environment). Their chemical stability—or lack thereof—and vulnerability to external insult are also key attributes.

For the purposes of this chapter 'explosives' is used as a catch-all term for any high-energy material, including propellants (which have the capacity to explode under certain conditions), 'ammunition' or 'munition' is used to describe an article, designed for a military purpose, containing one or more explosive material(s). For simplicity, military pyrotechnics such as smoke grenades and illuminating flares will also be covered by these terms. The abbreviation 'AE' will be used to collectively describe both ammunition and explosives.

3.1.1 Scope

Large-scale AE storage—in the order of tens, hundreds, or thousands of tonnes of net explosive mass (NEM)—is widely explored in openly-available literature. Although in theory such locations should carry the lowest degree of associated risk when well-managed, they nonetheless have the capacity to generate the most hazardous consequences.

It is acknowledged there are many more instances where military explosives might require unique storage arrangements, storage on a smaller scale, or on a less permanent basis. Some examples include:

1. Holdings of minor quantities of low risk, low consequence AE such as small arms ammunition for a guard post.
2. Storage of 'ready use' ammunition like that held aboard vehicles, ships, or aircraft when laden for combat.
3. Circumstances where ammunition is subject to routine maintenance or inspection and it is necessary for it to be stored on a temporary basis in a processing building.
4. Nuclear and chemical weapons storage.

The nature of a given item of military AE—and its purpose—often define the requirements of that AE's storage solution, along with the appetite to accept the risks presented by that AE. Tolerance of risk may be potentially much greater during combat operations than peace time, for example. As such, there are innumerable possible storage permutations and so discussing every aspect of military AE storage is beyond the scope of a single chapter. Instead this chapter will focus almost exclusively on Risk Management in permanent storage facilities holding large quantities of conventional military AE over long periods. Temporary field-expedient AE storage, designed only for use over a few months, will also be considered where relevant.

3.1.2 Risk Assessment

Risk can be quantified as a product of the likelihood of a hazard being realised and its consequences:

$$\text{Risk} = \text{Likelihood} \times \text{Impact}$$

The process of Risk Assessment usually involves five core elements:

1. Gain an appreciation of existing or potential hazards for the proposed activity.
2. Determine the likelihood of those hazards being realised.
3. Evaluate the impact of the hazard's effects on vulnerable personnel or material.
4. Implement measures to reduce or eliminate the hazards, and/or mitigate their effects.
5. Decide if the resulting degree of risk is of an acceptable level for the activity to proceed.

Applying these elements to the storage of AE is a complicated business requiring expert input on a broad range of topics (UK Defence Safety Authority (DSA) 2019a). Therefore it is rarely, if ever, possible to produce a single Risk Assessment document for a large-scale storage scenario. Instead, many nations or international organisations have published regulations and guidelines on hazard reduction and mitigation that can be applied to the majority of AE storage situations. Although not necessarily explicit in stating their intended purpose, such publications often incorporate aspects of the Risk Assessment process.

The three most prominent AE storage publications are the United Nations (UN) Office for Disarmament Affairs (UNODA) *International Ammunition Technical Guidelines (IATG)* (UNODA 2021), the North Atlantic Treaty Organisation (NATO) *Allied Ammunition Storage and Transport Publication* 1 (AASTP-1) (NATO 2015a), and the UK *Defence Code of Practice for In-Service and Operational Safety Management of Ordnance, Munitions and Explosives* (UK DSA 2020); there is a considerable degree of crossover between them since their authors are often using the same primary sources.

An examination of the evidence and assumptions underpinning these publications, along with their practical applications, forms the basis of this chapter.

3.2 Ammunition and explosives hazards

Examining explosives Risk Assessment and management processes first requires some consideration of the hazards that are both presented to, and generated by, AE under normal storage conditions.

3.2.1 Qualifying tests

Various national and international agencies have developed test standards to determine if a given material qualifies as an explosive and, if that material is deemed to be an explosive, the characteristics of that material with regards to its sensitivity, sensitiveness, and/or response to a number of stimuli. These test standards go some way towards determining the risks posed by a particular explosive.

Small-scale tests are used in the earlier stages of qualifying potential energetic materials. These tests typically evaluate the reaction of milligram quantities of the material, in liquid or powdered form, to the following:

1. Impact—A solid mass of a known weight is dropped onto the material, placed on an anvil, from a pre-determined height to see if there is an explosive response. If a reaction occurs the resulting acoustic event or gas production can be measured. The Rotter test, *Bundesanstalt fir Materialprìfung* (BAM) Impact test, or other Fallhammer tests are typical examples.
2. Friction—The material under test is placed on a plate made from a suitably hard substance and an equally hard peg or wheel is used to grind the material against the plate. The response (if any) is then compared to a descriptive scale of possible outcomes. While manual friction test criteria exist, for enhanced safety this can be conducted remotely using BAM Friction or Rotary Friction machines.
3. Ease of Ignition—The chosen material is placed in a test tube and exposed to a short burst of flame from the end of a length of burning fuze. The response is matched to one of a list of possible outcomes.
4. Temperature of Ignition (ToI)—The sample material is placed in a heater or other small oven and gradually heated until it responds. The temperature at which a response occurs is recorded, along with the level of the response that is graded according to a descriptive scale. (Note: A defining characteristic of a primary explosive is a material that detonates at its ToI. Primary explosives tend to be sensitive, whereas secondary explosives are comparatively less sensitive.)
5. Shock—The susceptibility of an acceptor material to explosive shock and subsequent sympathetic detonation is measured by using a high explosive donor charge to try and stimulate a response, then measuring the outcome by observing the resulting damage to a metal witness plate. These tests can be conducted at different scales; an example intermediate scale test arrangement is shown at figure 3.1. This type of test is sometimes called a 'gap' test as the donor and acceptor charges are usually separated by an inert material such as water or plastic.
6. Electrostatic Discharge (ESD)—To test the vulnerability of an explosive to ESD the material is placed on or between a pair of electrical conductors and an electrical spark is applied (the exact criteria, in terms of apparatus, voltage, energy, current etc varies depending on the nation conducting the test). This simulates an ESD from, for example, a person or faulty electrical equipment coming into contact with the exposed material. As with many of the other tests, the response is graded against a list of descriptors.

Further small-scale tests, using several grams of material, are then conducted to quantify the compatibility, stability, and both the mechanical and thermal properties of the explosive. Various techniques are used including, but not limited to, thermogravimetric analysis (TGA) differential scanning calorimetry (DSC), along with vacuum and compression tests. Eventually thermal stability and drop tests are conducted on larger scales—that is, all-up rounds of ammunition or AE packaged for transport—to determine suitability for transport and acceptance for military use.

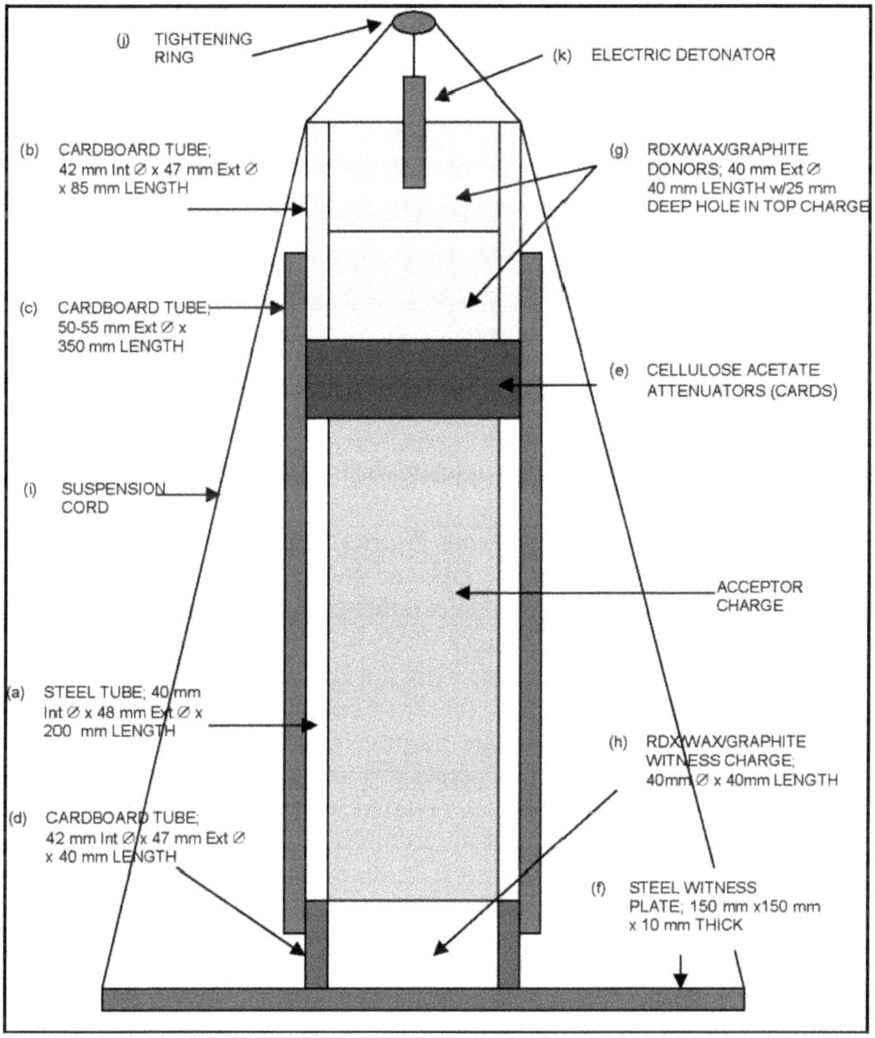

Figure 3.1. Intermediate-scale gap test setup (NATO 2009a). Reproduced courtesy of NATO.

Details of the explosives tests prescribed by the UN are found in Part 1 of their *Manual of Tests and Criteria* (UN 2019a). NATO requirements are published in *Allied Ordnance Publication 7 (AOP-7)* (NATO 2003) along with a variety of supporting Standardisation Agreements (STANAGs), which are listed in *STANAG 4170* (NATO 2008).

Despite the existence of the UN and NATO publications the specific methodology of a particular test can vary between country. This can give rise to different results even when the same material is tested (Williamson *et al* 2016), a problem which is exacerbated by the subjectivity allowed when the individual conducting a test must decide the result. For example, choosing between 'decomposition' and

Table 3.1. Result descriptors from small-scale ToI tests (NATO 2015b).

Small scale temperature of ignition (ToI) test	
Test 1. Test tube and heating block test: result descriptors	*Test 2.* Wood's metal bath test: result descriptors
No reaction	Ignition
Decomposition	Decomposition
Ignition	Deflagration
Explosion	Explosion
Detonation	

'ignition' of a material during a friction test depends how much smoke was observed and the relative size of any smear left on the friction surface (NATO 2009b).

Both the UN and NATO have attempted to harmonise many of the tests for their participating member states. Despite this there are a number of areas where disparities still exist (European Defence Agency 2010); one of the fundamental barriers to harmonisation is that the UN and NATO test publications are in the form of optional guidelines or agreements rather than strictly enforced regulations. Nor do they cover all tests (for instance, there is no standardised small-scale ESD test). It may be that total harmonisation will not be possible in the near future since there continues to be disagreement across the scientific community with regards to which tests are the best for a particular purpose (Gresshoff 2018).

The potential for differing test outcomes has implications for explosives storage safety. A scenario could arise where a material which ought to be classified as a sensitive, primary explosive is instead classified as a secondary explosive. An example of this is illustrated at table 3.1. Here, only one of the two tests must be completed; a material can be qualified differently depending on the (subjective) perception of the outcome—the various descriptors are not defined and so are open to interpretation. If a material 'explodes' in Test 1 then it would be classed as a secondary explosive, whereas the same reaction in Test 2 would make it a primary explosive.

Primary and secondary explosives are treated very differently in storage, with primary explosives being afforded considerably more protection from potential sources of insult; they are also often physically separated from other AE.

In general, countries participating in UN or NATO will accept the qualification results produced by other participating states since, in each case, there is an expected mandatory minimum data set which must be produced. Nonetheless, this acceptance must be tempered by an awareness that the final result may have been reached using different methods.

3.2.2 Hazards to ammunition and explosives

AE are susceptible to accidental initiation by a number of external environmental stimuli. Of these it is excessive heat—in particular fire—which is the most easily realised environmental hazard and a significantly frequent root cause of explosives storage accidents (Small Arms Survey 2014).

Nearly all explosives, by nature or design, can be initiated by an increase in temperature. As with most chemical reactions, transferring thermal energy into an explosive will bring its internal energy closer to its activation energy; once this point is reached the explosive begins to decompose resulting in burning or detonation.

The timescale over which the temperature rise occurs can dictate the violence of the reaction. A rapid temperature increase, as in an especially intense fire, invariably leads to deflagration or detonation even in low explosive compositions, potentially after only a matter minutes, seconds, or less. The fuel fire aboard the USS Forrestal is often cited as an example of this type of behaviour (Munitions Safety Information Analysis Centre (MSIAC) 2019).

Pressure is also a factor in the violence of an explosion. An explosive material which initiates under confinement will tend towards a much faster rise in temperature, which in turn leads to an increase in pressure, further raising the temperature and so on until the containment fails. This effect is called thermal runaway, and is the operating principle of improvised explosive devices such as pipe bombs.

Slower heating of an explosive from exposure to fire for a long period at distance (for example a blaze in an adjacent storage compartment) can ultimately result in a similar outcome to faster temperatures increases. This phenomenon is referred to as 'cook-off' and also occurs on a smaller scale inside weapon systems when a round of ammunition is left for too long in a hot chamber. In this instance the result can be the unintended firing of the weapon. To an extent the likelihood of weapon cook-off can be reduced through weapon design (by inclusion of a means of cooling the chamber between rounds) or by user training aimed at regulating rates of fire and preventing ammunition being left in a weapon unnecessarily.

In the longer term—over days, months or years—smaller temperature increases, below the activation energy threshold, can still lead to dangers in storage. The explosive 2,4,6-trinitrotoluene (TNT) is present in many older munitions and is still used to a lesser degree in modern ammunition. The melting temperature of TNT is around 80 °C and this can easily be reached in hot climates when ammunition is stored or transported in ISO shipping containers, as is an often the case during military deployments. As well as melting, the TNT will inevitably expand; in ammunition items like artillery shells this can lead to TNT exuding through joints and screw threads. When the TNT later cools it can leave behind sensitive dinitrotoluene (DNT) crystals which render the ammunition much more liable to accidental detonation if those crystals are subject to shock, for example from being dropped or fired.

Although not necessarily a problem during storage, the relatively low melting point of TNT (and other chemicals used in munitions; white phosphorous (WP)—with its 44 °C melting point—being one example) can lead to issues for the user if associated ammunition items are not stored appropriately. WP-filled mortar bombs are often stored on their side, however, in hot climates the combination of molten fill and the effect of gravity can result in a mass imbalance inside such bombs when the fill changes position. If fired, the imbalance will generate instability in flight that could lead to missing the intended target or, in the worst case, hitting friendly forces or non-combatants.

Prolonged exposure to heat can also accelerate the decomposition of nitrocellulose-based propellants. While such propellants naturally decompose gradually over time the process occurs more quickly with increased temperature since the extra thermal energy catalyses the evolution of nitrous oxides and the subsequent production of nitric acid, ultimately resulting in autoignition. A significant proportion of ammunition storage accidents are attributed to propellant autoignition, especially in developing countries or those holding ageing stockpiles (Carapic *et al* 2018).

Diurnal temperature cycling, or cycling caused by other wide variations in ambient temperature (for example, an aircraft carrying a munition from ground level to high altitude and back), can also be problematic. Rapid thermal expansion and contraction of energetic compositions can result in them becoming cracked and brittle, leading to sensitisation of those compositions to mechanical shock. This increases the risk of accidental initiation through poor handling practices, and once again poses a risk to the user where other sudden accelerations (like a rocket launching) could cause a premature explosion. Thermal cycling of non-energetic components can also render ammunition more hazardous, as when rubbers and plastics are used to provide environmental seals, particularly in complex ammunition items such as guided missiles. Failure of these seals can allow dust and moisture into the munition leading to damage of mechanical and electronic systems, as well as potential detrimental effects to any energetic fillings.

Electrical effects—be they from lightning, static, or mains supplies—are a hazard to AE for a variety of reasons. Foremost among these is the capacity for stray electric currents to produce heat and therefore start a fire or ignite an energetic material. Faulty, poorly maintained, or inappropriately used electrical equipment can be to blame, along with lightning strikes. For ammunition in the open this may occur as the result of a direct strike, but can also affect explosives stored indoors. Lightning which strikes a building has the potential to arc between the walls or roof into explosives inside that building. Alternatively, lightning can also cause combustion in flammable materials which may be co-located with AE.

Lightning strikes also produce high pressure blast waves in their immediate vicinity (Blumenthal 2016). However, the possibility of these shocks leading to sympathetic detonation in explosives has not been studied, perhaps because of the difficulties of disassociating the shock effect of a lightning strike from its electrical and heating effects.

Some explosives are inherently sensitive to electrical stimuli, as determined by the ESD tests detailed previously, and so must be protected from undesirable electrical inputs. Important electrical components in ammunition items—like control circuits in a guided missile—must be similarly protected or they may malfunction. There exist ammunition items that are sensitive to electrical initiation by design, as is the case for electrical detonators and other so-called electro-explosive devices (EEDs). Exposure of an EED to an unwanted current could result in that item functioning.

An electrical current need not be directly applied to be hazardous—in environments of powerful, frequent or close-proximity radiofrequency (RF) radiation it is possible to induce a current inside an EED through electromagnetic (EM) induction, leading to the same outcome as deliberate initiation.

Unless specifically designed or engineered to resist moisture most explosives and munitions do not react well to exposure to water, whether in large amounts over a short period (as in a flood) or smaller quantities over a prolonged period (condensation, for instance, building up during long-term storage in an environment where there are regular temperature variations). Low explosives, such as gunpowder, and pyrotechnic compositions are particularly vulnerable to moisture ingress, with some being soluble in water. They can be all but rendered inert if heavily wetted—a property which can occasionally be useful, as when guncotton requires transport over significant distances. The performance of explosives containing ammonium nitrate, which is hygroscopic, can be severely impaired by the absorption of water. However, it is worth noting that modern military explosives do not generally contain ammonium nitrate. Water can also detrimentally affect non-energetic components in munitions. Ferrous components can rust, risking mechanical failure, and circuitry can fail through either poor connections or short circuits.

Sand, grit, and other fine particulates can be an additional problem; explosives such as 1,3,5-trinitro-1,3,5-triazacyclohexane (RDX) can be sensitised if they become contaminated with these kinds of material. Mechanical components can also become fouled if fine material is able to accumulate inside munitions.

While most AE should be designed to withstand relatively robust handling by the end user, impact is nonetheless a potential hazard that must be considered. It is common for forklifts, cranes, and other mechanical handling equipment to be used to move palletised ammunition; a moment of inattention can easily result in an ammunition package being pierced by a fork or dropped from considerable height. In more extreme circumstances the collapse of a building or aircraft crash will also likely result in catastrophic levels of impact. For storage buildings situated too close to one another the possibility of an explosion in one building producing a crater which undermines its neighbour(s), causing it/them to collapse, is a realistic one. Close proximity storage means ground shock must also be considered lest it lead to sympathetic detonation and damage to building foundations.

Although AE are never entirely safe, and many of the hazards described above can be realised spontaneously as a result of circumstances beyond anyone's control, usually human error or lack of foresight is the ultimate cause of an explosives accident. Forgetting that AE are dangerous, through ignorance and complacency, are factors which must be addressed if the risks from explosives storage are to be as low as reasonably practical.

3.2.3 Hazards from ammunition and explosives

Ideally, every AE item would be stored completely independently of all other AE and a long way from any people; in reality this is rarely possible for practical and economic reasons. It is much easier to manage and control an AE stockpile in the large, purpose-built storage facilities and areas that are the focus of this chapter.

However, bulk storage brings with it the risk of a given item of AE accidentally initiating and the subsequent effect(s) presenting a hazard to the remainder of the stored AE, as well as anyone who is nearby.

A significant proportion of military ammunition contains high explosives and is designed to produce blast, high-velocity (often metal) primary fragmentation, or both. Blasts can also generate secondary fragmentation should they occur inside buildings or vehicles made from anything but the lightest of materials. In some cases a particularly focused and destructive effect is desired, as in munitions which use the Munroe effect to produce a shaped-charge jet (SCJ). High-energy propellants—used in propelling charges for shells or in missile and rocket motors—create intense thermal effects if ignited and, under confinement, can readily contribute to the conditions necessary for detonation to occur. Depending on their proximity and intensity all of the above effects can have a severely detrimental influence on other AE. In the best case damaged ammunition can be rendered unusable, with the associated economic loss and/or reduction in operational capability. In the worst case surrounding AE can also detonate or begin to burn, leading to a 'domino effect' of exploding munitions with potentially catastrophic consequences.

In addition to munitions that contain especially energetic materials stockpiles typically contain a great many military pyrotechnics which, during normal use, are not especially hazardous. However, in storage these items can become problematic. For example, a smoke grenade functioning in the open will produce a relatively harmless cloud of smoke that will disperse in a matter of minutes. But the same grenade functioning inside a small storage room, where the smoke cannot dissipate, will produce a toxic environment. Likewise, a flare used correctly outdoors will produce heat and light but little else; the same flare igniting inside a building can readily start a fire.

The need for a universally- and easily-understood system of indicating the potential hazards posed by AE during both storage and transport resulted in the UN defining a series of Hazard Divisions (HD), with each HD describing the likely effects of ammunition or explosives which fall into that HD. There are six HD in total, and two of those are further divided into subcategories specifically for storage (by NATO and some non-NATO countries, rather than the UN); these are known as Storage sub-Divisions, or SsD. Table 3.2 summarises the characteristics of each HD and SsD.

Those nations which subscribe to the UN HD system are expected to use signs and labels on ammunition packages and storage buildings to notify personnel, mainly emergency services responders, of the hazards they might encounter should AE become involved in an emergency situation, allowing those responders to react in an informed manner. HD and SsD are also used to determine quantity distances (QD)—discussed further at section 3.3.2, below—and so the allocation of the appropriate HD or SsD can have significant implications in terms of Risk Management and hazard mitigation.

Assessing the HD of a particular AE item is achieved through a series of tests prescribed by the UN. Whereas qualification tests are conducted on relatively small scales, for the most part HD tests are carried out on complete AE articles. These tests are described in appendix A. It is worth noting that the most UN tests require that the substance or article under test is presented as it would be for transport.

Table 3.2. Summary of HD and SsD (UK DSA 2019b; UN 2019b).

HD/SsD	Hazardous effect(s)
1.1	Mass explosion; virtually instantaneous explosion of an entire transport load or stack of AE.
1.2	Projections but no mass explosion.
1.2.1	Large fragments projected over an extended range.
1.2.2	Smaller fragments projected over a limited range.
1.2.3	Insensitive Munitions (IM)—effects are as per acceptable IM responses (see figure 3.2)
1.3	Fire, be that considerable radiant heat or progressive burning; minor blast, minor projection (or both). No mass explosion.
1.3.3	Mass fire hazard and considerable thermal radiation.
1.3.4	Sporadic burning.
1.4	No significant hazard. Any hazardous effects are contained within the package or limited to the immediate vicinity.
1.5	Very insensitive substances which could produce a mass explosion.
1.6	Extremely insensitive articles with no mass explosion hazard.

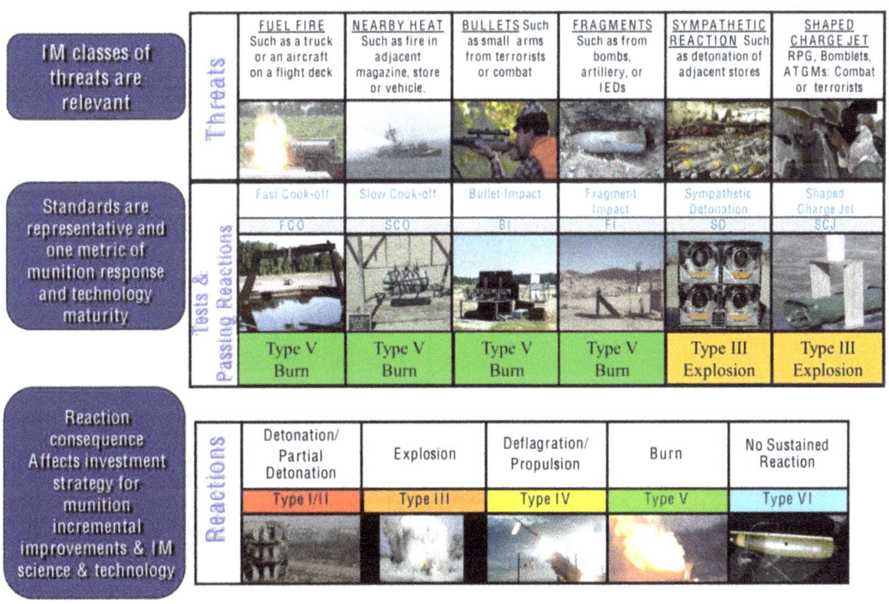

Figure 3.2. IM reactions and test pass criteria (Swierk 2013). Reproduced courtesy of NATO/ Thomas Swierk.

Other configurations, for example an unpackaged article, are not necessarily considered by the UN—however, in NATO due consideration is given to AE in scenarios beyond logistic transport, such as when carried unpackaged in combat conditions (NATO 2009c).

For simplicity it is preferable to store items from different HD separately, but for practical reasons it can be desirable to mix HDs in storage, i.e. to more effectively use the space available in a particular storage building. In such instances the numerically lowest HD of the stored munitions is usually adopted (though there are some exceptions) since this represents the worst case in terms of hazardous effects.

The UN has also produced a series of tests to decide if a material can be accepted into Dangerous Goods Class 1 (Explosives) in the first instance, and while there is some overlap with the qualification test criteria discussed at section 3.2.1, above, the UN tests seek to answer slightly different questions such as whether a material is too unstable for safe transportation.

There is currently a drive among NATO nations to design and procure so-called insensitive munitions (IM). These are ammunition items that are engineered to be less susceptible to accidental initiation and/or react less violently if inadvertently initiated. This is achieved through use of a combination of:

1. Low vulnerability energetic materials
2. Special munition features to prevent a response to threat stimuli or else reduce the violence of a response if one occurs
3. Packaging which provides improved protection from, and/or mitigation of, hazardous effects.

The reaction of an IM article or package to external insult must meet the requirements of figure 3.2.

IM present a number of advantages in terms of safety in storage since they would be expected to fall into HDs 1.5 or 1.6, with an associated reduction in quantity distance.

NATO has devised a series of standardisation tests to determine if a munition may be considered 'insensitive' taking into account not only accidental but also combat stimuli. The benefits are touted as enhanced survivability of logistic and tactical systems, minimisation of risk to personnel, and improved cost effectiveness in storage transport and handling (NATO 2010). However, none of these benefits can be quantified without also classifying IM as belonging to HDs 1.5 and 1.6.

It is therefore curious that the NATO standardisation tests for IM bear a superficial similarity to the UN criteria for placing a munition in HDs 1.5 and 1.6, begging the question of why go to the expense and effort of conducting the NATO tests if there is no measurable benefit without the need to also conduct the UN tests? The answer potentially lies in the differences between NATO and UN tests. Often the NATO tests are more stringent, detailed and realistic. The NATO tests are summarised in table 3.3; also highlighted are the areas where the NATO and UN tests diverge.

Clearly there are opportunities to harmonise several of the NATO and UN tests to avoid duplication of effort when assessing IM. Interestingly, the UN does not include a SCJ test in its HD 1.5 and 1.6 criteria. This seems to be an oversight since it is feasible that an IM in storage could be subject to a SCJ attack if a neighbouring item of ammunition (capable of forming a SCJ) should function accidentally. Alternatively, it may be the case the UN recognises that the majority of

Table 3.3. NATO IM standardisation tests.

STANAG	Test name	Description of test	UN equivalent test	Significant differences
4240	External fire	Article is placed in a bonfire that must reach 800 °C.	7g	The NATO test specifies the temperature rise time of the bonfire whereas the UN test does not. Liquid fuel must be used in the NATO test, wood can be used in the UN test.
4241	Bullet impact	3 × 12.7 mm projectile fired at the article.	7j	Exact target areas on the test article differ between the UN and NATO.
4382	Slow heating	Article is placed in an oven and temperature slowly raised.	7h	The UN test specifies a temperature range for the test; the NATO does not. Thermal preconditioning is mandatory for the NATO test and is optional for the UN test.
4396	Sympathetic reaction	One packaged AE article is initiated in a stack of the same packaged articles.	7k	The UN test uses an initiator appropriate to the article. The NATO test uses the credible threat which will likely lead to the most violent reaction.
4496	Fragment impact	A high-velocity simulated fragment is fired at the article.	7l	Unlike the UN test, the NATO test does not require the most sensitive component to be tested if a fragment threat to that component is not credible.
4526	Shaped charge jet (SCJ)	A SCJ is fired into the energetic components within the article.	N/A	No UN test for this reaction.

munitions—even those which show a good set of IM responses to other stimuli—cannot be expected to pass a SCJ test and therefore do not require a test that is a foregone conclusion.

While the safety benefits of IM during storage and use are clear, they pose an increased risk in terms of disposal. Traditional methods of Explosive Ordnance

Disposal (EOD) rely on sympathetically detonating or burning unsafe munitions. However, if future munitions are specifically designed to resist these stimuli then other, increasingly energetic or complicated methods must be employed, placing EOD personnel in potentially greater danger from their own render safe procedures than from the target munition (Pope and Baker 2018). Sadly, disposal is often close to the bottom of the list of priorities for IM designers and so it is unlikely that this issue will be addressed in the near future.

3.3 Risk reduction and hazard mitigation

3.3.1 Determining risk

The reasons for an unplanned explosives incident often fall into one of four categories (Greene *et al* 2005):

1. Physical or chemical deterioration of the ammunition or explosives.
2. Negligence, i.e. unsafe storage and transport practices, or poorly maintained infrastructure.
3. Accident, such as mishandling, non-malicious tampering, or a vehicle crash.
4. Deliberate attack (including sabotage).

Reducing the likelihood of an explosives accident is therefore a matter of preventing one or more of the above situations arising. UN, NATO, and various national regulations exist to address exactly this need. Correct application of these regulations facilitates the identification of risk and, where a risk cannot be eliminated, provides methods to minimise that risk.

Monitoring the physical and chemical condition a munition or explosive through its service life requires several frameworks and processes to be in place, often at a national level. This is especially important where large stockpiles—potentially consisting of many millions of ammunition items—must be managed across varying climatic environments.

Tracking the storage history of a single item of ammunition, or batch of items, can be particularly important. A munition which spends its entire service life inside a temperature-controlled storage building is likely to be in a very different condition to one which has been transported long distances by sea, road or air, and been stored in desert, arctic, tropical and maritime climates.

Determining the effects of different storage conditions on the life of a munition requires implementation of one or more surveillance programmes. Such programmes will usually aim to check the condition of a munition, or batch, through representative sampling at key stages in its life or when it experiences a significant change in environment. Both non-destructive techniques (visual inspections, electrical testing, and radiography for example) and destructive techniques (such as munition teardown and subsequent analysis of physical and chemical integrity) are available, along with monitoring of reliability during training and operations (Carapic *et al* 2018).

Since fire resulting from propellant autoignition is a frequent cause of ammunition accidents (Small Arms Survey 2014) regular sampling of stabiliser content is critical, especially for ageing stockpiles or those stored for long periods in hot conditions. During acceptance tests the suitability of propellant for long-term storage (10 years or more) can be assessed by artificially ageing the propellent (NATO 2006, 2007) and methods for sampling prior to transport are also available (UN 2019a, appendix 10).

Ensuring AE are physically protected from the environment and unwanted stimuli is a key part of storage safety, which can be achieved on several levels. While IM are designed with risk reduction in mind, many non-IM items of ammunition also incorporate intrinsic tried-and-tested safety measures. Examples include out-of-line fuzing, and the inclusion of mechanisms that prevent arming or firing unless a very specific set of criteria are met; for instance an artillery shell fuze might contain spring-loaded pins to hold a detonator away from a main charge and which cannot move unless many thousands of axial and radial *g* are experienced by the parent shell. Modern or complex munitions also tend to incorporate electronic safety features, such as timers and accelerometers, to prevent unintended initiation.

Packaging plays an important role in ammunition safety during storage and handling. As a minimum, packages should protect the ammunition contained therein from environmental exposure along with the wear and tear of routine handling, while also providing a means to move the ammunition safely (handles if to be moved by hand, latching points if the package will be craned, and so on). Well-designed packaging will protect from robust handling and resist chemical, biological and possibly even radiological attack. In the case of big, valuable, or complicated munitions desiccant packs and temperature data loggers may also be added to the package to protect from condensation and aid with surveillance, respectively.

Use of metallic packaging provides protection from RF induction through the 'Faraday's cage' principle, though the use of packages made entirely from metal becomes impractical for larger items due to weight implications. In these instances, surrounding the munition in a metallic mesh (embedded in the walls of a lightweight package made from plastic or fibreglass, for example), or hardening of individual sensitive components within the munition against RF induction, can be a more viable solution.

A means of improving the response of an IM to threat stimuli is inclusion of packaging features which will limit or mitigate that response. These can take the form of additional barriers inside the package to reduce the kinetic energy of projectile attack, or venting panels that allow pressure build-up to be released under control during a fire. The simple yet effective approach of rearranging the geometry of a package, and the spatial relationship of the associated ammunition to surrounding munitions, is also enough to enhance the IM qualities of a given item. The Swedish NLAW is an example of this approach; the rocket has a downward-facing SCJ warhead but in storage the design of its container is such that the rocket is canted so it will not directly fire its main charge at a neighbouring rocket.

At times it may be necessary to store ammunition outdoors for short periods, however, it is more common for munitions and explosive to be stored indoors. Ideally this takes place in purpose-built ammunition storage buildings, though adequate storage conditions can be realised using re-purposed structures, even ones as basic as ISO shipping containers. There are numerous benefits to indoor storage in terms security as well as protection from the environment and external stimuli.

Much greater control can be exercised over the environment experienced by stored ammunition when it is inside a building. To be deemed suitable a storage structure should be weatherproof, have fire detection and firefighting measures installed, along with a lightning protection system. Any electrical fittings should be designed to be compatible with the munitions that are stored. In the most sophisticated storage buildings it is possible to control and monitor the temperature in real time. Reinforced structures can be designed to resist nearby external explosions; the most heavily hardened can withstand deliberate direct attack and also contain internal explosions. With greater protection, however, comes greater cost and there is a balance to be struck between affordability and the value of the ammunition which is being stored.

For continued effectiveness it is vital to have a robust inspection and maintenance regime for explosives storage buildings since, if they are allowed to decay significantly, they can become more of a hazard to ammunition than a help. Management of the areas surrounding a storage building is an additional key consideration. Other dangerous goods and flammable materials must be kept away from ammunition stores to prevent any mutually hazardous situations arising. Vegetation in the immediate vicinity of a store must be removed or kept short to act as a firebreak, especially in climates where wildfires are prevalent. Vermin must be eliminated to protect both the ammunition and the infrastructure from damage caused by burrowing or gnawing. Aircraft must be diverted around large storage areas to avoid the possibility of a crash or emergency landing having an impact on a store. In coastal or riverine regions the provision of flood defence is a must. The list goes on.

To eliminate negligence comprehensive training must be given to personnel working in an explosives storage area, regardless of their role. If correctly delivered good training can greatly reduce the potential for accidents—errors in handling and unsafe practices resulting from ignorance ought to become rarities. Simple measures, such as restricting the presence of 'contraband' items in an explosives area, should be considered the norm.

Appropriate training must be supported by a culture of safety awareness and continuous improvement. As in any high-risk industry, enabling and encouraging good working practices is a management function which cannot be overlooked, and any failings must be addressed as soon as they arise. National and international organisations have produced competency frameworks which align training standards with levels of qualification; these frameworks often have a foundation in health and safety law. High levels of training can, however, be both expensive and time consuming to complete and personnel who reach expert degrees of competence can be highly sought after in the competitive international job market.

Security of ammunition in storage and transport cannot be overlooked, even in peace time. Good security will not only prevent theft and sabotage, it will also deter the curious but uninformed from inadvertently causing an accident. In times of conflict security and protection become even more important since the capacity of any modern fighting force to conduct operations is reliant on the availability of ammunition supplies.

The level of security required will be highly dependent on the value and quantity of the items which are being protected, and the situation in the surrounding area. At the most basic level a simple locked box may be sufficient to protect minor quantities of small arms ammunition, whereas the protection of nuclear weaponry will dictate the use of multiple layers of physical security, a dedicated guard force, remote sensors, and so on (UNODA 2021, volume 9).

If the factors described in this section are taken into account by personnel responsible for ammunition storage then the majority of accidents can be avoided by reducing the inherent associated risks to a practicable minimum. This is clearly beneficial for those working in, or living near, an ammunition storage site. The cost of replacing ammunition which has been destroyed or damaged, along with repairs to buildings which are involved in an accident, can be considerable, and so prevention is also better than a cure in financial terms. Avoiding accidental loss of munition stocks also maximises the availability of ammunition at any given point in time, so that training and operational munition needs can be met in the short-, medium-, and long-terms.

3.3.2 Consequence analysis

Even where the risk of an unplanned explosion has been reduced to the lowest possible degree there will always remain the chance of an accident occurring. Therefore, in addition to risk evaluation, ammunition storage locations must also be subject to consequence analysis examining the likely effects of an explosion. Knowledge of these effects can then be used to dictate what mitigation measures are required to limit the impact of an accident.

The effects of an explosion on people and buildings are discussed in depth in various publications (USA Federal Emergency Management Agency (FEMA) 2003, Weggel 2010, Jorolemon *et al* 2021) but are nonetheless summarised at figures 3.3 and 3.4 for convenience.

The proximity of personnel and structures to explosives stores is a necessary evil. Workers can be more effective if their accommodation is nearby to cut down time lost to commuting. If a 24-h guard force is needed then their presence close to the explosives is essential. It is much simpler to secure and administer multiple storage buildings when they are situated close to one another, and nearby transport links make it much easier to move ammunition to where it is needed. These practicalities mean that, in some circumstances, exposing people and buildings to explosive danger is difficult to avoid entirely.

Certain materials and installations, sited near AE storage locations, can create a mutual hazard. Bulk fuels containers, electricity transmission lines, high power radar, and so on, are not only placed at risk by the close proximity of AE but also present a risk to the same AE. However, as with the presence of people, the co-location of these mutually hazardous material and installations is occasionally unavoidable.

Category	Characteristics	Body Part Affected	Types of Injuries
Primary	Unique to HE, results from the impact of the over-pressurization wave with body surfaces.	Gas filled structures are most susceptible - lungs, GI tract, and middle ear	- Blast lung (pulmonary barotrauma) - TM rupture and middle ear damage - Abdominal hemorrhage and perforation - Globe (eye) rupture - Concussion (TBI without physical signs of head injury)
Secondary	Results from flying debris and bomb fragments	Any body part may be affected	- Penetrating ballistic (fragmentation) or blunt injuries -Eye penetration (can be occult)
Tertiary	Results from individuals being thrown by the blast wind	Any body part may be affected	- Fracture and traumatic amputation - Closed and open brain injury
Quaternary	- All explosion-related injuries, illnesses, or diseases not due to primary, secondary, or tertiary mechanisms. - Includes exacerbation or complications of existing conditions.	Any body part may be affected	- Burns (flash, partial, and full thickness) - Crush injuries - Closed and open brain injury - Asthma, COPD, or other breathing problems from dust, smoke, or toxic fumes - Angina - Hyperglycemia, hypertension

Figure 3.3. Explosion-related injuries (USA Centers for Disease Control and Prevention (CDC) 2003).

DAMAGE	INCIDENT OVERPRESSURE (kPa)
Typical Glass Window Damage	1.03-1.52
Minor Damage to Some Buildings	3.45-7.60
Panels of Sheet Metal Buckled	7.58-12.41
Failure of Concrete Blockwork	12.41-19.99
Collapse of Wood Framed Buildings	over 34.47
Serious Damage to Steel Framed Buildings	27.58-48.26
Severe Damage to Reinforced Concrete Structures	41.37-62.05
Probable Total Destruction of Most Buildings	**68.95-82.74**

Figure 3.4. Explosion damage to buildings resulting from peak incident pressure (Samali *et al* 2018, reproduced with permission © Springer).

Nonetheless, there are measures that can be taken to safeguard people and buildings from explosion effects, and to protect AE from mutually hazardous materials and installations. The foremost and simplest of which is putting distance between the two. Blast, fragmentation and thermal effects are all reduced at increasing distances from the source and at long range eventually become inconsequential.

The application of quantity distances (QD) is a commonly used method of calculating the appropriate separation between stored ammunition and any potentially endangered people or buildings. A QD is essentially the distance—which is derived from the net explosive mass (NEM) of AE at a given site—at which the hazardous effect(s) from the accidental initiation of the contents of that site are acceptably low. It is important to note that a QD is not a safety distance; at a given QD there still exists a risk of hazard exposure, however small. For instance, current regulations for the distance between an ammunition store (located above ground) containing HD 1.1 munitions and the closest inhabited building are based on the acceptance of 5 kPa peak side-on blast overpressure (enough to damage windows, door frames, and chimneys; for some building types partial collapse may occur) and debris which produces one lethal fragment, with kinetic energy of greater than 80 J/56 m^2 (UK DSA 2019c).

True safety distances, especially when it comes to fragmentation throw, are often so large as to be prohibitive. Safety distances would only usually be applied in circumstances when explosives will be deliberately initiated, such as a demolitions area or live firing impact area; since the likelihood of an explosive event occurring in such areas is guaranteed the associated mitigation measures must be of proportionately greater effectiveness.

QD for different HD are most pertinent when applied to above ground storage where blast, fragment, and thermal effects are all able to develop freely. In underground storage—defined as storage where there is sufficient overhead cover that an explosion within the storage chamber will not cause failure of the cover (UK DSA 2019d)—it is expected that all HD 1.2 and 1.4 effects would be completely contained, leaving only HD 1.1 and 1.3 effects to consider. For HD 1.1 the predominant effects dictating QD are ground shock and the throw of crater debris, along with directional blast out through entrances (these, tending to be the weakest part of an underground store, are expected to fail before the rest of the store and so will act as routes for pressure relief). Venting of HD 1.3 thermal effects through entrances is also a potential hazard, and the volume of a given storage chamber must be compared to the NEM of HD 1.3 within that chamber since heavy confinement can lead to rapid pressure rises during a fire, which in turn may cause large quantities of HD 1.3 material to behave like HD 1.1 materials.

Work to determine QD began in the 1930s and has been ongoing ever since. QD have been added to and revised through analysis of wartime building damage data, practical small- and large-scale trials, computer modelling, and knowledge exchanged between NATO members and other countries. A summary of the most influential trials can be found at appendix B.

QD for HD 1.1 are fundamentally based on the principle of cube root scaling. Scaled distance (Z, with units of m $kg^{-1/3}$) for peak blast overpressure in free air, can be calculated as:

$Z = R/W^{1/3}$ where R is range in metres (from a spherical or hemispherical charge) and W is the TNT-equivalent charge mass in kilograms (Hopkinson 1915, Cranz 1925).

In QD calculations R is replaced by D (for distance), W becomes Q (for quantity) and Z becomes a constant of proportionality which depends on the peak incident pressure accepted for each category of exposed building or personnel. Some commonly used QD formulae, along with the reasoning behind their use, are summarised in table 3.4.

HD 1.2 QD are dependent on a combination of logarithmic calculations and measurements of fragment throw during practical trials. For HD 1.3 the acceptable levels of thermal radiation density at a given distance are used to

Table 3.4. HD 1.1 QD constants of proportionality (UK DSA 2019e).

QD	Exposed buildings/personnel at this QD	Expected damage/injury
$44.4Q^{1/3}$	– Large buildings of vulnerable construction. – Critical infrastructure. – Venues used for mass gatherings.	– Superficial damage to normal structures. – Vulnerable structures may collapse. – 50%+ of large glass panels facing the blast will break. – Direct injuries unlikely; injuries will result from glass or debris from vulnerable structures.
$22.2Q^{1/3}$	– Normal residential buildings of brick construction. – High volume transport links.	– Normal buildings will suffer damage to windows, doors and chimneys. Partial collapse possible. – Fatalities unlikely; injuries may arise from secondary debris such as glass.
$14.8Q^{1/3}$	– Medium volume transport links. – Outdoor public spaces (i.e. sports fields, car parks).	– Normal buildings will suffer damage requiring some repair. – Injuries from primary fragmentation are possible for people in the open.
$9.6Q^{1/3}$	– Military training areas. – Outdoor spaces used by military personnel.	– Significant structural damage to normal buildings. – Cars, aircraft and ships will suffer minor damage. – Temporary hearing loss, injury from fragments, debris and translation.
$8Q^{1/3}$	– AE processing buildings. – AE packing/transit storage areas.	– Significant structural damage to normal buildings. – Serious injuries and fatalities likely.
$3.6Q^{1/3}$	– Temporary, AE-laden vehicle holding areas (barriers must be present).	– Normal buildings could be completely demolished. – Severe injuries or death are expected.
$2.4Q^{1/3}$	– Personnel conducting minor, short-term AE storage and shipping tasks.	– Normal buildings will be demolished. – Fatalities and severe injuries highly likely.

determine QD. HD 1.4 does not attract a QD since the hazardous effects of munitions in this category should be contained to their packaging. The effects of HD 1.5 explosives are considered equivalent to HD 1.1, although the risk of accidental initiation ought to be much lower for items in this HD. HD 1.6 is not well understood because there are so few items that fall into this HD; currently they are treated as either HD 1.1 or 1.3 depending on the particular storage circumstances and the attributes of the munition in question.

Aside from distance, the next most effective method of protection from the hazardous effects of an explosive accident is placing barriers between the likely source of an explosion and the buildings or people that need safeguarding. In situations where space is limited the use of barriers may be the preferred safety option. Barriers can resist or redirect blast and thermal effects, and can absorb or slow fragmentation. However, constructing an entirely explosion-resistant barrier is usually impractical and not cost effective, especially for NEQs in the thousands of kilos (though there are exceptions; the *Klotz* underground blast closure valve being one example, see Weerheijm and Conway 2018; use of this type of barrier was only possible by the very heavy confinement achieved through underground storage). Instead, it is often sufficient for a barrier to arrest the immediate effects of an explosion and thereby prevent simultaneous propagation from one stack or storehouse of explosives to another. This aspect is important to ensure operational capability by avoiding the potential loss of an entire stockpile during an accident. While it may be necessary to accept the possible loss of one storage building, the loss of surrounding buildings or an entire site to a mass accident is less palatable and can have far-reaching consequences.

Effective barriers can be made in a variety of geometries depending on the effects against which they are designed to protect, and can be built from materials such as wood, steel, concrete, earth, sand and even water. Due consideration must be given to the chosen material's ability to resist environmental damage, and whether it will add to the hazard of an explosion by forming secondary fragmentation. Some examples are shown at figure 3.5.

Storage buildings can also incorporate design features to reduce or delay the hazardous effects of an unintended explosion. In the case of buildings built from reinforced concrete the thickness of the walls can match the geometry of barriers made from the same material; the walls effectively act as barriers in their own right with the potential to completely contain any hazardous effects. For an explosion inside a heavily constructed store where total containment is unachievable, deliberate structural weak points, in the form of frangible blow-out panels, can relieve and re-direct blast pressure before the remainder of the structure fails and forms fragmentation or debris. These are best placed in the roof so that blast is directed upwards, the direction in which the effects is least likely to affect any nearby buildings or people. However, the use of deliberate weak points must be weighed against the need for security; additional physical security measures may be needed to prevent a blow-out panel from becoming an easy access route into a storage building.

One or more doors are a necessary feature of a storage building and are usually the weaker part of the structure. As an alternative to the extra cost of including blow-out panels the doors may be designed to resist external blast but fail if loaded by internal blast. A barrier can then be placed in front of the doors—in larger storage areas this barrier can be the rear of another storage building as per the arrangement seen in figure 3.6.

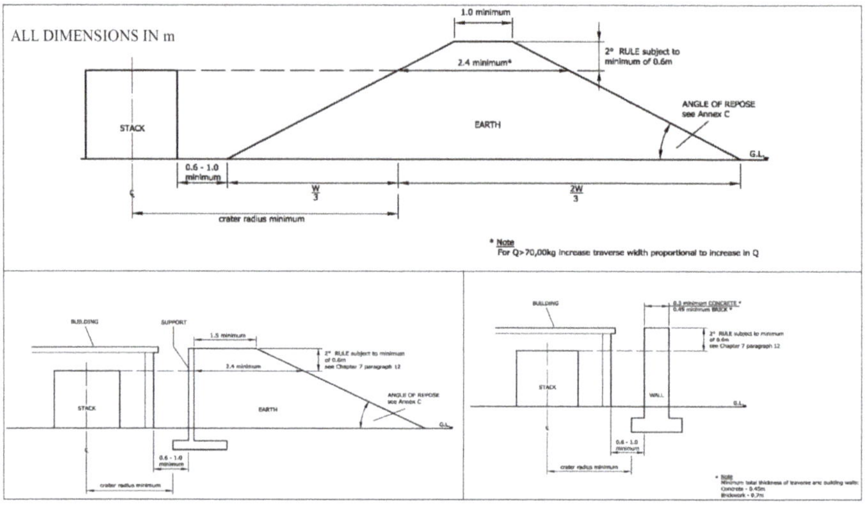

Figure 3.5. Example barrier geometries used in explosives storage areas (UK DSA 2019f).

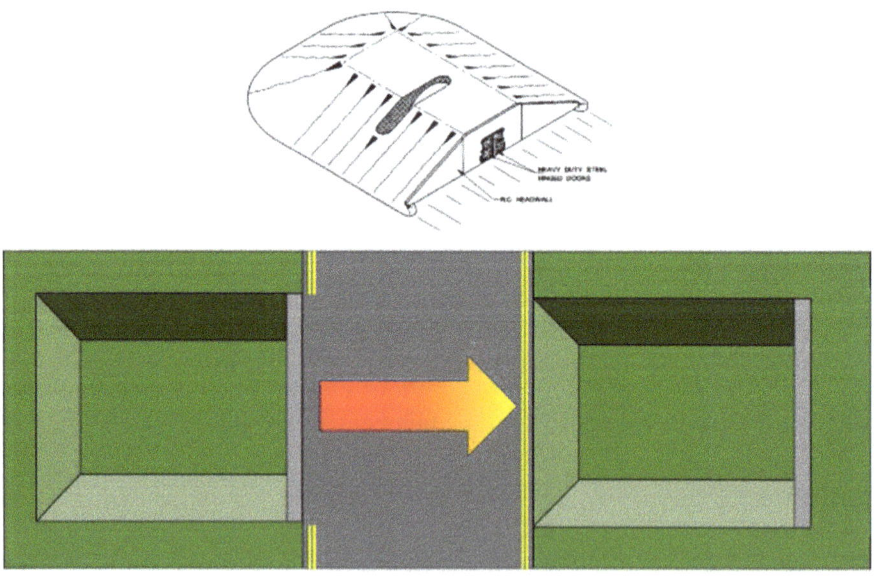

Figure 3.6. Storage building spatial arrangements for hazardous effect reduction. Top: Isometric perspective diagram of an earth-covered ammunition store (UK DSA 2019g). Bottom: Plan view of earth-covered stores (not to scale); explosive effects from the store on the left will be intercepted by the rear of the store on the right. Contains public sector information licensed under the Open Government Licence v3.0. See http://www.nationalarchives.gov.uk/doc/open-government-licence/version/3/ (Accessed: 28 July 2021).

Good storage practices also play a role in reducing the impact of an explosives accident. Where sensitive AE items, like detonators and fuzes, must be stored with their parent munitions, simple distance separation of the two—even if just a few metres—can prevent a relatively minor event, involving only the sensitive items, from becoming a much larger event. The same separation principle can be applied to stacks of ammunition, regardless of sensitivity, and will limit the tendency of the sum total of ammunition in a building to explode simultaneously.

An extension of this practice is the use of Compatibility Groups (CG). As well as a UN Hazard Division (HD), AE are routinely allocated a CG—in combination the HD and CG form a Hazard Classification Code (HCC). AE in the same CG can be stored and transported together without adversely enhancing any hazardous effects in the event of an accident. Definitions of the CG are show at figure 3.7. Some CG can be safely mixed with AE from other CG, whereas others cannot—for example storing detonators of CG B with high explosives of CG D would be unacceptable since the initiation of the detonators (which on their own would be

Description of Substance or Article to be Classified (a)	Compatibility Group (b)
Primary explosive substance.	A
Article containing a primary explosive substance and not containing two or more effective protective features.	B
Propellant explosive substance or other deflagrating explosive substance or article containing such explosive substance.	C
Secondary detonating explosive substance or black powder or article containing a secondary detonating explosive substance, in each case without means of initiation and without a propelling charge, or article containing a primary explosive substance and containing two or more effective protective features.	D
Article containing a secondary detonating explosive substance without means of initiation with a propelling charge (other than one containing a flammable or hypergolic liquid).	E
Article containing a secondary detonating explosive substance with its own means of initiation, with a propelling charge (other than one containing a flammable or hypergolic liquid) or without a propelling charge.	F
Pyrotechnic substance, or article containing a pyrotechnic substance, or article containing both an explosive substance and an illuminating, incendiary, lachrymatory or smoke-producing substance (other than a water-activated article or one containing white phosphorus, phosphide or a flammable liquid or gel).	G
Article containing both an explosive substance and white phosphorus.	H
Article containing both an explosive substance and a flammable liquid or gel.	J
Article containing both an explosive substance and a toxic chemical agent.	K
Explosive substance or article containing an explosive substance and presenting a special risk needing isolation of each type.	L
Article containing only extremely insensitive detonating substances.	N
Substance or article so packed or designed that any hazardous effects arising from accidental functioning are confined within the package unless the package has been degraded by fire. In this case all blast or projection effects are limited to the extent that they do not significantly hinder or prohibit fire fighting or other emergency response efforts in the immediate vicinity of the package.	S

Figure 3.7. Compatibility group definitions (UK DSA 2019b).

a relatively minor event) could lead to the sympathetic initiation of the high explosives (a much more significant event).

In a similar vein, NATO has developed the concept of Sensitivity Groups (SG) for HD 1.1 and 1.2 AE. The purpose of SG are to define the robustness of a given AE item and its susceptibility to sympathetic detonation. AE deemed to be 'robust' can then be stored more closely with other AE, whereas 'sensitive' AE must be kept further apart; the SG of an AE item therefore feeds into QD between stacks of ammunition and/or storage buildings (NATO 2009d).

Robust emergency response procedures can help limit the consequences of an AE accident. Regular evacuation training for personnel working in an AE storage area, the installation of hardened emergency shelters, and the ability to quickly close roads, shipping lanes and airspace can all prevent or reduce injuries and fatalities.

Timely fire detection and a speedy firefighting response are also very important mitigation measures. The early stages of an ammunition accident are often less violent than later stages; it is rare for an entire stockpile to accidentally initiate at once. A fire usually takes time to become established even when AE are stored in close proximity to one another. Effective firefighting, from a safe distance, can limit the consequences of such an incident to the initial area. If left untended a fire can spread, leading to explosions which in turn start further fires and explosions in potentially rapid succession. Left unchecked the situation can quickly escalate to the point it becomes too dangerous to contain, leaving evacuation as the only option and the possible loss of the entire storage area must be anticipated.

3.3.3 Case studies

Accidents involving explosives have been occurring since the invention of gunpowder. However, the growing proliferation of personal recording devices and social media means first-hand AE accident information is much more readily available in the modern era.

It is therefore possible to examine three recent large-scale explosives accidents through open-source material, exploring the risks prior to the accident and the resultant consequences.

3.3.3.1 Balakliya (Балаклія), Ukraine (NB: English spelling can vary)

Background: Between 2014 and 2022 Ukrainian armed forces were engaged in a conflict with separatists in the Donbas region of Ukraine. The Balakliya ammunition depot, located on the northern edge of the town, is located relatively close to the contested region and in 2017 additional ammunition stocks—beyond the designed capacity of the depot—were being stored there to support the front line. Those munitions consisted of (at least) high explosive tank and artillery shells, artillery rockets, anti-tank missiles, and large surface to air missiles (SAM), i.e. ammunition in HDs 1.1 and 1.2 (and likely HD 1.3 propellant for the tank and artillery shells). The depot is easily recognisable as an ammunition storage location from aerial imagery; there were dozens of clearly separated storage sheds each surrounded by earthwork barriers, along with stacks of unsheltered ammunition containers also encompassed by barriers. The sheds consisted of lightweight barn-like structures with some open sides. There appear to also have been a pair of earth-covered storage buildings in the middle of the depot.

Risks: Open source reporting (Narwhal8915 2017) shows images supposedly from inside the Balakliya depot of large-calibre propellant cartridge cases stacked from floor to ceiling inside the storage sheds, and of other shells and AE boxes stored out in the open. The validity of these reports is questionable—they may be images from other depots in Ukraine, and some images show inert or expended items. It is nonetheless clear that storage conditions were not ideal, and ammunition stocks included ageing Soviet-era munitions. Estimates put the total quantity held at 72 000 tonnes.

Aerial images of the depot from 2015 show several dozen large containers stored together in the open. The size and shape of these containers corresponds to the dimensions of the 3M9 missiles which are part of the 2K12 'Kub' SAM system (the NATO designation is the SA-6 'Gainful') used by the Ukrainian armed forces. These missiles weigh approximately 600 kg, with a 57 kg warhead and 239 kg of propellant. Their design range is 24 km with a maximum velocity of Mach 2.8.

Consequence analysis: While some of the 3M9 missile containers were oriented towards earth barriers, many were not. Even so, the Earth barriers had sloped sides; UN and NATO regulations recommend that rockets and missiles, or other self-propulsive ammunition, be stored oriented towards a vertically-faced barrier to arrest their flight should they launch by accident. Alternatively, they can be stored inside a building designed to resist their effects, or specific devices can be installed to prevent their flight. In cases like Balakliya, where missiles were stored without any of these mitigation measures, QDs ought to have taken into account the flight range of those munitions.

QD appear to have been applied to protect personnel and buildings outside the depot perimeter, with a border of at least 400 m surrounding most of the site. Except on the southwest of the site, closest to the town, there were no roads or buildings within this border, which is the minimum QD for lightweight storage structures (without barriers) storing HD 1.1 of up to nearly 6000 kg NEM.

To the southwest were buildings within 400 m, although they were still at least 270 m away—this distance is acceptable for lightweight structures surrounded by a barrier for a HD 1.1 NEM of nearly 2500 kg. However, if the reporting of shells stacked as high as the roof of the storage sheds is correct then the Earth barriers, the tops of which were considerably lower than the top of the sheds, would have been rendered ineffective and the 400 m QD would apply.

The distances between storage buildings or stacks of ammunition were relatively small, as little as 50 m in some places. While these distances are practical for small quantities of HD 1.1 ammunition, or when using hardened structures for storage, there is almost no protection for lightweight stores against secondary fires and explosions caused by blast, flame, firebrands, projections and lobbed munitions originating in a neighbouring lightweight store holding HD 1.2 munitions.

The accident: In the early hours of 23 March, 2017 a fire started in the storage area followed by a series of explosions. The fire, and further explosions, continued until 25 March. Self-propelled munitions could be seen launching and flying erratically throughout. In the interim 20 000 inhabitants of the surrounding area were evacuated and an initial 50 km exclusion zone was enforced. Video footage taken by bystanders during the first 24 h showed large fires and blast events occurring sporadically; other footage from a drone used to monitor the situation showed the point at which at least one stack of missiles transitioned from burning to detonation (Citizen Truth 2017).

Aftermath: The accident resulted two fatalities and several injuries. A large portion of the storage sheds were completely destroyed. Damage to buildings in the town was extensive, especially those closest to the depot. An estimated 32 000 tonnes of ammunition was lost and the disposal of unexploded ordnance (UXO) in the immediate area took until at least 6 April, involving around 700 personnel. Inhabitants of Balkliya were allowed back into their homes from the start of April, though at the time it was expected to take two to three years until the entire surrounding region was free from UXO; some missiles were found many kilometres from the depot and buried several metres below ground level—it is likely such items will be found over time as the resulting of farming or building activity.

The cause of the fire is still to be determined, with sabotage being one avenue of investigation (Interfax Ukraine 2018, Boiko 2020). Due to the presence of ageing ammunition stocks autoignition also seems a possibility.

Observations: The design and layout of the ammunition depot at Balakliya conforms to many of the conventions, regulations and guidelines that are used by UN and NATO for large ammunition storage sites; an aerial overview is shown at figure 3.8. Although it is not possible to know with any certainty the NEM of ammunition in a given storage shed, the QD used between the sheds and the surrounding town seem to have been mostly appropriate since the damage to buildings outside the depot is consistent with the expected effects of different HD at those QD.

Open source imagery shows the majority of UXO in the town consisted of 3M9 missile parts (in one especially worrying case, a missile body is seen adjacent to a fuel station forecourt (Replyanchuk 2017)). This is evidence that the QD used for those missiles was not correct, mainly as a result of their having no barriers or other mitigation measures in place to arrest their flight if accidentally launched. It is perhaps worth considering that, even if the missiles had been pointing away from the town or towards an appropriate barrier, the explosion of one missile in a stack would almost certainly change the orientation of any nearby missiles, potentially rendering any barriers irrelevant. Therefore, the storage of the 3M9 missiles in the open, stacked one atop the other and adjacent to other stacks, also contributed to the involvement of so many missiles in the accident.

During the incident, blasts in some storage sheds projected many hundreds of artillery and tank shells from their initial locations into other parts of the depot and surrounding fields, some travelling 600 m or more. Those shells either landed intact, deflagrated, or detonated. It is highly likely the majority of the sporadic explosions witnessed during the incident occurred when burning shells (and indeed missiles) were thrown clear of their original location and landed on a neighbouring storage

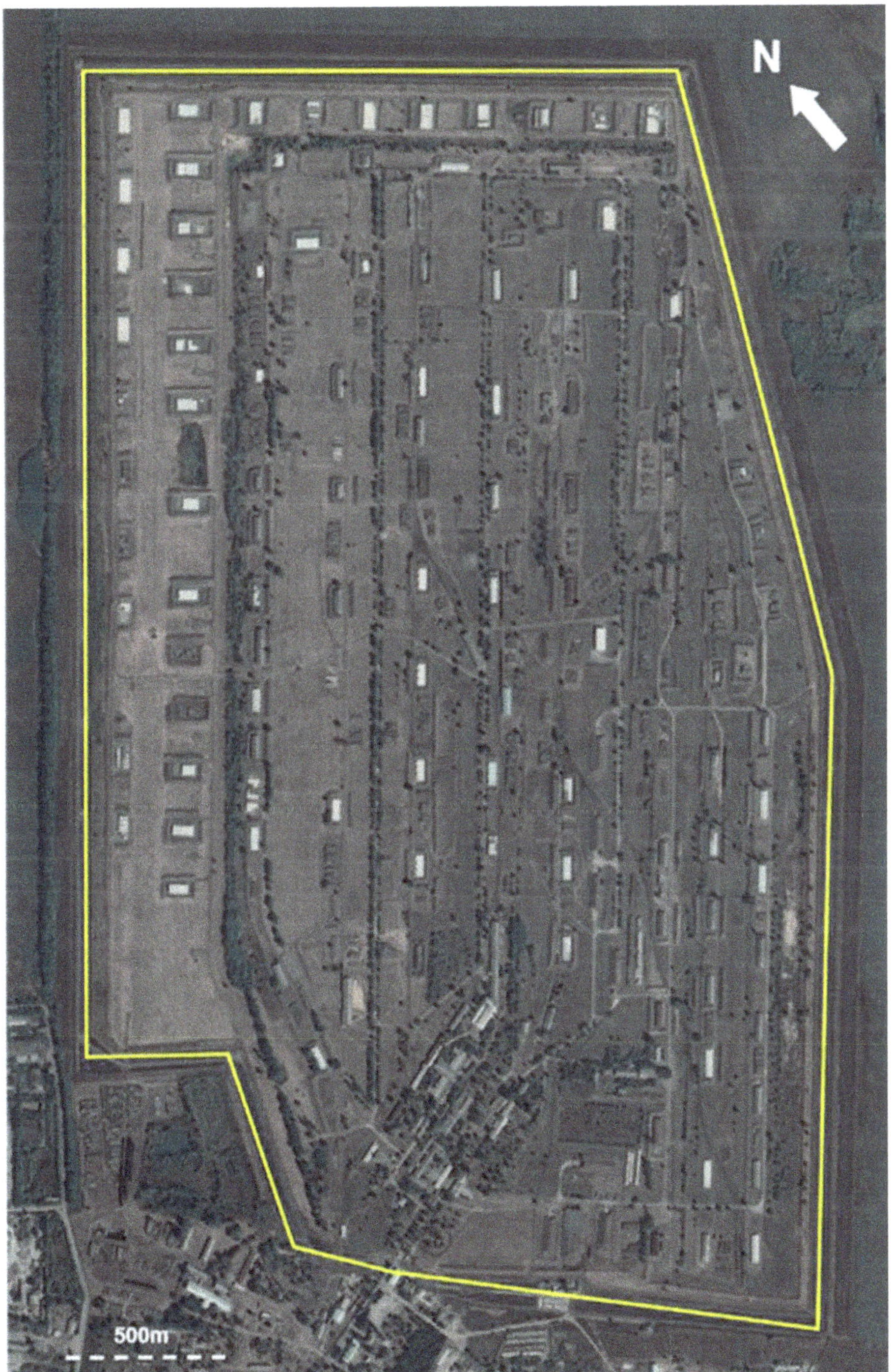

Map Data: Google, © 2019 CNES/Airbus

Figure 3.8. Balakliya ammunition depot circa 2015 (Google Earth 2019a). The yellow line defines the depot perimeter. The town begins immediately to the southwest of the depot. Map Data: Google, (c) 2019 CNES/Airbus.

shed before starting a fire there, or exploding and sympathetically detonating the stock therein (multiple small fires can be seen during the drone video). This behaviour is evident from damage to the ground where the storage sheds once stood; in some cases the soil has a peppered and scorched appearance from shell impacts and subsequent burning, in other cases there is a large crater indicating a mass explosion event (see figure 3.9).

This is the type of 'domino effect' that appropriate QD between storage buildings is supposed to prevent or limit. If, however, the storage sheds at Balakliya contained more than their designed capacity of ammunition, and the QD had not been

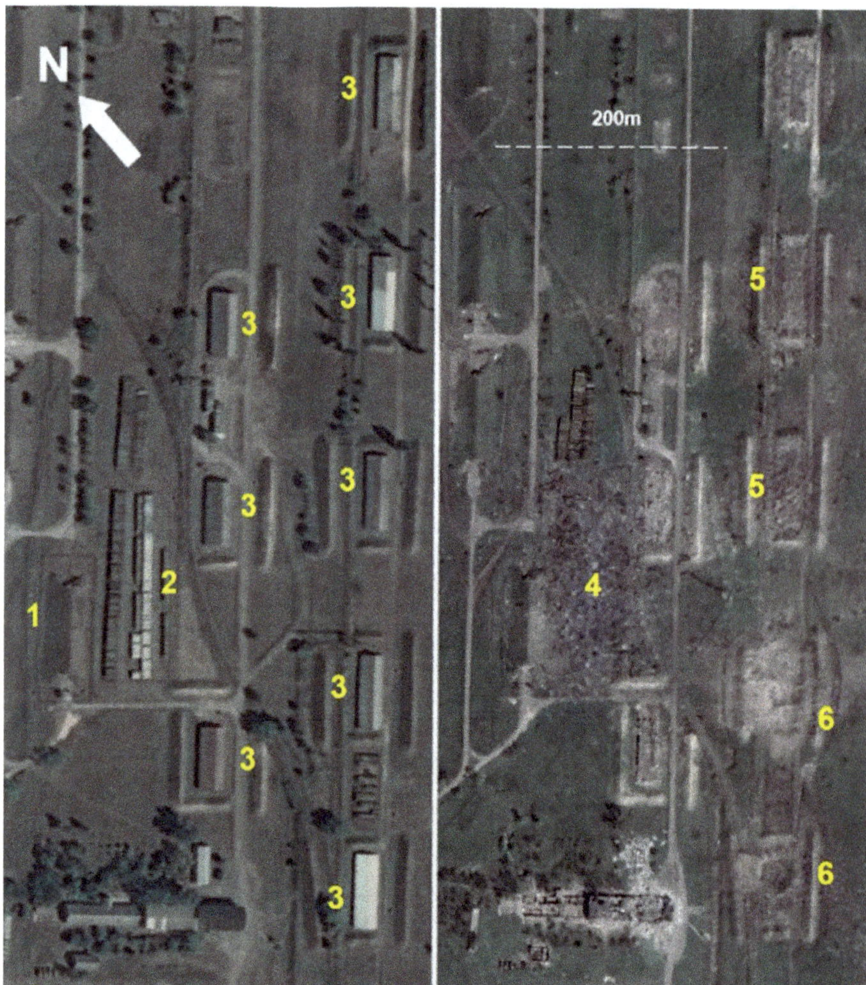

Figure 3.9. Post-blast damage (right) to depot compared to original condition (left) (Google Earth 2019a, 2019b). (1—Earth mound barrier. 2—Large stack of 3M9 missiles. 3—Storage shed. 4—Remains of 3M9 missile stack. 5—Small craters from sporadic explosions. 6—Large craters from mass explosions.) Map Data: Google (c) 2019 CNES/Airbus/Digital Globe.

adjusted accordingly, then the destruction of so many sheds was to be expected. This limited degree of inter-building protection requires the implementation of other mitigation measures, such as improved fire detection and firefighting methods.

Despite the loss of so much of the depot infrastructure and stockpile, the QD being used at the time were sufficient to prevent the simultaneous mass detonation of the entire contents of the depot, which undoubtedly afforded the local authorities the time needed to complete an admirably comprehensive evacuation.

3.3.3.2 Beirut, Lebanon

Although the explosion which occurred in the Port of Beirut in August 2020 did not involve military AE, many of the same storage Risk Management and mitigation principles are normally applied in the civilian environment. In fact, the tolerance for risk in a non-military context ought to be lower, so safety standards should be more stringent than those accepted for use by armed forces.

Background: On 27 September 2013 the cargo ship MV Rhosus was detained in the Port of Beirut by Lebanese authorities. The ship was carrying 2750 tonnes of ammonium nitrate (AN). After legal proceedings in 2014 the AN was transferred to warehouse number 12 in the port in Beirut where it remained until the accident in 2020.

Photographs of the warehouse 12 interior from before the accident show multiple 1000 kg bags of AN stacked up inside (Dalatrm 2020, DimaSadek 2020). The warehouse was a lightweight fully-enclosed structure made from thin metal sheeting over a steel girder frame.

Immediately to the west of warehouse 12 was a group of 48 concrete grain silos containing Lebanon's strategic 120 000 tonne grain reserve. Other warehouses in the port were all of similar construction to number 12; these warehouses occupied most of the space to the south of warehouse 12, a distance of at least 410 m to the limit of the port.

Risks: AN is not inherently explosive but is instead an oxidising agent. However, in the right conditions large quantities can be made to detonate either through shock initiation or by burning to detonation (UK Agricultural Industries Federation 2015). Over time, when stored in bulk, AN has a propensity to pressure-set. This occurs when AN prills absorb moisture and, under pressure from surrounding AN masses, set to form a solid conglomerate which increases the likelihood of a deflagration to detonation transition (DDT) during a fire. Given the age and way in which AN was stored is warehouse 12 it is likely that some pressure-setting had occurred.

Contamination of AN by combustible contaminants can also increase the susceptibility of AN to detonation. Some reports state AN prills in warehouse 12 had 'darkened' indicating the possible presence of contamination.

As the result of multiple large-scale accidents involving AN during the 20th century, many national and international regulations regarding AN storage have been produced. The most common key regulations are summarised below (UK Health and Safety Executive 2004, Australia Department of Mines and Petroleum 2013, Galassi 2014, Safex International 2014):

- AN should be stored away from other combustible materials in an area that can be well-ventilated in the event of a fire.
- Where stacks of AN exceed 250 tonnes a sprinkler should to be fitted.

- Stacks should not exceed 300 tonnes, or take up a space greater than 2×3 m depending on the density of the AN.
- There should be at least 1 m between stacks of AN (depending on the AN density).
- Pressure-setting is to be monitored/prevented.
- Local firefighting agencies are to be informed of bulk AN storage locations.
- Sources of fire are not to be allowed near AN stores.

It is not clear if any of the practices shown above were adopted in warehouse 12. It seems likely that few, if any, of those measures were in place since the risks posed by the AN in the Port of Beirut were highlighted to government authorities on multiple occasions between 2015 and 2020 (Nakhoul and Bassam 2020).

Reports indicate that, in addition to the AN, there was also 23 tonnes of fireworks, a small quantity of detonating cord, 100 tonnes of vehicle tyres, and 55 tonnes of other combustible materials inside warehouse 12.

Consequence analysis: QD for AN are based on a TNT equivalence of 25% and take into account the largest single stack of AN. The size of stacks in warehouse 12 is not known, though a comprehensive image analysis has estimated the entire 2750 tonnes of AN was stored in warehouse 12 as a single haphazard stack (Forensic Architecture 2020).

If correct, application of civil QD (military QD are also included for comparison) should have resulted in the separation distances shown in table 3.5. (It is interesting to note that that civil QD are more permissive than similar military QD in relation to structure types. There is no apparent reason for these disparities, except for different recommendations for the acceptable peak pressures experienced by structures at the proposed distances.)

The closest buildings not directly associated with port activities were approximately 410 m from warehouse 12. These consisted of various commercial and residential buildings, vulnerable structures including glass-fronted tower blocks, along with a hospital.

The accident: Just before 6 pm on the evening of 4 August 2020 smoke was seen rising from the northeast corner of warehouse 12, a fire having apparently been started accidentally by welding activities. The smoke quickly thickened and grew in volume; it rapidly became evident the corner of the warehouse was ablaze. After a few minutes a relatively small explosion occurred in the northwest corner of the warehouse and fireworks were seen launching out over the port. A few seconds after this a detonation

Table 3.5. Military and civil QD for 2750 tonnes AN (based on TNT equivalence of 25% NEM) (UK DSA 2019e, Australia Department of Mines and Petroleum 2013).

Building type—QD	QD calculation	Resultant QD (m)
Military QD:		
Vulnerable—$44.4Q^{1/3}$	$44.4 \times 687\ 500^{1/3}$	3919
Residential—$22.2Q^{1/3}$	$22.2 \times 687\ 500^{1/3}$	1959
Civil QD:		
Vulnerable—$22.2Q^{1/3}$	$22.2 \times 687\ 500^{1/3}$	1959
Residential—$17.8Q^{1/3}$	$17.8 \times 687\ 500^{1/3}$	1571

occurred which completely destroyed warehouse 12, heavily damaged the neighbouring grain silos, and caused extensive damage to the port and areas around the port. Various analyses of the resultant seismic data put the explosion as being equivalent to 0.5–1.1 kilotonnes of TNT, which is consistent with most, if not all, of the 2750 tonnes of AN being involved in a simultaneous detonation.

Aftermath: The explosion resulted in 207 fatalities, and approximately 6500–7500 people were recorded as injured. In reality it is likely the number of injured was much higher; hospitals at the time of the incident became so overwhelmed with casualties it was not possible to record all admissions. This problem was compounded by damage to the three hospitals nearest the port being so extensive they could not accept patients and had to be evacuated (Mansour *et al* 2021).

Traumatic blast injuries to brain and lung accounted for the majority of those who required intensive care, whereas fragmentation injuries—mostly from flying glass—made up the most common reason for casualties needing immediate surgical intervention. Other surgeries were required for patients with compartment syndrome caused by blunt trauma (from displacement or debris impact). Hearing loss, temporary and permanent was also a frequent injury. Over time psychological injuries came to light, with reports of anger, depression, and PTSD symptoms becoming apparent.

At the seat of the explosion was left a crater measuring approximately 130 m at its widest point and 40 m deep (see figure 3.10). The port was devastated, with many warehouses razed. 34 of the 48 nearby grain silos were destroyed, with the associated

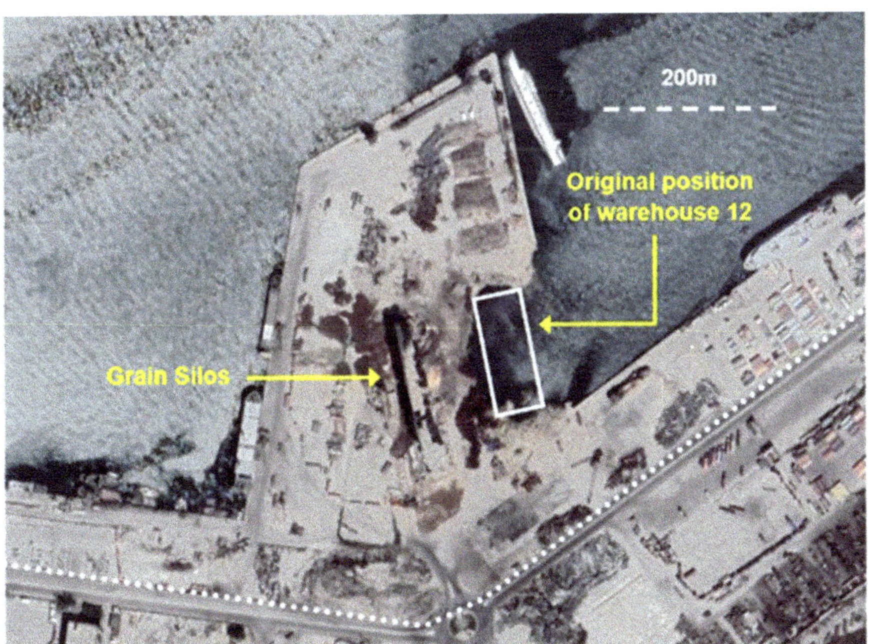

Figure 3.10. Port of Beirut after the explosion (dotted line indicates port perimeter) (Google Maps 2021a). Map Data: Google (c) 2021 CNES/Airbus/Maxar Technologies.

loss of around 15 000 tonnes of the grain reserve. Buildings within 600 m suffered serious structural damage, with some partially collapsing, while buildings out to 10 km suffered some degree of minor damage. Innumerable glass windows in the city were blown out by the blast. A map of the damage can be seen at figure 3.11 The total cost of the damage is believed to be in the region USD15 billion. An estimated 30 000 people were left homeless.

At the time of the accident Lebanon was both in the midst of an economic crisis and suffering the impact of the COVID-19 pandemic. COVID-19 transmission increased in the immediate aftermath of the explosion as a result of hospital overcrowding. Nearly a year later the political and economic ramifications of the incident were still being felt after the resignation of the Lebanese government in the face of public outcry over perceived state negligence (BBC News 2021a).

Observations: A combination of factors created the conditions for the Port of Beirut explosion. These were:

1. Storage of different UN dangerous goods (DG) class materials in close proximity to one another (fireworks of DG class 1 (explosives) and AN of DG class 5.1 (oxidising agent)). Mixing of different DG is normally prohibited without additional specific safety measures. The presence of other combustible materials also contributed to the spread of the initial fire.
2. Flame-producing activities (welding) being conducted in and around warehouse 12.
3. Poor ventilation of warehouse 12 which allowed the build-up of both pressure and the combustion products of AN, further catalysing the fire.
4. Inappropriate storage of the AN; it was not separated into stacks of the recommended size and so the mass involved in the explosion consisted of the entire 2750 tonnes simultaneously.

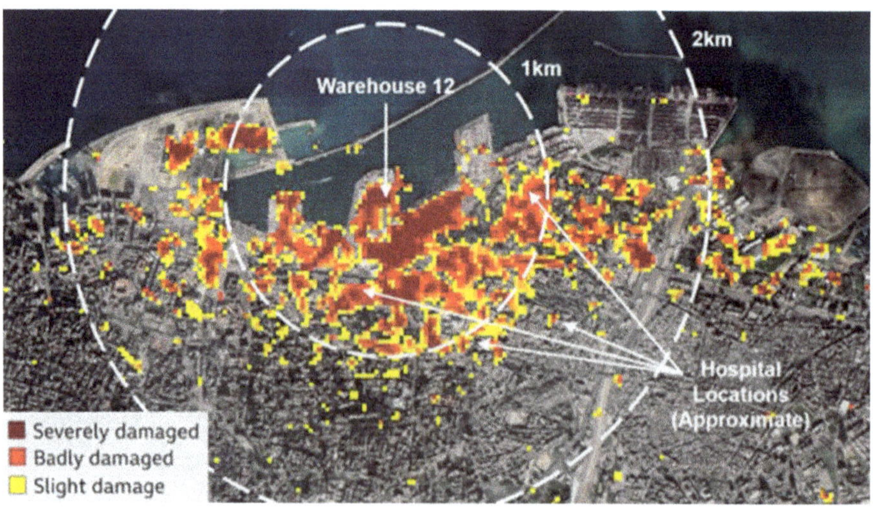

Figure 3.11. Assessed damage to buildings surrounding the Port of Beirut (USA National Aeronautics and Space Administration (NASA) Jet Propulsion Laboratory (JPL) 2020).

5. The AN was stored at least 1 km too close to the city.
6. The age and apparent contamination of the AN is likely to have made it to more susceptible to DDT.
7. The warehouse was not fitted with any sort of fire suppression system.

The presence of the fireworks was arguably the single biggest contributing factor to the explosion. A team of firefighters arrived within 4 min of receiving a report of the fire—at this time it may have still been possible to quench the combustion reaction and vent the structure before the AN detonated. However, their arrival coincided with the involvement of the fireworks in the fire; once a portion of the fireworks exploded the reaction quickly ran out of control and the AN detonated shortly thereafter. Sadly the firefighters were among the first casualties of the explosion.

3.3.3.3 Bata, Equatorial Guinea; 7 March 2021

Background: A military base was established in Nkoantoma, on the eastern edge of Bata, in 2004. At that time the base was surrounded by forest and farmland, with very few inhabited buildings nearby. As of 2021 the area had become much more densely populated with residential and commercial properties in close proximity to the base (BBC News 2021b), i.e. less than 400 m. The base itself included several accommodation buildings housing soldiers and their families.

Risks: It is not known exactly what AE were being stored at the base. UXO photographed after the accident show tank shells, recoilless rifle rounds, mortar bombs and hand grenades. The age of the munitions could not be accurately determined, though many were of Soviet and Chinese designs which have been in production for several decades. Furthermore, the government of Equatorial Guinea stated the accident involved dynamite. Dynamite is notoriously unstable in long-term storage if not correctly packaged, carefully handled and closely monitored due to its sensitivity to mechanical shock and capacity to sweat highly-sensitive nitroglycerine.

Comparison of aerial images taken before and after the explosions show 30–40 ISO shipping containers which are later replaced by craters (Maxar Technologies 2021); the craters are assumed to be the seats of explosions and therefore the ISO containers are deemed to have been the storage facilities for at least some of the AE at the base (see figure 3.12). Map Data: Google, Imagery © 2021 Maxar Technologies, Map Data © 2021.

Consequence analysis: The ISO containers were situated side-by-side in an area measuring approximately 100 ×50 m and were 100 m from the closest buildings on the base. Due to their close proximity to one another the NEM of explosives in the ISO containers would be aggregated together since practically instantaneous propagation could be expected between AE in those containers.

Dynamite is a HD 1.1 explosive. Assuming at least some of the ISO containers had dynamite inside, and given the lack of any barriers either between containers or around them, for HD 1.1 explosives a minimum QD of 400 m should normally be applied between the site and any surrounding residential buildings regardless of the NEM stored.

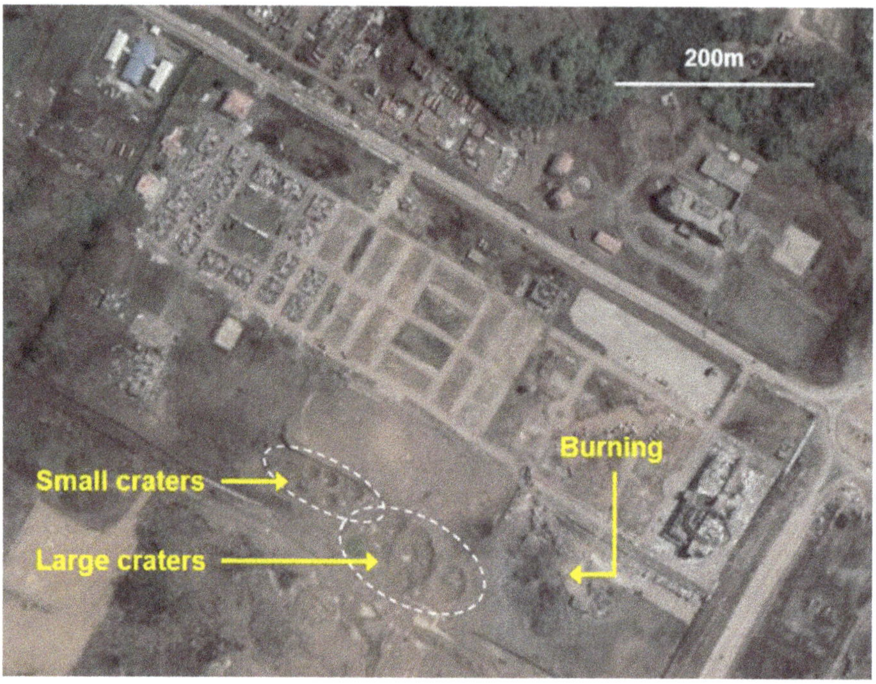

Figure 3.12. Nkoantoma base showing aftermath of the explosions (original position of ISO containers indicated by dash circles) (Google Maps 2021b). Map Data: Google (c) 2021 Maxar Technologies.

The accident: Three explosions occurred in quick succession at the Nkoantoma base on the afternoon of 7 March 2021, with a fourth explosion some 2 h later. Multiple buildings on the base were immediately destroyed along with vegetation within hundreds of metres. Two large overlapping craters, produced by explosions in the middle of the area where the ISO containers had been, each measured approximately 60–65 m in apparent diameter. Fires were reported at the site, and imagery from eyewitnesses shows a large column of thick black smoke above the base for a period of time after the accident.

Aftermath: At least 107 people died as a result of the explosion, with a further 600 or more injured. Local hospitals were quickly inundated with casualties, prompting the government to declare a state of emergency and request international aid.

Hundreds of buildings were either destroyed or rendered uninhabitable by heavy damage. Social media showed images of local inhabitants posing with UXO, prompting the launch of an educational campaign warning of the dangers of handling UXO.

The government of Equatorial Guinea stated the accident had been caused by local farmers using fire to clear vegetation, though this is disputed by other agencies. Negligence on behalf of the personnel charged with managing the AE was also put forward as a cause of the explosions (BBC News 2021c).

Observations: Detailed information about the AE at Nkoantoma, along with their storage conditions, is scarce. As such, it is only possible to infer what happened prior to and during the accident from analysis of open source imagery.

Post-accident photographs of buildings, vehicles and ISO containers in other parts of the base show evidence of a significant blast overpressure event (BBC News 2021d); building roofs were displaced, and there was some structural collapse of walls facing the blast site. Cars and containers had a 'bowed in' appearance, as though crushed by a gigantic blunt force. Images containing lamp posts showed them leaning away from the direction of the explosion—since they were leaning over rather than totally collapsed it appears as though there was single blast loading event, as multiple significant events would likely have knocked the posts over.

There was little to no sign of fragmentation damage, though it may be the case that fragmentation receptors close to the seat of the explosion, such as walls, were destroyed by the blast wave. Thermal effects, specifically scorched earth, appeared to be limited to an area some 130 m east of the craters. It is likely this was the result of a fire started after the initial explosion and may account for the large smoke plume and the explosion which occurred 2 h after the initial event; for example, from one or more burning items of heavy-cased ammunition projected away from the original site which later burned to detonation.

A formula for estimating crater size in soil can be used in an effort to evaluate the possible NEM of explosives held in the ISO containers (Ambrosini *et al* 2003):

$D = 0.8\ W^{1/3}$ where D is the crater size in m and W is the TNT equivalent NEM in kg. Rearranging for NEM and applying the crater size dimensions from Bata give the following results (note that the crater size has been reduced from its apparent diameter to an estimate of its true diameter):

$(D/0.8)^3 = W\ (55\ \text{m}/0.8)^3 = 324\ 951$ kg.

Adjusting TNT to dynamite (a factor of 1.25):

$324\ 951 \times 1.25 = 406\ 188$ kg.

There is undoubtedly a margin of error in these values since the effect of the soil is not taken into account and it is unlikely the entire stockpile consisted of dynamite. Likewise, the formula was derived from detonation of unconfined explosives; it is possible the AE in Nkoantoma were subject to some confinement from the ISO containers. Nonetheless it would appear that tens, if not hundreds of tonnes of AE were involved in the largest explosions. These values are reasonable given the storage volume and weight capacity available in 30–40 ISO containers.

In any case, current regulations recommend that ISO containers used for HD 1.1 explosive storage without barriers should be located at least 400 m from the nearest inhabited buildings, and then only storing a NEM of 5–6 tonnes across the entire site. The use of barriers, and separation between the ISO containers, could have dramatically reduced the size of the Nkoantoma explosions. Preventing the encroachment of civilian buildings into the QD could also have limited the damage, injury and loss of life.

3.4 Opportunities for harmonisation and standardisation

There are several aspects of explosives storage risk evaluation and consequence analysis which differ slightly between regulatory organisations. Earlier in the chapter differences in explosives qualification methodology were highlighted that could have an indirect impact on storage safety if sensitive explosives, or components containing sensitive explosives, are not correctly classified.

Harmonising the small disparities between the UN HD1.5/1.6 and NATO IM tests could be advantageous for both bodies in the long term; the UN could endorse tests which are both more stringent and more realistic, while NATO would avoid unnecessary repetition of tests to gain the benefits of IM QD.

QD values used by the UN and NATO nations are useful, but close examination reveals a number of underlying flaws. The NATO Munitions Safety Analysis Information Centre (MSIAC) is conducting ongoing work to challenge the status quo and address these issues.

The foremost problem is that the data upon which QD lookup tables were originally based dates back to the late 1940s, and incorporates information from trials using outdated bulk explosives such as cordite (which is no longer used *en masse*). Building damage from explosions was included in the early data underpinning QD, but building design and construction has changed significantly in the intervening 70+ years. In some cases, the evidence supporting QD have been lost to time and so those QD are currently accepted at face value for want of any better alternative (van der Voort *et al* 2016).

QD for HD 1.1 have traditionally been based on peak pressure values from a TNT blast. In recent years it has been realised that impulse is an equally, if not more relevant measurement and as a result some QD may have been overestimated. For small NEM storage, impulse appears to be the key criteria (see figure 3.13). However, this realisation brings with it further complications in terms of TNT equivalence since a particular explosive may have different TNT equivalencies for peak pressure and impulse. When we consider that many modern munitions do not

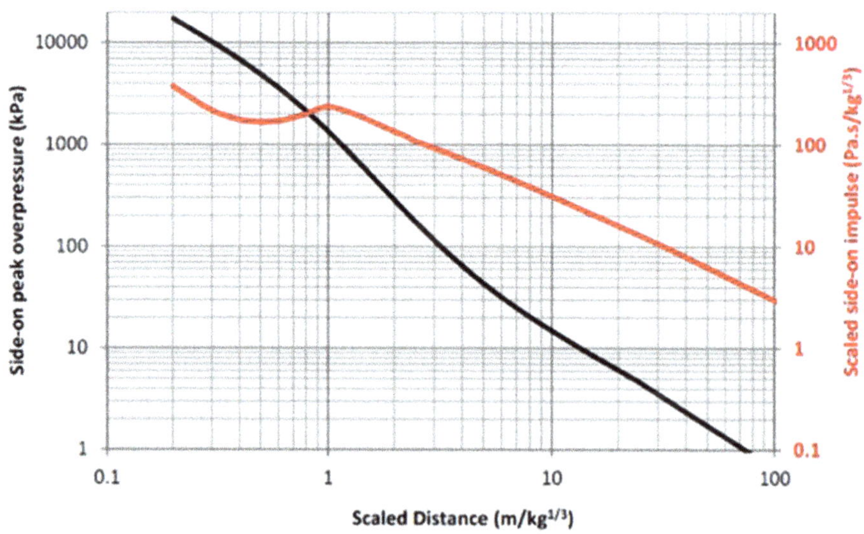

Figure 3.13. Side-on peak overpressure and scaled side-one impulse as a function of scaled distance for a hemispherical surface burst (van der Voort *et al* 2016).

contain TNT—instead containing RDX, HMX, Polymer Bonded Explosive (PBX) and so on—the accurate determination of appropriate QD therefore becomes an ever more complicated process.

Debris and fragmentation throw of building remains from HD 1.1 explosions inside brick and reinforced concrete (RC) structures has been studied more closely in the 21st century. It has been found that existing QD for these types of buildings have previously been underestimated, as per figure 3.14. It appears that blast is not always the predominant hazard for HD 1.1 AE when stored inside an RC structure; for small NEM it is debris and secondary fragments which present a greater danger.

However, QD for most brick and reinforced concrete storage buildings have always been applied with the assumption that debris throw is uniform in all directions. Studies have revealed this is not the case; debris from surfaces normal to the blast wave (such as walls and the roof) travels much further than debris from corners (see figure 3.15). There may therefore be opportunities to use reduced QD depending on the orientation relationship between such storage buildings and surrounding structures or personnel. This is already the case for so-called 'igloo' earth-covered stores, where the sides and rear of the building are more heavily built than the front.

Launch of debris is dependent on several factors including not only NEM and building material, but also internal volume. The internal volume is not currently considered during QD calculations. To bridge this gap The Klotz group has developed—and continues to develop—a computer model to predict debris throw (Weerheijm and Conway 2018).

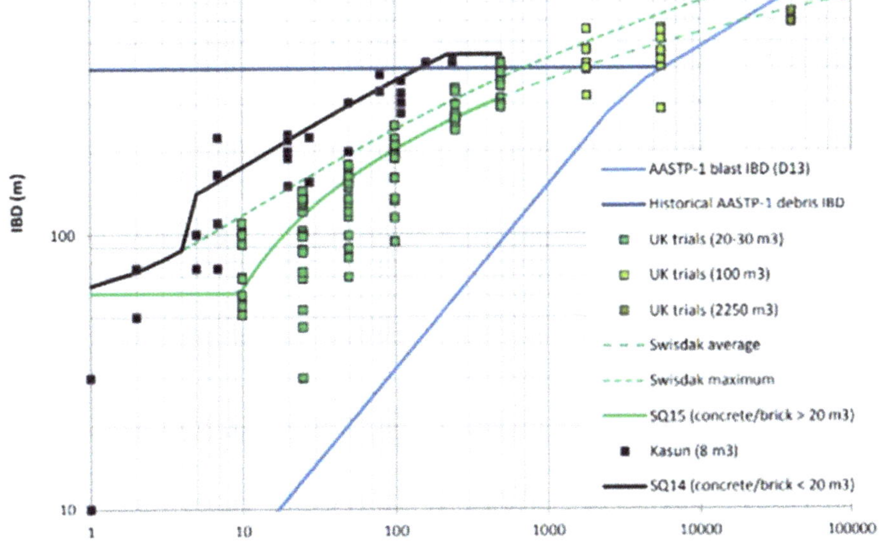

Figure 3.14. Debris inhabited (residential) building distance (IBD) plotted against results of practical trials (NEQ is used in place of NEM) (van der Voort *et al* 2016).

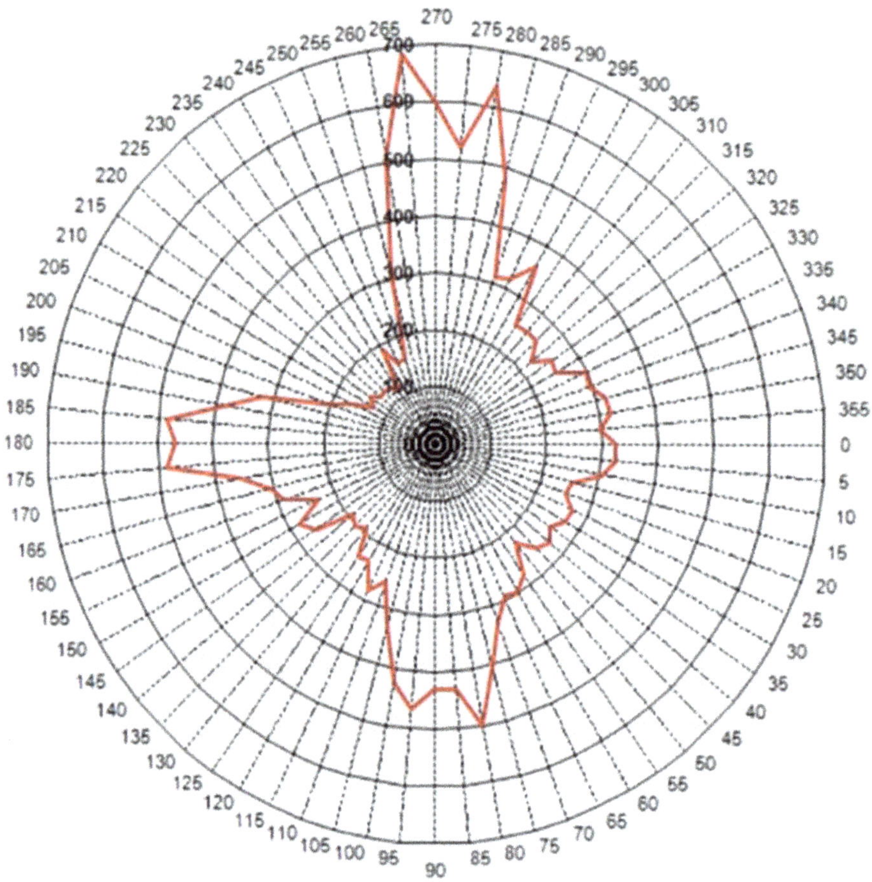

Figure 3.15. Plan view of pseudo-trajectory normal debris field from SciPan test 4–1000 kg TNT inside a reinforced concrete/concrete masonry unit building. (Figures in bold are distance in metres; remaining Figures are bearing in degrees). Walls faced 0, 90, 180 and 270 degrees (Conway *et al* 2010). DDESB. See: https://discover.dtic.mil/privacy-security-copyright/ (Accessed: 28 July 2021).

MSIAC have also examined methods to apply QD which are more relevant to the predominant likely hazardous effect of the ammunition being stored, be that blast, debris or fragmentation, UXO throw, or thermal radiation. Also incorporated is adoption of the 'Maximum Credible Event' (MCE) concept for ammunition in HD 1.2. Ammunition in HD 1.2 tends to be heavy-cased, with a relatively low charge: case ratio. As such an ammunition stack in this HD is unlikely to explode simultaneously since propagation between items is less probable. Instead, practical trials show that only a single container of HD 1.2 items tend to be destroyed at a time, and so the NEM of a single container can be used as the MCE (Q) value when calculating QD.

However, the concept of MCE does not account for the likelihood of packaging being compromised during an accident. Packages involved in an explosion may be split open, spilling their contents, thereby invalidating any separation or other protection afforded by the packaging. In these circumstances the potential grows for larger credible events to occur as an incident progresses. Similarly, any rockets, missiles, or SCJ warheads could change orientation during an accident and so may no longer be pointing in a safe direction after an initial explosion, again raising the odds of a larger than anticipated credible event.

The consequences of an explosives accident over time are rarely considered with regards to explosives storage safety. The implied assumption of most regulations is that the hazardous effects of an accident occur simultaneously, and that those effects only happen once. Examining accounts of past large-scale accidents reveals that in reality it is more common for several explosions to occur over a period of minutes or hours, often with larger events occurring some time after the initial, smaller ones. This has implications in terms of consequence analysis; for example, while an inhabited building at the appropriate QD may withstand a single 5 kPa side-on pressure pulse with minimal damage, the question of whether the same building would withstand several similar pulses has not been addressed. Similarly, it has been assumed that people who might be exposed to debris or fragmentation will seek cover. However, in truth evacuation and rescue efforts often begin almost immediately, so for progressive events fragmentation density over time becomes relevant. The effects of an explosives accident over time were considered as far back as the 1960s, but seem to have been overlooked in modern publications.

3.5 Conclusion

Evaluating the risks posed by military AE storage and assessing the consequences of accidents involving those munitions is a complicated process requiring expert multidisciplinary input. This means that achieving harmonisation, standardisation, and rationalisation across organisations and nations is a difficult prospect. MSIAC have made headway in redressing some of the problems relating to storage safety but it will take time for their findings to be accepted by the likes of the UN and national organisations since their respective bureaucratic apparatus must ratify the proposed changes.

The move towards considering each hazardous effect of an item of ammunition separately, and the use of MCE, are admirable aspirations but make an already convoluted process increasingly complex; as QD tables become more intricate the likelihood of potentially fatal errors in their application increases. To overcome human error MSIAC have mooted increased reliance on physics-based computer modelling to conduct QD work. However, this becomes increasingly difficult in conflict-afflicted or low-capacity environments where AE risks are high and the ability to quickly apply simple, effective QD—with the minimum of training—is highly desirable.

High explosives have existed for over 150 years yet their effects, particularly at large scales, are still not fully understood. As new discoveries are made and understanding changes so too do requirements for safe storage. AE storage Risk Assessment is therefore a living process, making it very difficult to establish a universal standard.

The common goal of safety organisations concerned with explosives is ever safer standards of storage, but many of these organisations are working independently using slightly differing methods. As such, true harmonisation may never be achieved in this field without a single governing body to act as the authoritative explosives safety entity.

Appendix A United Nations hazard division tests (UN 2019a)

Test code	Test name	Description
Test Series 5—Is it a very insensitive explosive substance with a mass explosion hazard (i.e. HD 1.5)?		
5a	Cap sensitivity test.	Tests the sensitiveness of a substance to initiation using a standard blasting cap.
5b	Deflagration to detonation (DDT) test.	Tests the tendency of the substance to burn to detonation under confinement when ignited by a low explosive charge
5c	External fire test for HD 1.5.	A packaged item is placed on a bonfire for at least 30 min to see if it reacts to external fire.
Test series 6—Which HD (of 1.1 to 1.4) corresponds most closely to the behaviour of a product if involved in a fire or an internal explosion?		
6a	Single package test.	A packaged AE item (or items) is confined by other empty containers and deliberately initiated using the most appropriate means of initiation for the AE type. The reaction (if any) is observed.
6b	Stack test.	As per 6a, but instead of empty containers a stack of the same packaged AE surrounds the item which is initiated to see if propagation occurs between packages.
6c	External fire (bonfire) test.	One or more packages (depending on total volume) are placed in a bonfire for at least 30 min
6d	Unconfined package test.	As per 6a, but the AE package is no longer surrounded by other containers. The hazardous effects (if any) which escape the package are observed.
Test Series 7—Is it any extremely insensitive explosive article (i.e. HD 1.6)? (7a–7f test the explosive substance in the articles to see if it is an Extremely Insensitive Substance (EIS); 7g–7l test the complete AE article)		
7a	EIS cap test.	As per 5a.
7b	EIS gap test.	A gap test using a standard detonator to initiate the substance. RDX is used as the acceptor charge separated by a PMMA rod from the test substance. The outcome is measured on a witness plate.

7c	Friability test.	A small quantity of the substance is launched at hundreds of m/s towards a hard surface to see if the impact causes a reaction.
7d	EIS bullet impact test.	A 12.7mm projectile is fired at the substance to see if the impact causes a reaction.
7e	EIS external fire test.	As per 6c using small pipes filled with the test substance.
7f	EIS slow cook-off test.	Samples of the substance are placed in small pipes which are in turn placed in an oven and gradually heated to 365 °C to see if a reaction takes place.
7g	1.6 article external fire test.	As per 6c except a complete packaged AE article is used.
7h	1.6 article slow cook-off test.	As per 7f using a complete packaged AE article.
7j	1.6 article bullet impact test.	A 3-round burst of 12.7mm projectiles is fired at the packaged AE article to see if a reaction occurs.
7k	1.6 article stack test.	As per 6b.
7l	1.6 article fragment impact test.	Simulated fragments are fired at the packaged article at very high speeds to see if a reaction occurs.

Appendix B Explosives trials

Trial name/location	Date	Organisation
France	c.1930	Eugene Burlot

Summary

- Explosives trials which led to the origination of many of the QD still in use today.

Trial name/location	Date	Organisation
Soltau	1946	UK; Explosives Storage and Transport Committee (ESTC)

Summary

- Investigation into sympathetic detonation of munitions from blast and fragmentation effects.
- Identified that fragmentation was a much greater hazard to munitions than blast.
- Found that barriers were vital in preventing propagation between munition stacks.
- Double slope earth traverses were found to be especially effective at small QD in terms of blast mitigation.

Trial name/location	Date	Organisation
Dune/Heligoland	1947	UK; ESTC

Summary

- An extension of the Soltau trials, including ignition of bulk cordite underground.
- As expected, heavy confinement of cordite led to a very rapid pressure increase; in some cases this led to deflagration bordering on detonation.
- It was found that cordite could be made to detonate if initiated with a high explosive.

Trial name/location	Date	Organisation
Big Papa	1968	US: USAF Weapons Lab

Summary

- Trials to investigate appropriate distances between bomb stacks, determine optimum barrier geometry, and acquire blast and ground shock data.
- Found that QD could be smaller for ammunition stored in the open, i.e. which was not stored in buildings which gave rise to secondary debris and fragmentation.
- Standard earth barriers have a limited influence on blast effects, but are very effective at containing fragmentation.
- Blast propagation between stacks of heavy-cased munitions is unlikely even at small QD.
- Barriers should not be made of materials which can give rise to secondary fragmentation.

Trial name/location	Date	Organisation
ESKIMO Series	1970s–1980s	US; Department of Defense Explosives Safety Board (DDESB)

Summary

- ESKIMO is an acronym for Explosives Safety Knowledge Improvement Operation, a series of explosives trials conducted to gather blast and debris data from a variety of munition/structure combinations, mainly focused on earth-covered stores and intermediate to large NEMs (tens to hundreds of tonnes).
- Seven tests (I–VII) were conducted in total.
- The data was used to develop current safe separation criteria for earth-covered storage buildings.

Trial name/location	Date	Organisation
Woomera	1990s–2000s	Australia; ESTC & UK; ESTC

Summary

- A series of independent Australian and joint Australian/UK trials to examine various structure/barrier responses to explosive events. In particular the debris and blast effects of small NEM (tens of kg) explosions inside brick structures.
- Data was used to inform and validate computer modelling.

Trial name	Date	Organisation
Kasun series	2000s	Norway; Forsvarsbygg (and Sweden; also funded by Klotz group).

Summary

- A series of tests using 'Kasun' structures (relatively small reinforced concrete cubes) containing different munitions/small NEM (tens of kg). Structure breakup, blast pressure, debris formation and debris throw data was gathered.
- Data used to further develop the Klotz Group Engineering Tool (KG-ET) for debris throw prediction.

Trial Name	Date	Organisation
'SciPan' series	2000s–2010s	US; Department of Defense Explosives Safety Board (DDESB)

Summary

- 'SciPan' is an abbreviation of Science Panel; the SciPan tests are one of a series of tests forming an extension of the ESKIMO programme.
- Tests used to determine debris throw from reinforced concrete structures containing intermediate NEM (tonnes) of high explosives.
- Demonstrated that current QD for residential buildings are not adequate for storage buildings made form reinforced concrete.

References

Akhavan J 2011 *The Chemistry of Explosives* 3rd edn (Cambridge: RSC Publishing)

Ambrosini D, Luccioni B and Danesi R 2003 Craters produced by explosions on the soil surface *Mecánica Comput.* **22** 678–92

Australia Department of Mines and Petroleum 2013 *Safe Storage of Solid Ammonium Nitrate—Code of Practice* 3rd edn (Perth: Department of Mines and Petroleum) https://dmp.wa.gov.au/Documents/Dangerous-Goods/DGS_COP_StorageSolidAmmoniumNitrate.pdf

BBC News 2021a Lebanon crisis deepens as PM-designate quits over cabinet deadlock *BBC News* 15 July https://bbc.co.uk/news/world-middle-east-57854141

BBC News 2021b Bata blast: satellite images show equatorial Guinea destruction *BBC News* 10 March https://bbc.co.uk/news/world-africa-56337856

BBC News 2021c Bata explosion: equatorial Guinea blasts kill 31 *BBC News* 8 March https://bbc.co.uk/news/world-africa-56311677

BBC News 2021d Bata explosion: equatorial Guinea death toll rises to 98 *BBC News* 8 March https://bbc.co.uk/news/world-africa-56300088

Blumenthal R 2016 The explosive effects of lightning: what are the risks *Acad. Forensic Pathol.* **6** 89–95

Boiko V 2020 Arms warehouse in Balakliya blew up again. This time it's political, *Naspravdi. Today* 26 April https://naspravdi.today/en/2020/04/26/arms-warehouse-in-balakliya-blew-up-again-this-time-it-s-political/

Carapic J, Deschambault E, Holtom P and King B 2018 *A Practical Guide to Life-Cycle Management of Ammunition.* (Geneva: Small Arms Survey) http://smallarmssurvey.org/resources/publications/by-type/handbooks/lcma-handbook.html

Citizen Truth 2017 *Explosions in Balakleya, Ukraine—Footage from Drone* [Video] https://youtu.be/YcvwjnAn59g

Conway R, Tatom J and Swisdak M 2010 *SciPan 4: Program Description and Test Results* https://apps.dtic.mil/sti/citations/ADA532254

Cranz C 1925 *Lehrbuch der ballistik* (Berlin: Springer)

Dalatrm 2020 *First photos emerging from #Beirut Port's #Hangar12 with workers stacking what seems to be the salvaged Nitroprill Ammonium Nitrate which caused the #Beirut Blast* [Tweet]. 4 August https://mobile.twitter.com/Dalatrm/status/1290780414114779137/photo/1

DimaSadek 2020 *This is how the bags of death were stacked. The photos were taken days before the accident when the State Security Service reported the presence of ammonium nitrate in warehouse 12* [Tweet]. 6 August https://twitter.com/DimaSadek/status/1291486030458224640

European Defence Agency (EDA) 2010 *Harmonization of Ammunition Qualification* (Brussels: EDA) https://eda.europa.eu/docs/default-source/projects/20150715_dteb_ennsa.pdf

Forensic Architecture 2020 *The Port of Beirut Explosion* https://forensic-architecture.org/investigation/beirut-port-explosion

Galassi T 2014 *Guidance on the Ammonium Nitrate Storage Requirements in 29 CFR 1910.109(i)* https://osha.gov/laws-regs/standardinterpretations/2014-12-03

Google Earth 2019a Balakliya 2015 [Map] 49°28'57'N 36°52'36'E (Accessed in 2019 via Google Earth Pro 'Historical Images' function)

Google Earth 2019b Balakliya 2017 [Map] 49°28'57'N 36°52'36'E (Accessed in 2019 via Google Earth Pro 'Historical Images' function)

Google Maps 2021a Port of Beirut [Map] 33°54'04'N 35°31'08'E https://google.com/maps/place/33°54'04'N+35°31'08'E/

Google Maps 2021b Nkoantoma, Bata [Map] 1°49'06'N 9°48'32'E https://google.com/maps/place/1°49'06.0'N+9°48'32.0'E/

Greene O, Holt S and Wilkinson A 2005 *Ammunition Stocks: Promoting Safe and Secure Storage and Disposal* (Belgrade: South Eastern and Eastern Europe Clearinghouse for the Control of Small Arms and Light Weapons (SEESAC)) https://seesac.org/f/docs/SALW-Destruction-2/Biting-the-Bullet-EN.pdf

Gresshoff M 2018 Insensitive high explosive shock-to-detonation transition criteria *16th Int. Detonation Symp.* (Cambridge, MD, 15–20 July 2018) (Livermore: Lawrence Livermore National Laboratory) https://osti.gov/servlets/purl/1465274

Hopkinson B 1915 *British Ordnance Board Minutes 13565* (Kew: The National Archives)

Interfax Ukraine 2018 Sabotage in Balakliya ammo depot in 2017 was carried out by combat drone—matios *Interfax Ukraine* 31 May https://en.interfax.com.ua/news/general/508903.html

Jorolemon M, Lopez R and Krywko D 2021 *Blast Injuries* https://ncbi.nlm.nih.gov/books/NBK430914/

Klapötke T 2019 *Chemistry of High-Energy Materials* 5th edn (Berlin: De Gruyter)

Mansour H *et al* 2021 The Beirut Port explosion: injury trends from a mass survey of emergency admissions *Lancet* **398** 21–2

Maxar Technologies 2021 Open data response to the explosions in Equatorial Guinea, [Blog] *Maxar For a Better World blog* 15 March https://blog.maxar.com/for-a-better-world/2021/open-data-response-to-the-explosions-in-equatorial-guinea

Munitions Safety Information Analysis Centre (MSIAC) 2019 *USS Forrestal, Gulf of Tonkin, Vietnam—29 Jul 1967* (Brussels: MSIAC) https://msiac.nato.int/sites/default/files/media/accident_posters/msiac_accident_poster_13_-_uss_forrestal_vietnam_29_jul_1967.pdf

Nakhoul S and Bassam L 2020 Exclusive: Lebanon's leaders were warned in July about explosives at port—documents *Reuters* 10 August https://reuters.com/article/us-lebanon-security-blast-documents-excl-idUSKCN2562L7

Narwhal8915 2017 *#Balakleya After viewing photos provided by @colonelcassad its a wonder this ammo dump didnt blow up years ago* [Tweet] 23 March https://mobile.twitter.com/Narwhal8915/status/844815738732896257/photo/3

North Atlantic Treaty Organisation (NATO) 2003 *Allied Ordnance Publication 7 (AOP-7)—Manual of Data Requirements and Tests for the Qualification of Explosives Materials for Military Use* 2nd edn (Brussels: NATO) https://nso.nato.int/nso/nsdd/main/standards

NATO 2006 *Standardization Agreement 4581—Explosives, Assessment of Ageing Characteristics of Composite Propellants Containing an Inert Binder* 1st edn (Brussels: NATO Standardization Office) https://nso.nato.int/nso/nsdd/main/standards

NATO 2007 *Standardization Agreement 4582—Explosives, Nitrocellulose Based Propellants Stability Test Procedure and Requirements Using Heat Flow Calorimetry* 1st edn (Brussels: NATO Standardization Office) https://nso.nato.int/nso/nsdd/main/standards

NATO 2008 *Standardization Agreement* 4170—*Principles and Methodology for the Qualification of Explosive Materials for Military Use* 3rd edn (Brussels: NATO Standardization Office) https://nso.nato.int/nso/nsdd/main/standards

NATO 2009a *Standardization Agreement 4488—Explosives, Shock Sensitivity Tests* 2nd edn (Brussels: NATO Standardization Office) https://nso.nato.int/nso/nsdd/main/standards

NATO 2009b *Standardization Agreement 4487—Explosives, Friction Sensitivity Tests* 2nd edn (Brussels: NATO Standardization Office) https://nso.nato.int/nso/nsdd/main/standards

NATO 2009c *Allied Ordnance Publication 15 (AOP-15)—Guidance on the Assessment of the Safety and Suitability for Service on Non-Nuclear Munitions for NATO Armed Forces.* 3rd edn (Brussels: NATO) https://nso.nato.int/nso/nsdd/main/standards

NATO 2009d *Allied Ammunition Storage and Transport Publication 3 (AASTP-3)—Manual of NATO Safety Principles for the Hazard Classification of Military Ammunition and Explosives* (Brussels: NATO)

NATO 2010 *Allied Ordnance Publication 39 (AOP-39)—Guidance on the Assessment and Development of Insensitive Munitions (IM)* 3rd edn (Brussels: NATO) https://imemg.org/wp-content/uploads/AOP-39E(3).pdf

NATO 2015a Allied Ammunition Storage and Transport Publication 1 (AASTP-1)—NATO Guidelines for the Storage of Military Ammunition and Explosives (Brussels: NATO)

NATO 2015b *Standardization Agreement 4491—Explosives, Thermal Sensitiveness and Explosiveness Tests* 2nd edn (Brussels: NATO Standardization Office) https://nso.nato.int/nso/nsdd/main/standards

Pope M and Baker E 2018 Insensitive munitions explosive ordnance disposal challenges *Insensitive Munitions and Energetic Materials Technology Symposium* Portland (Brussels: MSIAC) pp 23–26

Replyanchuk D 2017 *Unnamed Image* [Photograph] (Kiev: Hromadske International) https://en.hromadske.ua/posts/balakliya-special-report

Safex International 2014 *Good Practice Guide: Storage of Solid Technical Grade Ammonium Nitrate* (Blonay: Safex International) https://ime.org/uploads/public/COVID19/SAFEXGPGStorageofSolidTGAN.pdf

Samali B, McKenzie G, Zhang C and Ancich E 2018 Review of the basics of state of the art of blast loading *Asian J. Civil Eng.* **19** 775–91

Small Arms Survey 2014 *Unplanned Explosions at Munitions Sites (UEMS)—Excess Stockpiles as Liabilities rather than Assets* ed E Berman, P Reina (Geneva: Small Arms Survey) http://smallarmssurvey.org/?uems-handbook

Swierk T 2013 *IM Testing and Assessments* (Brussels: NATO Science and Technology Organization) https://doi.org/10.14339/STO-EN-AVT-214-02-pdf

UK Agricultural Industries Confederation (AIC) 2015 *The Storage, Handling and Transportation of Ammonium Nitrate-Based Fertilisers 2015* Peterborough: AIC) https://aictradeassurance.org.uk/latest-documents/storage-handling-transport-ammonium-nitrate-based-fertiliser/

UK Defence Safety Authority (DSA) 2019a *DSA 03—Ordnance, Munitions and Explosives (OME), Part 1—Defence Code of Practice and Guidance Notes for OME Acquisition, Chapter 8—Risk Management* (Bristol: Defence Ordnance Safety Regulator (DOSR)) https://gov.uk/government/publications/dsa03-ome-part-1-ome-acquisition

UK DSA 2019b Storage and Handling of Explosives *DSA 03—Ordnance, Munitions and Explosives (OME), Part 2—Defence Code of Practice for In-Service and Operational Safety Management of OME—Chapter 13* (Bristol: DOSR) https://gov.uk/government/publications/dsa03-ome-part-2-in-service-and-operational-safety-management-of-ome

UK DSA 2019c Quantity distances and other explosives licensing criteria for above ground storage *DSA 03—Ordnance, Munitions and Explosives (OME), Part 2—Defence Code of Practice for In-Service and Operational Safety Management of OME—Chapter 10*, section 2 Bristol: DOSR https://gov.uk/government/publications/dsa03-ome-part-2-in-service-and-operational-safety-management-of-ome

UK DSA 2019d Quantity distances and other explosives licensing criteria for underground storage *DSA 03—Ordnance, Munitions and Explosives (OME), Part 2—Defence Code of Practice for In-Service and Operational Safety Management of OME—Chapter 10*, section 6 (Bristol: DOSR) https://gov.uk/government/publications/dsa03-ome-part-2-in-service-and-operational-safety-management-of-ome

UK DSA 2019e Quantity distances and licensing criteria *DSA 03—Ordnance, Munitions and Explosives (OME), Part 2—Defence Code of Practice for In-Service and Operational Safety Management of OME—Chapter 10*, section 1 Bristol: DOSR) https://gov.uk/government/publications/dsa03-ome-part-2-in-service-and-operational-safety-management-of-ome

UK DSA 2019f Traverses *DSA 03—Ordnance, Munitions and Explosives (OME), Part 2—Defence Code of Practice for In-Service and Operational Safety Management of OME—*

Chapter 7 (Bristol: DOSR https://gov.uk/government/publications/dsa03-ome-part-2-in-service-and-operational-safety-management-of-ome

UK DSA 2019g Buildings associated with military explosives *DSA 03—Ordnance, Munitions and Explosives (OME), Part 2—Defence Code of Practice for In-Service and Operational Safety Management of OME—Chapter 6* (Bristol: DOSR) https://gov.uk/government/publications/dsa03-ome-part-2-in-service-and-operational-safety-management-of-ome

UK DSA 2020 *DSA 03—Ordnance, Munitions and Explosives (OME), Part 2—Defence Code of Practice for In-Service and Operational Safety Management of OME* (Bristol: DOSR https://gov.uk/government/publications/dsa03-ome-part-2-in-service-and-operational-safety-management-of-ome

UK Health and Safety Executive (HSE) 2004 *Storing and Handling Ammonium Nitrate* (London: HSE) https://hse.gov.uk/explosives/ammonium/index.htm

United Nations (UN) 2019a *Manual of Tests and Criteria* 7th edn, (New York, Geneva: UN) https://unece.org/rev7-files

United Nations (UN) 2019b *Recommendations on the Transport of Dangerous Goods—Model Regulations—Volume I.* 21st edn (New York, Geneva: UN) https://unece.org/rev-21-2019

UN Office of Disarmament Affairs (UNODA) 2021 *International Ammunition Technical Guidelines (IATG)* 3rd edn (New York: UNODA) https://unsaferguard.org/un-saferguard/guide-lines

USA Centers for Disease Control and Prevention (CDC) 2003 *Explosions and Blast Injuries—A Primer for Clinicians* (Atlanta: CDC) https://cdc.gov/masstrauma/preparedness/primer.pdf

USA Federal Emergency Management Agency (FEMA) 2003 *Reference Manual to Mitigate Potential Terrorist Attacks Against Buildings, Chapter 4—Explosive Blast* (Washington, DC: FEMA https://fema.gov/sites/default/files/2020-08/fema426_0.pdf

USA National Aeronautics and Space Administration (NASA) Jet Propulsion Laboratory (JPL) 2020 *ARIA Damage Map: Beirut Explosion Aftermath* https://jpl.nasa.gov/images/aria-damage-map-beirut-explosion-aftermath

van der Voort M, Deschambault E, de Roos J and Taylor T 2016 Experimental and theoretical basis of current NATO standards for safe storage of ammunition and explosives *Military Aspects of Blast and Shock (MABS) Symposium 24* Halifax, 2016 (Brussels: MSIAC) https://mabs.ch/spiez-base/mabs-21-to-25/mabs-24/

Weerheijm J and Conway B 2018 *The Klotz Group Contribution to Explosives Safety* https://ndiastorage.blob.core.usgovcloudapi.net/ndia/2018/intexpsafety/WeerheijmPaper2.pdf

Weggel D 2010 Blast threats and blast loading ed N Uddin *Blast Protection of Civil Infrastructures and Vehicles Using Composites.* (Cambridge: CRC Press) 3–43 https://doi.org/10.1533/9781845698034.1.3

Williamson D, Gymer S, Nicholas E, Taylor S, Walley M, Jardine P, Glauser A, French S and Wortley S 2016 Characterisation of the impact response of energetic materials: observation of a low-level reaction in 2,6-diamino-3,5-dinitropyrazine-1-oxide (LLM-105 *RSC Adv.* **6** 27896–900

IOP Publishing

Challenges in Risk Analysis for Science and Engineering
Development of a common language
Tracey J Temple and Melissa K Ladyman

Chapter 4

The application of systems thinking to Risk Assessment: left shifting safety

D Holley, P Gill, N Mai and R Vrcelj

Failure Modes and Effects Analysis (FMEA) has previously been used successfully as part of Quality Control for production processes. The development of the FMEA systems thinking model for identification of quality issues automatically lends itself to viewing a production process from a safety perspective.

In this chapter, the concept of using the FMEA approach to analysing the safety and thus assessing the risks of a process are described. Two examples are given, the first for a seemingly simplistic home-based hobby, incorporating easily accessible concepts and processes. This is then extended to an apparently more complex example, where processing of energetic materials is considered. Both approaches show that with good practice and a well-defined FMEA approach, not only is Quality Control enhanced, but also a more robust and reliable consideration of the Safety Features can be made, leading to a better assessment of the Risk for each process.

The FMEA approach to Risk Assessment implies that more time will be spent at the start of any process, however, this extra time is more than compensated for, when consideration is given to the time and cost spent modifying an industrial process that has already been designed, built and commissioned. The earlier that possible safety issues can be identified, the easier it is to design appropriate mitigations for the critical safety issues. Thus minimising the risk for both the process as a whole and the individual parts of the process.

4.1 Introduction

A safety management system (SMS) is essential to ensure that all aspects of work are conducted in a safe manner. The UK Health and Safety Executive (HSE) recommends using the 'Plan, Do, Check, Act' cycle to achieve a balance between the systems and behavioural aspects of safety management [1]. It discusses treating

doi:10.1088/978-0-7503-3643-7ch4 © IOP Publishing Ltd 2022

health and safety as an integral part of good management generally, rather than as a stand-alone system. It also emphasises that the ‘Plan, Do, Check, Act’ cycle should be repeated as frequently as required.

The implementation may vary, depending on the industry or sector but for the ‘Do’ step of the cycle they all in part follow the same basic 5-step process: (i) identify hazards; (ii) assess the risks; (iii) control the risks; (iv) records findings; and (v) review the controls. Some industries, such as the explosive industry provides extensive guidance on hazard identification and evaluation and the assessment of explosives risk [2].

The UK MODs Acquisition Safety and Environmental Management System (ASEMS) [3] outlines 16 different techniques/tools for profiling and assessing risk. Examples include Bow Tie, Risk Matrices, Fault Tree Analysis and Failure Modes and Effects Analysis (FMEA). They are a mixture of qualitive and quantitative processes and all have different strengths and weaknesses.

In her book ‘Engineering a Safer World’, Leveson [4] discusses that event-based models can easily miss subtle and complex interactions involving organisational and social factors, human decisions and software design. Traditional hazard analysis techniques such as fault tree analysis do not work well for human errors, software errors and other design errors. Leveson further argues that Risk Assessments that are firmly rooted in probabilistic analysis of failure events have proven disappointing when attempting to incorporate complex factors such as software, human and organisational factors. Systems Theory provides a much better foundation for safety engineering than the classic reduction approach underlying event-based models of accidents. Combining systems theoretic approach to safety with system engineering processes will design safety into the system as it is being developed. Systems engineering can merely be viewed as good engineering or common sense and therefore represents more of a shift in emphasis than a change in content. A shift to move the traditional ‘Plan, Do, Check, Act’ cycle to the left and focus more on the Plan with a ‘Plan, Check, Confirm, Do’ process.

Two common tools used for product or process development are Failure Modes and Effects analysis (FMEA) and Quality Function Deployment (QFD). Ginn *et al* [1, 2] discuss how FMEA has traditionally only been used to support the later stages of product development. In contrast, QFD being a four-phase technique that supports the whole product development cycle, in reality is normally only implement in the early stages of product development. They conclude that the use of both QFD and FMEA drives performance and quality by strengthening the link between the requirements and the design.

Failure modes and effects analysis (FMEA) was one of the first systematic techniques used for failure analysis. It was developed in the United States military [3] as a reliability evaluation technique to determine the effect of system and equipment failures. Failures were classified according to their impact on mission success and personnel, equipment and safety. Subsequently it has been used by NASA during the Apollo program, the automobile industry and many others as a tool to complement their design processes [4].

FMEA is a qualitative technique focusing on identify failure potential in a design or process i.e. to foresee failure before it actually happens. After defining the boundary of the system, the following steps are addressed [4]:

(i) How can each component fail?
(ii) What might cause these modes of failure?
(iii) What could the effects be if these failures did occur?
(iv) How serious are these failure modes?
(v) How is each failure mode detected?
(vi) What are the safeguards in place to protect against accidents resulting from the failure modes?

The addition of a numerical assessment of severity of effect, probability of occurrence and ability to detect introduced a critical analysis (CA) stage. Hence, conducting an FMEA and following it with a CA gives a Failure Modes and Effects Critical Analysis (FMECA).

Despite FMEA being a relatively simple technique it is difficult to successfully implement because of a lack of understanding of failure [5]. The concept of linearity in cause and effect leads the process towards the assumption that an effect has one cause, but in reality, there are normally multiple, complex causes and effect chains. The FMEA process is often open to interpretation and dependent upon an individuals or groups personal view and experience.

Burge proposes that there is a need to adopt a systems approach to FMEA as a precursor to a Design-FMEA or Process-FMEA called a Functional-FMEA to identify issues from the beginning and design them out. Applying the concept and principles from systems thinking to create a clear and structure approach to the application of FMEA would provide a more consistent and complete understanding of failure [5].

Quality function deployment (QFD) methodology was developed in the late 1960s in Japan, to improve products at the design stage. The QFD process is widely applied in various industries around the world [6] to assess customer requirements for a system/service and translate them into measurable technical characteristics using a set of matrices [7]. A classic QFD process for product development generally involves four requirements translation phases (figure 4.1).

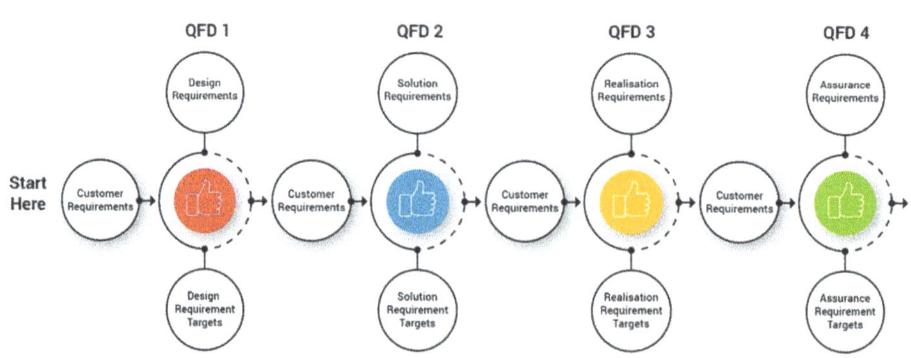

Figure 4.1. The four quality function deployment (QFD) requirements translation phases.

The first phase, which is fundamental in product development, consists of collecting requirements and translating them into measurable design requirements. The second phase, which corresponds to the product concept, translates the design requirements into design solution requirements which are then translated into realisation system targets in the third phase. In the final phase, realisation system requirements are translated into a set of delivery requirements [8]. With this systematic process, each output generated from the inputs of the previous phase are converted into the next phase as its inputs [9].

The QFD methodology has been extended to safety Risk Management (table 4.1) where the holistic analysis of hazards considers the interactions between the system, the users and the working environment but also the preventive measures [10, 11].

The application of QFD to systems safety allows moving the traditional 'Plan, Do, Check, Act' cycle to the left and focus more on the Plan with a 'Plan, Check, Confirm, Do' process. The combination of QFD and FMEA provides a systems toolset that allows identification of many of the possible pitfalls before you get to the 'Do' stage. Using both QFD and FMEA allows a shift in the thinking so that the fault trees and failure modes are available, in detail pre-accident.

There might be many types of FMEA but they are following the same procedure to detect, eliminate and/or mitigate failures resulting from product/system/process/service failures. The two common types include Design-FMEA (D-FMEA) and Process-FMEA

Table 4.1. The four QFD phases.

	Quality focused	Safety focused
QFD Phase 1	Collecting requirements and translating them into measurable technical requirements	An examination of the relationship between non-functional and functional requirements. Identification of missing requirements and requirements needing more detail.
QFD Phase 2	Translates the technical requirements into design solution requirements	An examination of the relationships between the solution and the functional requirements; identifying the key design features that enable the solution to meet the requirements. Identification of the key design features and possible system architectures.
QFD Phase 3	Translates design solution requirements into realisation system requirements	Defines the relationship between the key features and the solution chosen to achieve the process function in a safe fashion.
QFD Phase 4	Realisation system requirements are translated into a set of delivery requirements	Looks at the relevance of proposed control measures with respect to key design or safety features

(P-FMEA). As the name suggests, D-FMEA focuses on assessing the potential risks and their causes and effects, occurring in the early stage of the product design and development, which is more cost effective. This methodology has been widely applied to improve performance but also safety. For example, Li *et al* [12] implemented this technique to identify and mitigate the risks associated with a lithium-ion battery energy storage power station. The other type of FMEA process, called P-FMEA, investigates the potential failure modes related to the production process. This procedure has been used in many sectors. Recently, P-FMEA was employed to define the existing defects observed in the manufacture of roller bearing for the automotive industry and to mitigate them [13]. While for many years P-FMEA has been mainly used in the manufacturing process, it has been applied more recently to maritime operations. The hazards during the transfer of petroleum products from one ship to another, their causes and possible effects have been evaluated [14] and the results showed that P-FMEA can be implemented to other sectors than production.

The application of Functional-FMEA (F-FMEA) to Risk Assessment has not been openly reported in the literature. Russomanno *et al* [15] reported the use of the F-FMEA approach to describe how a system works with a focus directed toward expected goals and determining the effects of a system's failure modes. The functional perspective had the advantage that the language used was more general than any one system being modelled. Ideal for the application of collecting and storing data for computer aided FMEA.

A Functional-FMEA examines the potential design options early; allowing the selection of a safer and more resilient solution before it becomes a D-FMEA or a P-FMEA. This chapter discusses the application for F-FMEA for the Risk Assessment of a process or design.

4.2 Application of functional-failure modes and effect analysis (F-FMEA)

Historically, process development has been a matter of the 'try it and see if it works' means of identifying problems. This might make any issues easy to spot, but when it goes wrong, as it commonly does, time and money will have already been spent to get to the stage of being able to realise a design and subsequently a lot of money, time and effort may now be required to put it right.

Although identifying and analysing problems in design and manufacture early on can be difficult and requires more thought, it is more cost and process effective, both from a quality and a safety perspective. However, the design takes more time to appear 'in the flesh'. This often goes against project culture, where only tangible 'in the flesh' designs are seen as progress and this requires ongoing education of the powers that be (Senior Managers, Project Managers, etc) to understand that a different initial approach is required, not just to 'get on with it'.

A basic outline of what is needed when the life of the device is considered can be readily listed as below and each point can be considered a 'sub-life' of the device:

1. It has to be manufactured safely
2. It has to be tested appropriately to confirm the 'goodness' of the manufacture

3. It has to be delivered
4. Stored
5. Used (and function properly on use)
6. Disposed of at the end of its allotted span.

Of course, for any device this list can be more complex and any sub-life can be broken down into further sub-lives.

Each of the sub-lives will have performance and safety aspects that need to be considered and the approach will be similar for any sub-life:

- Collate and analyse the requirements placed on the design and process
- Examine the functionality of the design and process
- Analyse the functional flow and predict the consequences of failure
- Choose how the functions are going to be achieved in the design
- Perform a trade-off study on those options and select the best one (or two)
- Analyse design for critical factors
- Repeat the above process for the required design process, taking a functional model for the process and looking at the consequences of failure
- Work out where in the process the critical features are to be manufactured
- Look at what in-process controls that can be implemented to deliver the critical features
- Study the noises that can cause unwanted variation in the process
- Devising control measures that will make sure the process controls always hit the target value and that help to eliminate noise and thus reduce variation.

Whilst this reads as a long and complex list of tasks to perform, this is performed through thinking, discussion and co-operation with colleagues. This paper exercise is not as time-consuming or expensive a process as procuring expensive machinery or using trial and error. This process will provide a blueprint for a well-planned safe process.

The above process will be iterative and there will be design and process changes with each forward step. However, the consequences of those changes will be that a better working solution will be provided on completion of the analytical phase of the article and process design. This apparently complex approach is not what it seems, much of what has been discussed already occurs, but this approach moves the whole approach to the start of the whole cycle. Design errors may be more difficult to spot, but inexpensive to correct. To aid in this, a toolkit has been developed.

4.3 The F-FMEA toolkit

In developing a Systems Thinking FMEA approach, it is useful to have a toolkit that can aid and inform any downstream process, from manufacture through to usage/disposal. However, as each process is different, there will be no ‘tick-list’ for any process, but the toolkit is a way of thinking and approaching such a process. The Systems Thinking FMEA approach will also be a dynamic too, so this toolkit will always be changing and developing. In developing the Toolkit, it is important to have a number of concepts regarding the process; the design rationale, the system requirements and the functionality. In addition, a number of methods exist that can give qualifiable and quantifiable support for any particular choice.

The first consideration is for a holistic view of the design, requirements and functionality of the device. These three points are intertwined and cannot be considered separately and this leads naturally to a Systems Thinking approach, which automatically leads to a broader overview of safety and quality. Whilst design, requirements and functionality are to be considered holistically, from a linear narrative perspective (this chapter), they will be discussed slightly separately in the following section for clarity.

These requirements have traditionally been the drivers to design and are defined by the variety of stakeholders that exist, ranging from the customer, through mandatory or legislator requirements, through quality and internal management requirements. As each stakeholder has different priorities, any final design (process and device) has always been a compromise between conflicting requirements.

The question remains how are the conflicts resolved to provide the best solution for the design and process? The answer lies with functional requirements. Functional requirements are focused on what the system actually does. In a processing perspective 'mix the composition' is a functional requirement, it has a verb/noun construction and a direct object. Once it is understood what has to be done, the means can be selected and a choice can be made in how they are done. Prior to that candidate, selection of the functions can be arranged into a logical flow and the consequences of failing to deliver a function can be analyzed.

Function flow should be logical. The nodes adequately describing the function with the input and output arrows described appropriately. It is important to understand the required inputs and desired output for each of the functions.

Inputs such as components or heat and outputs such as waste of solvent vapour can be considered as 'Stores'. Functions may be decomposed into a lower level as required. 'Press material' could be decomposed into a series of functions such as; lower drift, apply load, maintain load, release load, extract pellet.

It important at this stage to keep the descriptions of the functions as generic as possible. Even the act of pressing material describes a solutionised action. Phrases such as 'Change powder form' can be considered as a function. This opens up process choices and so it is important to revisit the functional flow as the design of the article and process proceeds. The basic functional set will be added to as progress is made.

4.3.1 Functionality

Functionality is a description of the functional aspects of the process or device in a broader form and these can be broken down into four main groups (table 4.2).

This more general approach to Functionality allows an early appreciation of impending problems in the design and in the process. There will exist a functional flow diagram, which describes the behaviour of the device, and also a similar diagram, which describes the functions that must be performed to realise the design. All with inputs and outputs and the required withdrawals and dumps to storage.

Table 4.2. The four main functionality aspects.

Solution Dependent Functionality: A particular generic function solution may require further solution specific functionality.	
Legislation Dependent Functionality: Legislation requires incorporation of additional functionality.	
Architectural Functionality: At the system level the collected functional solutions require further functionality typically to support, align, contain, and protect.	
Emergent Functionality: Functionality aimed at managing Unwanted Behaviour. Usually relates to informing the user of failure situations	

So, what happens when things go wrong? The design or process fails and does not meet the requirement set or this could be expressed in terms of a quality failure, a defect. It could also be considered in terms of an accident, a safety issue, resulting in injury to the user, unintended targets or bystanders and the public at large. This should be put under an FMEA. This can occur either on the design (D-FMEA) of the process (P-FMEA), or as in how later examples will be examined, at a functional level (FD-FMEA and FP-FMEA).

There is some confusion and imprecision around what a failure mode is, for example, 'Poor composition homogeneity' has been described as a failure mode, whereas it should be defined as a failure mechanism. The following definitions are offered, table 4.3.

In the simple case of a paper clip, the function is to hold paper. Repeated bending will cause the paper clip to break, so it can no longer function (hold paper). The failure mode is 'No hold'. The mechanism by which this occurs is metal fatigue and the cause is cyclical loading (repeated bending, causing metal fatigue). To help remove the failure mode 'No hold', a design or means of achieving the function that will resist cyclical loading must be looked for. Alternatively, the design could be left as it is and put in place restrictions on use so that the cyclical loading is reduced. This may involve introducing other functions that will inevitably have associated failure mode. If the emergent functionality is not controlled then the original failure mode will result. The failure modes can then be ranked according to severity, probability and detection.

This sounds complex, but leads to the following generalised flow diagram (figure 4.2) to identifying, documenting, and prioritising potential functional-failure modes, effects and causes, which helps identify additional functionality or design features *ab initio* that will make the system more robust in the hands of the customer.

To successfully perform this F-FMEA, it works best when drawing upon the expertise and knowledge of a multi-disciplinary team that considers the system functions and how they might fail, identifying potential: Failure mode; Effects of failure and the Causes of failure.

Ratings are assigned to the probability of occurrence (preferably from previously collected data, but general experience as well), severity of an occurrence and probability of detection, and these are then used to determine an overall risk value.

Table 4.3. Failure and failure mode definitions, reproduced from [5].

FAILURE: The termination of the ability of an item to perform a required function [16].
FUNCTION: what the item does, its purpose—should be written as a verb-noun phrase [8]. (See functional requirements earlier in this chapter.)
FAILURE MODE: the manner in which the item fails to meet its intended function [17]—and therefore must be described in terms of the function. Failure modes are used in preference to failures, since a given failure may have several failure modes, but there exist only five failure modes:
(1) The function could simply not be delivered or (2) there could be too much, (3) too little, (4) intermittent delivery and (5) unintended delivery. Precise definitions for each particular scenario must be carefully documented and in some cases one or more of the five principal failure modes may be difficult to imagine however they operate very well for routine guidance.
FAILURE MECHANISM: The physical, chemical or other process that results in failure [16].
EFFECT: what the CUSTOMER/USER of the system might experience as a result of the failure mode [3].
CAUSE: the underlying event that leads to a failure mode, it initiates the failure mechanism [8].

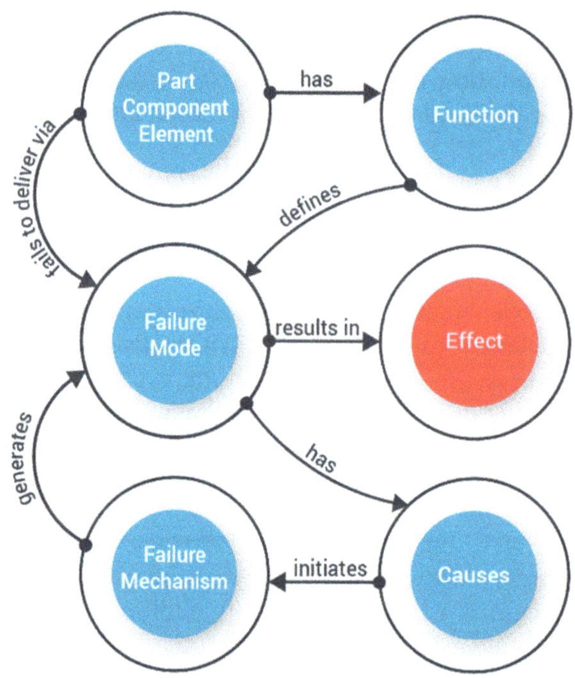

Figure 4.2. Generalised failure flow diagram, adapted from [5].

If an analysis of the potential failures can be conducted, their effects on the community and the probability and severity of the occurrence (risk) can be established and used as a guide to establish the most hazardous areas. This enables design suggestions which may guide choices in the next phases of design.

Historically, much emphasis has been placed on the quality or goodness of the design and process from the perspective of manufacturing a reliable and safe product. The section of the process that is sometimes disregarded is the process safety aspect, as it is considered that there are different tools for that. However, the holistic view proposes that the safety of the manufacturing process will depend on the nature of the design choices made, for example:

- Is there a less sensitive formulation one could use?
- What are the safety consequences of getting the process wrong?
- Where do the most hazardous components lie in the design?
- Where in the manufacturing process are these assembled?
- What controls will be imposed to lessen the risk of an incident in these process steps?
- What noises will interfere with these controls?
- Can design changes be implemented and additional control measures engaged to present a safer AND more reliable means of manufacturing?

Thus, in parallel to the understanding of where the performance critical functions are achieved at the design stage, a similar set of tools can be used to analyse the safety of the system and the associated manufacturing system. Again, the emphasis is on left shifting the design process and looking at the implications of design and process choices and the implications of choice and change before the process is physically constructed. Finding a problem whilst the design and process are still at the drawing and planning stage is much more economical to fix than when equipment, perhaps even the wrong equipment is installed.

While this is more economical, it has already been stated that this is a more difficult process, which is why the application of a toolkit that promotes logical thinking (Systems Thinking rather than Systems Engineering) is important to demonstrate control over the hazards presented. The process, like the design itself, will comprise a series of functions. Each with a desired input and output, each with its own inputs from store and 'dumps' such as waste, vapour and excess heat. The designer will again be able to choose the means for achieving the functions.

4.3.2 Functional MAP illustration

Just as it is important to look at perhaps three or more designs, it is also important to look at three or more process designs and compare the fit against the requirements set both for performance (how many in a week, reject rate etc) and safety, the probability and severity of a potential incident.

Functionality and functional requirements have been discussed mostly as a mechanism for exploring the choices for achieving design function so that the design meets the non-functional requirement set. The process functionality can be seen in a similar way and just as the generic functionality in the design will give rise to the four design functional dependencies (table 4.2). These additional functional natures will become apparent in the process.

These process functionalities parallel that of the argument in design. Additional functionality has been added into the process and as failure was defined as the inability of a component of the system to deliver its assigned function, by adding more functions, more opportunities for failure have also been added.

These added functions will have additional interactions with the requirement set and these may not just be with the primary target requirement, others may be involved. They may interfere with the ability to complete other functions to the required specification and also add complexity in the manner in which higher level functions are delivered.

Of these four functional types the solution dependent functionality and the legislation dependent functionality are the raison d'être of the designer and the safety person respectively. They are perhaps sometimes overlooked, but easier to spot and provide sound design choices for in the 'function/means' stage of process.

It is the architectural and emergent functional types that cause more issues, as they are often 'patches' that are included in the process system to compensate for a design inadequacy. To put right a fault in the design that was not spotted earlier on or simply to remove side effects of a process that could be better managed.

The Systems Thinking FMEA approach toolkit is present below (figure 4.3), with each step explained and discussed.

Step 1. Requirements vs Function. The correlation between requirements and functionality is determined. QFD1 will indicate which of the design or process functions are responsible for delivering the key requirements. It can be used to identify critical functions in the design or the process with respect to safety or performance.

Step 2. Design/Process Concept. The design concept is generated by choosing how each of the functions is to be performed. It is best to generate three or more

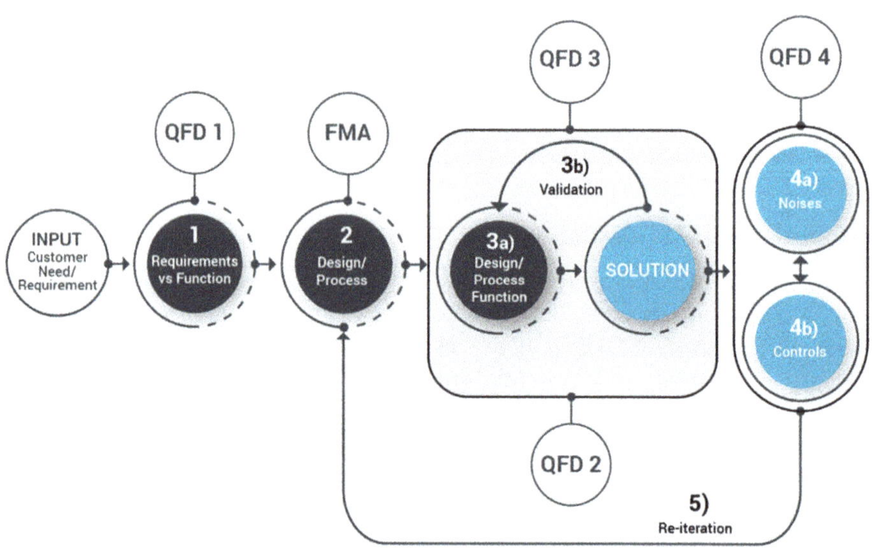

Figure 4.3. Systems thinking FMEA approach toolkit to Risk Assessment.

concepts and perform a multi-criteria analysis against the requirements set, both from safety and design perspectives. A Function Means Analysis (FMA) is a tool that will help understand how the choice of how functions are performed may have an impact on the more sensitive parts of the process. Solution dependent functionality may be generated that may impinge adversely on other parts of the process.

Step 3a. Design/Process Function. It is important to know which of the selected design solutions are responsible for delivering the critical design and process functions. This may be a one-to-one relationship, but this is rare. QFD 2 is a tool to better understand the interactions across the functional map. It is as well to know this before engaging on actually physically putting the process together. A function may dependent upon a series of solutions and conversely a solution may be responsible for delivering more than one function. The consequences of a change in solution must therefore be carefully considered on safety and performance aspects.

Step 3b. Validation. QFD 3 identifies where the critical or sensitive materials or devices are made in the process (if any). Critical steps should be subjected to further 'What/Why' analysis in order to establish the controls that are susceptible to noise and what the most prominent noises are (safety and design).

Step 4. Noises and controls. P-diagrams will help to examine noises and QFD 4 will examine the effectiveness of the controls on the noises. What are you trying to control in the design or process operation and what noises affect that control? A control that is affected by many noises needs control measures to make it work and a noise that affects multiple controls requires further examination. For many processes that involve chemical processing, the biggest two noises are humans (dexterity, aptitude, fatigue and boredom) and the environment (temperature and humidity).

The final QFD puts the control/safety measures into place. Initially this should then lead to a re-examination and re-iteration of the whole system, to ensure that no further emergent functionality arises. To show how this can be achieved, two examples will be given, the first a very generic mechanical example, the second a more detailed and complex example from the Defence sector.

4.4 Example 1: small scale woodworking

This example examines how the toolkit (figure 4.3) can be applied to wood handling for making model ship hulls (figure 4.4). Whilst this initially sounds overly simple, it gives clear examples of how an iterative process can be used to guide both safety and quality.

Step 1. Requirements vs Function. The hull is a plank and frame construction where a spine is created, ribs are placed across the spine and blanks, or fillets are nailed to the ribs to form the hull shape. Hulls are not square and so the planks have to be bent to shape and held in place whilst the adhesive sets to hold them in place.

Functionally, for the whole process is:

1. Prepare plank—The planks are usually soaked to allow them to be bent especially at the bow.

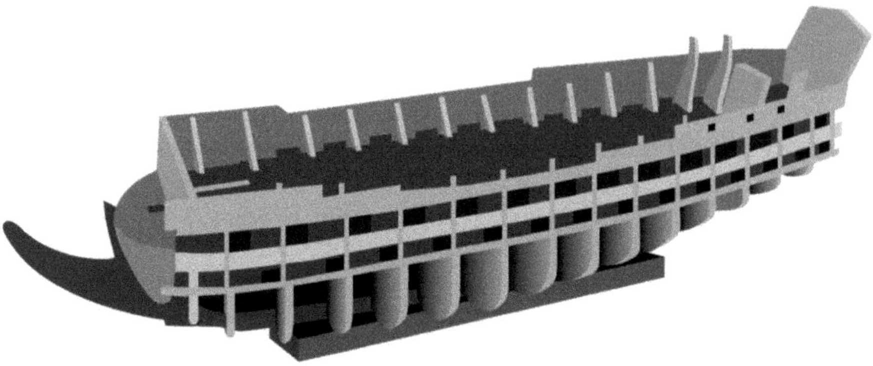

Figure 4.4. Model ship hull under construction.

2. Shape plank—The steaming, bending, cutting and sanding operations that make the plank fit the work.
3. Fix plank—The process of fixing the plank to the frame.
4. Retain plank—The process of retaining the plank in place whilst the adhesive sets.
5. Trim plank—Final shaping and sanding to the required 'boat like' shape.

As with all functional models there are several ways of completing each of the functions, some will be more hazardous than others, some will be more effective at producing quality work than others. What is desired is a reliable means of completing the function, meeting the quality requirements and preventing injury.

Step 2. Design/Process Concept. At this point the design of the process can be considered and written out as a function against means table, where the choices can be listed to winnow out obvious losers and highlight potential winners. A list of reasons can also be added as to why each choice was made, aiding design history and lessons learned. It is also easy to add a hazard identification step at this point as well.

Step 3a. Design/Process Function. Now there is an outline of the process for attaching the wooden boards to form the hull of the model ship. This outline is:

- Prepare the planks by hot soaking the planks to bend them into shape
- To shape the plank, hot soak and then use a modelling knife to rough cut them into shape and abrasives to fine adjust
- Apply a strong PVA adhesive and use brass pins to fix the plank to the frame and retain it in place until the adhesive dries and
- To finally trim the plank, rough trim using the Dremel and fine adjust with abrasives.

A process can now be written based on the above outline and then perform a P-FMEA on it, which can be considered from both a quality control perspective and a Risk Assessment viewpoint too.

A first draft of the salient functions would give a list comprising of:

- Unintended shaping. Shaping of the finger rather than the wood.
- Unintended Fixing. Pinning of the finger rather than the wood.
- No Fixing. Glue or pin in the wrong place.
- Too much fixing. Pin bent by hammer.
- Too little fixing. Pin and glue in the wrong place, loose wood.
- No retention. Work finishes up in the wrong place.
- Unintended retention. Finger pinned or glued to work.
- Unintended trimming. Finger trimmed.
- Too much trimming. Damage to work piece.
- Too little trimming. Work piece not finished.

From these examples it can be seen that there are both quality and safety implications for the choices of means made earlier. These can be mitigated by control measures, which can be identified by a severity, probability and detection analysis and propose control measures. These controls measures may well be additional processes and/or functions. Because of the emergent undesirable behaviour, more functionality has had to be put in place to control the process and as this is a function it too will be prone to failure and it too will need its control measures. It should be noted that there is a difference in terminology here, in Systems terminology, Controls are the things (variables) one tries to control, Control Measures are the methods put in place to make sure the existing controls stay in control and don't wander.

Even at this low level, it is evident from the process list above that there is also another form of emergence, that of solution-based emergence. In selecting to use a pin to fix the planks in place, there is an inherent assumption that it will be hammered in 'Too much fixing. Pin bent by hammer'. In the original analysis, 'too much fixing' was considered as too much force applied by hammer and there are safety considerations when you consider 'unintended hammering' could also hurt the thumb or fingers. The use of the hammer was assumed but has escaped the analysis.

Instead of 'hammer pin', a solution independent description should be used, e.g. 'Apply force to pin'. This may force finding some safer alternatives than wielding a hammer, but if a hammer is required, then suitable control measures can be examined and implemented.

As a first rough draft of the process, this is adequate, but it is clear that some of the functional detail has been missed, for example, to position the work, mark the pin position, hold the pin in place, and apply force to the pin. These need to be added into the functional analysis and then the analysis repeated with these emerging functions included. There will be more failure modes to cope with, frequently occurring ones such as 'Too little hold (of the pin) by the manual method chosen' results in dropping the pin during the process of applying force to the pin resulting unintended application of force to the finger!

In designing and thoroughly examining the process, it has been possible to identify quality and hazard issues. Not only that but in doing so, assumed and emergent functionalities emerged, which in their turn come up with yet more failure modes that need controlling and there are still problems. Sore thumbs, bent pins and pins in the wrong place (Unintended application of force, no hold and unintended marking) being some of the issues. Two more steps are required. Implementation of a more reliable, less hazardous way of performing some of these functions and examination of how to control the process against outside influences.

For the first of these steps, a dry run will be included. Position the plank in place and mark the positions for each of the pins. (Rather than just positioning the plank and fixing them in.) In this case the choice of marking is to use a drill to mark the position, doing this also gives a small pilot hole that will also hold the pin in place.

Again, this generates a new function 'Make hole' that needs a P-FMEA run on it. Too much hole making could result in an oversized hole that will result in a wobbly fix for the pin resulting in 'no hold'.

Some of the causes will come out of the P-FMEA analysis but the process will still be subject to variation and that variation could in turn be the cause of either poor quality work or an accident (table 4.4).

Step 4. Noises and controls. As ever, the first question is 'Control of what?'. It is an important concept which is often overlooked. Some things (variables) one will definitely be able to control and other things (variables) can only be monitored. For example, the temperature at which the model is made can be controlled by doing it indoors with the central heating set at a particular temperature. However, if the relative humidity is important for, say, glue setting, that can only be monitored (unless there is a domestic air conditioning unit available). The latter is a noise and a source of variation.

The processes that should be controlled are something like:

- Soaking time
- Soaking temperature
- Grit of abrasive
- Position of plank
- Pre-drill hole diameter
- Pre-drill hole depth
- Pre-drill hole position
- Pre-drill hole angle
- Pin drive in force
- Pin drive in depth
- Drill speed
- Amount of adhesive
- Type of adhesive
- Cure time for adhesive

The noises that emerge from the above list come in three types; Environmental (including the modeller/operative), deterioration and unit to unit variation and typical noises would be something like:

Environmental

- Modeller getting tired
- Competence of modeller

Table 4.4. Example 1 function against means table.

Function	Desired outcome	Means	Use	Hazards	Comment
Prepare plank	Plank bent and cut to the correct shape ready for fixing	Soak in cold water	No	Water spills	Good but soak times are long
		Soak in hot water	Yes	Water spills scalding	Better. Short soak times and wood works better hot
		Leave dry	No		Useless unless used with a bending tool. Risk of fracture on the bend.
Shape plank	Shape plank to fit required section of hull	Soldering iron	No	Hot surfaces on soldering iron.	OK but heat is too localised
		Steam clothes iron	No	Hot surfaces on steam iron	Better for longer bends but very cumbersome
		Hot soaking	Yes	Scald from hot work. Hot water. Water spills	Best process. Hot soaking puts heat into work evenly.
		Plank bending tool (Shallow slots cut on the reverse)	No	Sharp edges on bending tool. Splinters from work	Not a bad process but precision required. Can result in damaged work
		Abrasive	Yes	Sore fingers. Dust	Best used for fine adjustment
		Saw	No	Sharp edges. Dust	Difficult to control
		Cut with modelling knife	Yes	Sharp edges	Used for initial coarse adjustment
Fix plank	Fix plank to frames	PVA type adhesive	Yes		Takes a long time to set but allows slip and adjustment
		Superglue	No	Instant adhesive. Irritating vapour	Sets too quickly. Little time to adjust joint. Issues with vapours and noxious dust created on sanding
		Superglue 'Gel'	No	Instant adhesive. Irritating vapour	Better than standard superglue but still sets too quickly. Little time to adjust joint. Issues with vapours and noxious dust created on sanding

		Brass pin	Yes	Tools	Pins weak and bend sometimes. Risk of splitting work
		Steel pin	No	Tools. Sparks on grinding	Pins stronger but still bend sometimes. Risk of splitting work. Sparks created on sanding back
Retain plank	Make sure the work stays in place whilst the glue dries	Brass pin	Yes	Tools	Pins weak and bend sometimes. Risk of splitting work
		Steel pin	No	Tools. Sparks on grinding	Pins stronger but still bend sometimes. Risk of splitting work. Sparks created on sanding back
		Clamps	No	Moving parts on tools	Little and limited access for clamps on rib and plank hull
		Fingers	No	Instant adhesive. Chemical hazard	If you want to hold the piece of work all day, Carry on. Only really viable for use with superglue but there's the risk of being stuck to the work
		Removable pin	No	Tools. Sharp edges	Damage to work on removal. Adhesive not strong enough on its own to hold some joints
Trim plank	Trim plank to the shape of the boat after completion of fixing it and the glue has dried	Abrasive	Yes	Sore fingers. Dust	Best way for final shaping
		File	No	Tools. Sharp edges. Dust	OK for coarse shaping
		Dremel sanders	Yes	Tools. Sharp edges. Dust	OK for coarse shaping
		Dremel saw	Yes	Tools. Sharp edges. Dust	OK for coarse shaping

- Lighting in the room
- Comfort (Chair adjustment)
- Ability to hold the work
- Pin 'Fatigue' in storage

Deterioration
- Drill wear
- Tool damage

Unit to unit
- Drill speed
- Variation in hole position
- Variation in hole depth
- Pin type
- Pin strength
- Variation in direction of force
- Variation in quantity of force
- Variation in everything you are trying to control, but the above list is expected to be the most important

A P-diagram is a good way of laying this out, but one also needs to examine how the controls are affected. Application of a 'What/Why' table allows a noises and controls analysis to pinpoint which controls had the most hits from noises (Are you really in control?) and which noises are hitting the controls the most (What needs to stop happcning? What extra controls might be needed?)

In this case of positioning the pin, the choice to use an electric drill. The best choice for a fine positioning hole is with a fine drill bit, but this can break a lot of drill bits. For less breakages of drill bits, a slightly bigger bit can be used, but can lead to the drill skidding and the hole ending up in the wrong place. Also, the hole could end up too wide and too deep and the pin not grip.

With a smaller hole, more force was required to drive it home and hammering accurately becomes more important. Reversion to a hand drill and the use of a pin driver rather than a hammer better controlled the noises and, more importantly prevented a bruised thumb and pins through the fingers.

4.4.1 Example 1 summary

In step 1 (requirements vs functionality) the correlation between the requirements set and the performed functions were examined. The requirements for the hull shape were understood. There's a drawing, it needs to look good and the job needs to be done safely and accurately. But the functions that have to be performed need to be understood to do the task and how well they need to be done. An initial assessment of quality and hazard was performed and a plan to mitigate against them was drawn up. This is often where the process stops when dealing with energetic materials. Especially in the laboratory, one is only dealing with a little bit of it so what could possibly go wrong?

In step 2 (design/concept) the functions were examined and it was decided how they were going to be performed. A means to deliver the functions in the task has been selected. This is very much 'Take 1' and may need thinking about. Again, a necessary step that isn't always perform.

In step 3a (design/process function) QFD 2 was used to look at the interdependency of the solutions across the range of functions that need to be achieved. It helped identify the critical process steps. In this case the drilled hole and the pin are the critical solutions. The P-FMEA has examined the actions and what could happen if there is a failure to deliver the functions in the process or where could problems be caused in the safety of myself and others, and in the quality of the product and hence a possible safety issue for someone else further down the line, i.e. the user. This is in the knowledge that in a full manufacturing scenario, it is entirely possible to make dangerous materials or devices with very little chance of harm to one's self!

- Are extra steps needed? More functions?
- Why are extra steps needed? What has emerged?
- Could things be done differently? Revisit means and reselect.
- If more functionality has been added into the process or changes the way it operates, what happens if the new functions fail or the old ones fail when delivered by a new means?
- Quick fixes—these are elements that are often overlooked. When a 'quick fix' to a process is implemented it is often, if not always forgotten that there may be consequences.

In step 3b (validation) QFD 3 was used to look at where in the process the critical components are assembled. What are the critical process functions that are being performed that give rise to a successfully nailed plank that will contribute practically and aesthetically to the final design.

In step 4 (noises and controls) the process was then revised again and an examination made of the noises and controls that may be encountered. The results of this analysis protects against interference from the outside world, and very importantly looks at the role of Operators in the loop. Added control factors may now be required OR a revision of the operating procedure. The electric drill has been removed because of the control issues and a pin driver included instead of a hammer to modify the application of the force to the pin and avoid bruised thumbs.

Both factors served to remove the result of most noise factors interfering with the process, the effects of variation. The noises and controls were examined in the first selected process and it was discovered that there are things one is trying to control that are influenced by variation in other control factors. One may either add additional control measures or, better, look at the process and redesign it.

QFD 4 puts the additional control measures in place or, as in this case, prompts the designer to go back to the design and rethink things slightly! Adding more controls effectively adds more opportunities to fail, so the analysis must be repeated.

Many see this as additional work. In actual fact it is pre-work not re-work. Much better to spot the issues now, on paper than have to redesign a process after an accident. For this particular example, the method that eventually evolved came through trial and error, after a few bashed thumbs, split planks, missed pins, bent pins and sloppy fixings. All of which had to be treated and fixed. However, it can be seen how taking an overview of the whole design process from the start and simply considering what might be needed allows a clear outline of the way forward.

The process 'Evolved', but clearly thought could have been put into the design at an early stage, which does lead to opportunities to understand the implications of

- Hazards
- Hazardous operations
- Quality
- Consequences of Failure
- Control of 'Work rounds'
- The implications of change on a process
- The proper exploration of new means.

While this may appear to be a trivial example, it clearly shows how putting forethought into the process can improve quality and reduce risk. In more inherently hazardous areas, e.g. the energetic materials sector, the equivalent of the bruised thumb or the split plank is more dramatic, with injury, damage to property and even loss of life being possible outcomes. This will be investigated in example 2.

4.5 Example 2: ignition composition for tracer bullet

In this example from the Defence sector, an examination will be made of the seemingly simple matter of manufacturing by mixing, a tracer composition for use as a low visibility ignition composition for a small arms tracer round (both 5.56 mm and 7.62 mm calibre). This process has been selected as there has been a history of incidents that have happened during manufacture of the composition and several 'quick fixes' have been put in place that themselves have caused further problems.

The manufacture of the composition is in three stages; (i) creation of premix 1, (ii) creation of premix 2, and then (iii) blending of the two premixes to create a final composition that will be used as the low visibility ignition composition within the tracer. An outline of the components is given in table 4.5.

This would normally be approached from a design perspective, looking principally at performance and quality control but in this particular case, an examination will be made of controlling the process from a safety perspective. In addition, a number of requirements are effectively set in stone, the operational requirements for tracer and design of bullets will not change significantly in the foreseeable future.

Table 4.5. List of generalised component materials used in the tracer composition.

Premix 1		Premix 2	
Binder	Bitumen	Fuel	Metal
Binder solvent	Organic solvent	Antioxidant	Organic 'resin'
Oxidant	Metal oxide	Solvent	Organic solvent
Diluent	Metal salt		

From the design perspective the key functions are:

- To ignite from the propellant charge
- To burn for the correct time
- Transfer ignition to the next stage (main tracer).

Whilst this appears to be a relatively simple chain of events, it needs to be seen within the greater scope of the final usage. There are a number of environmental factors that could affect the final performance and working of the above chain. The four predominant environmental factors are:

(i) Large changes in pressure from ambient to high positive pressure within the gun barrel at launch
(ii) A transition to a negative pressure at barrel exit
(iii) Travelling at supersonic speed
(iv) Rotation (spin) of several hundred times a second.

This automatically leads to some emergent functionality in the design, e.g. that the composition should resist the high-pressure environment and provide sufficient strength and rigidity to survive the journey.

If this were to be examined from a critical quality perspective, the following features of the composition design are important:

- Oxidant purity—This is controlled to remove natural variation, pre-manufacture
- Mix homogeneity—The ratio of fuel and oxidant controls the safety (sensitiveness), intensity of light (strength) and the burn rate of the composition (speed)
- Granule size—The composition needs to be of the correct granule size to enable it to flow through the press tooling (for consolidation into a pellet) and giving the correct aliquots for filling.

These features are dependent upon the correct control of the means that have been chosen to produce the composition and would be mapped in a QFD 3 matrix (figures 4.7 and 4.8) and the noises and controls explored from a quality perspective via a P-Diagram (figure 4.5) and 'What/Why' table (figure 4.6).

For this current perspective of Risk Assessment, this raises the following questions:

- How is the risk altered if the process is explored from a safety critical perspective?
- Would this alter the findings of the Risk Assessment?

As previously mentioned, the mix homogeneity is still a key feature, as an oxidant rich mixture has been found to be much more sensitive than the balanced composition. Control of the granule size, archived from a quality perspective by controlling the solvent levels during the mixing, drying and granulating process also has safety implications because of the ease (or otherwise) of granulation, blocking of the sieve causing heat, hence a possible ignition and the amount of dust that could be readily ignited and spread the fire. This can be captured by a functional flow diagram, as shown in figure 4.9.

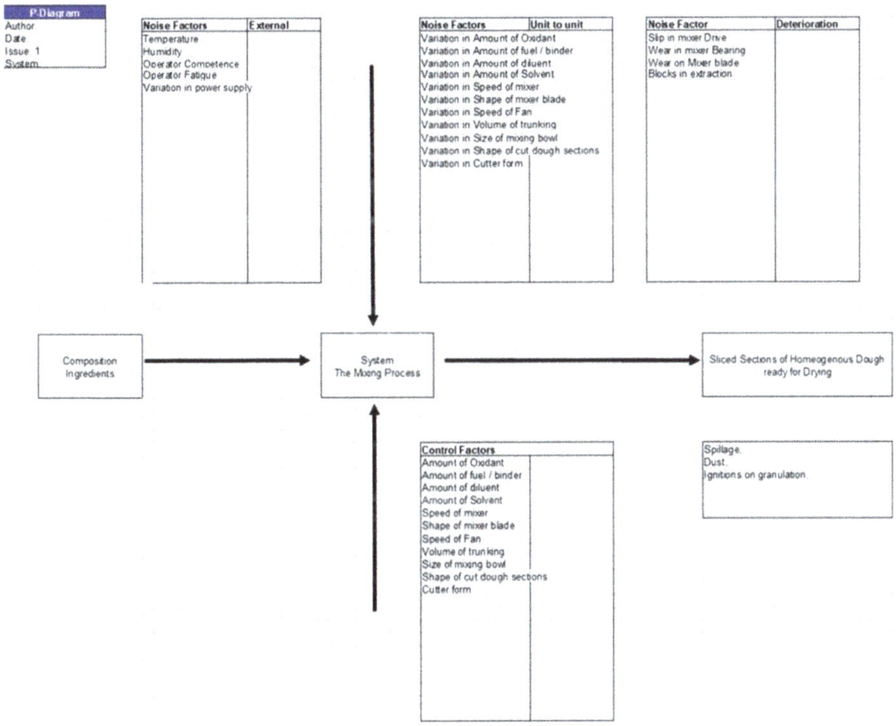

Figure 4.5. Parameter diagram (P-diagram) for the mixing process described in example 2: ignition composition for tracer bullet.

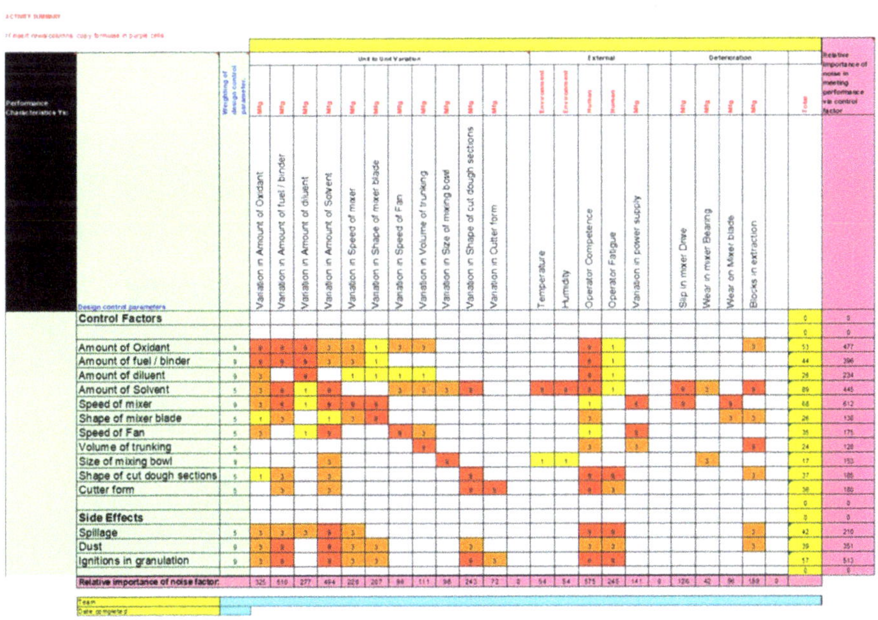

Figure 4.6. 'What/Why' table for the mixing process (example 2: ignition composition for tracer bullet).

		Weigh out ingredients . (Manual electric balance.)	Dispense solvent (Manual, measuring cylinder)	Dissolve Bitumen (Air powered stirrer)	Add ingredients to mixer (manual operation)	Mix (Hobart Mixer)	Extract fume (LEV fan)	Remove from Mixer (manual operation)	Cut dough (manual operation)		
Important design features											
Homogeneity of composition	5	9	3	9	3	9	3	1			
Rheological Properties of mix	5		3	3		9					
Purity of Peroxide	5	9									
Ratio of fuel and oxidant	5	9									
Granule size	5		9	9		9			9		
Dust levels	5					9			9		
		135	75	105	15	180	15	5	90		0
Target											
Benchmark											
Ideal Value											

Figure 4.7. QFD 3 table for example 2: ignition composition for tracer bullet.

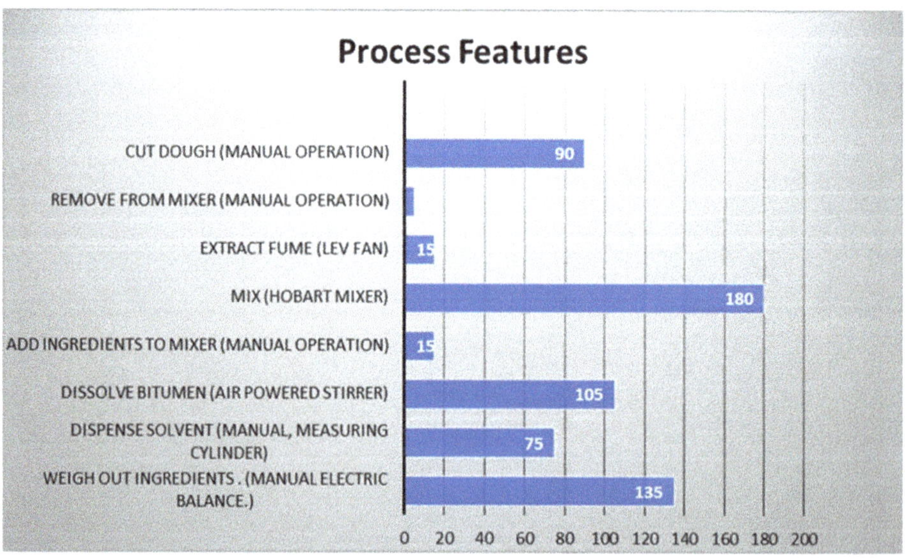

Figure 4.8. QFD 3 relative importance chart for example 2: ignition composition for tracer bullet.

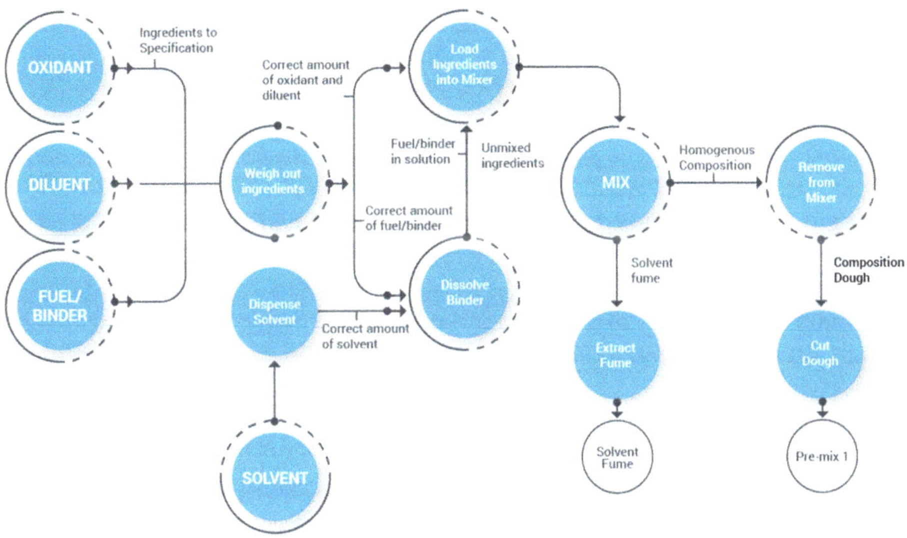

Figure 4.9. Function flow diagram for the pre-granulation phase of premix 1 for example 2: ignition composition for tracer bullet.

The failure modes that would increase the likelihood of an unwanted ignition can be examined via a Process Functional-FMEA and then examine the 'what why' table for key noises and controls. The final stage would be to compile the QFD 4 matrices to tie in the control measures proposed. (Control factor, noise and control measure definitions). Three important definitions at this stage are:

(i) Control factor—Something controllable in the process (e.g. mixer speed)
(ii) Control noise—Something not controllable that affects the process (Humans, Temperature, deterioration)
(iii) Control measure—Something put in place to mitigate against noise of tightened process control.

In constructing the P-FMEA a function is selected (mixing) and then examined to determine, describe and define the failure modes:

- **No mixing**—defined as no torque input to the mix via the mixer blade.
- **Too much mixing**—defined as mixing for more than the prescribed time set for the mix, or at higher than the prescribed speed.
- **Too little mixing**—the opposite action.
- **Intermittent mixing**—defined as a variation on the mix condition, deviation in input to the mix, during the appropriate mixing time, of the appropriate mixing torque.
- **Unintended mixing**—as mixing in of something that is not required or deviation from the approved recipe.

The hazardous effects of the failure modes may fall in two areas: (a) what might the effect be for the manufacturer and (b) for the end user. For (a) it is most likely that the most hazardous effect will be an ignition incident during the manufacturing process. For (b) the user, the hazard (in this case) is likely to be e.g. misfires, but more likely quality issues (e.g. no ignition, failure to work) and this is more likely to be monitored through quality control.

No mixing is possibly the simplest to consider. This is defined as no torque input to the mix via the mixer blade. Too much mixing may either be achieved by mixing for more than the prescribed time set for the mix, or at higher than the prescribed speed. Too little mixing is the opposite action. Intermittent mixing may be defined as a variation on the mix condition, deviation in input to the mix, during the appropriate mixing time, of the appropriate mixing torque. Unintended mixing is generally taken as mixing in of something that is not required or deviation from the approved recipe.

For the example of the tracer, the P-FMEA is one way of identifying the need for these controls and interventions in the process and the QFD process also identifies the need for further controls, both in the safety and quality of the product, so it is possible to return to the P-FMEA for the mixing process.

In conducting a P-FMEA for each function in the process, (ingredient preparation, mixing, granulating, pellet pressing and loading) the interaction between the processes is examined so that the desired output from one process can be understood, so that a successful transition to the next can be made. This is done from the perspective of eliminating rejects or eliminating incidents. In many instances there is a strong relationship between the two.

4.5.1 Individual process—premix 1

If the fabrication of one of the premixes is examined, it will be possible to see how an FMEA approach aids both the quality and the safety. For Premix 1 (table 4.5) the fabrication process (table 4.6) consists of stirring together bitumen (dissolved in the organic solvent) with a powdered oxidant, cut back (mixed) with the metal oxide (diluent). These components, when properly mixed, result in a product that will granulate to make a free-flowing powder (for the next mixing stage with the Premix 2). Examining this mixing process, each of the failure modes described above can be analyzed:

- No mixing leads to no blending of the ingredients and, if the mix is taken without checking to the next stage of the process (mixing with Premix 2), it will lead to an unsatisfactory composition which will not perform its function. However, there is little in the way of a safety risk. As the components are not intimately mixed, ignition of small amounts, is unlikely to proceed to a full ignition of the entire mix. There is a hazard here, in that the failed mix will then require removing from the manufacturing process and disposed of. But this becomes a separate hazard issue (in the context of this mixing example, but not of the whole process).
- Too little mixing will lead to a poorly mixed composition, which, although can cause ignitions, is most likely to result in poor performance. Aggregations of oxidant and/or diluent are likely to exist and areas which are either fuel

Table 4.6. Function against means table for premix 1.

Premix 1					
Function	Desired outcome	Means	Use	Hazards	Comment
Dissolve binder in solvent	Solution made with no excess solid binder	Stir binder in cold solvent	No	Solvent spills or evaporates	OK but may take time to dissolve up the binder (bitumen).
		Stir binder in warm solvent	Yes	Solvent spills or evaporates. Possible flash point fire.	Good, more bitumen can be dissolved up and initial dissolution is quicker. Evaporation rate is faster for warm solvents. May have to do it in enclosed vessel. If the temperature control incorrect, may cause a fire due to solvent reaching its flash point.
Addition of solid powder (oxidizer and diluent) and mixing	Powder completely coated with bitumen and evenly mixed to yield free-flowing powder	Pour powder mixture into binder solution and mix	No	Generation of dust powder. Splashing and spillage of binder solution	The powder is not free flowing at this stage, so will not pour well.
		Pour binder solution into powder mixture and mix	Yes	Spillage of binder solution	It's easier and more controllable to pour a liquid into a solid.

rich or fuel poor (compared to the desired fuel/oxidant ratio) will be common. For the binder, the solvent will not be able to evaporate at the same rate throughout the mixture, leading to possible areas where solvent is trapped in a bitumen coat. This is likely to then lead to a poorly flowing powder, which then does not mix well when taken to the next mixing stage. This will lead to quality issues, as the composition will have poor ignition and burning properties. The most likely hazard issues arise in the removal and disposal of this composition, similar to those faced in the case of 'No Mixing'.

- Intermittent mixing will not be properly mixed and the case is identical to that of 'Too Little Mixing'.
- Unintended mixing is a more problematic case. Unintended mixing may cover a wide variety of possibilities. Two probable cases are: (i) the addition of foreign bodies (e.g. metal shavings or splinters from process machinery) or (ii) the wrong amounts of the correct materials have been added. This will lead to quality issues and will have an impact upon the risks. The most obvious risk will be from metallic foreign bodies within the mix. The mix may proceed 'normally' and the appropriate intimately mixed composition achieved. But there will always be a possibility of frictional heating with the metallic objects; resulting in an ignition leading to a unintentional burning hazard.
- Too much mixing increases the work done on the product, which may cause attrition of the oxidant powder and cause over evaporation of the solvent resulting in a material which is too friable to granulate properly resulting in powdery, dusty product. This leads to a range of possibly damaging issues, e.g. ignitions during granulation, dust fires, problems in dosing the composition during loading and ignitions during pressing.

Studying the causes of 'too much mixing' allows a judgement to be made on the severity and probability (and hence risk) of the failure occurring. It is possible to factor in detectability, a measure of the ability to spot the problem arising before it causes an issue. In this case, detectability of a product that it is 'too dry and friable' to granulate is a matter of experience. So, there is a failure mode whose effect on the system is to introduce a product that has an increased risk to the safety of the process and is difficult to detect unless you are an experienced operator. This leads to the requirement of having have some controls in place to prevent this failure mode from occurring.

4.5.2 Individual process—premix 2

Premix 2 (table 4.5) is manufactured by stirring together the antioxidant resin (dissolved in the organic solvent) with the fuel (metal). When properly mixed (table 4.7), this should lead to a dry composition of metal powder coated in the antioxidant. Coating the metal prevents the metal reacting with atmospheric moisture and possibly resulting in the generation of hydrogen gas. Examining this mixing process, each of the failure modes described above can be analyzed:

Table 4.7. Function against means table for premix 2.

Premix 2

Function	Desired outcome	Means	Use	Hazards	Comment
Dissolve resin in solvent	Solution made with no excess solid binder	Stir resin in cold solvent	No	Solvent spills or evaporates	OK but may take time to dissolve up the bitumen.
		Stir resin in warm solvent	Yes	Solvent spills or evaporates. Possible flash point fire.	Good, more resin can be dissolved up and initial dissolution is quicker. Evaporation rate is faster for warm solvents. May have to do it in enclosed vessel. If the temperature control incorrect, may cause a fire due to solvent reaching its flash point.
Addition of solid powder (metal) and mixing	Powder completely coated with bitumen and evenly mixed	Pour metal powder into resin solution and mix	No	Generation of dust powder. Splashing and spillage of binder solution	The powder may not be free flowing at this stage, so may not pour well.
		Pour resin solution into metal powder and mix	Yes	Spillage of binder solution	It's easier and more controllable to pour a liquid into a solid.

- No mixing leads to no blending of the ingredients and, if the mix is taken without checking to the next stage of the process (mixing with Premix 2), it will, even if this second mix is performed, lead to an unsatisfactory composition which may not perform as required. However, there is a greater risk in this case, due to the unprotected metal possibly reacting with atmospheric moisture and generating hydrogen. From a production perspective, this is unlikely to be of major consequence, as the reaction rates are low. It is feasible that the performance at initial testing is adequate and the items shipped out to the user/customer and it is at this stage that the risk can manifest itself. For example, long periods of storage before use gives the timescales needed for hydrogen build up (and has been seen in previous systems). If this builds up, even for a small tracer round, there is a risk at both storage and operation. In storage, if there is an ignition source (spark, flame), this can ignite the hydrogen being emitted by the whole batch and lead to a major fire and unintended initiation of the tracers. For the operator, when the tracer is being fired, the hydrogen will again ignite and cause some form of misfire. Both can have major consequences for the personnel near the event.
- Too little mixing leads to a poorly coated metal, which will lead to an outcome similar to those of no mixing. In this case, the possible generation of hydrogen at later stages will be less, but still will generate a risk in the same way that No Mixing does.
- Intermittent mixing leads to the metal not properly coated and the case is identical to that of Too Little Mixing.
- Unintended mixing poses more questions than might be identified. Again, two very possible cases are those of the addition of foreign bodies (e.g. other metal shavings or splinters from process machinery) or that the wrong amounts of the correct materials have been added. In this case, the immediate effect of e.g. metal shavings is minimal (but may have an effect downstream), as they ought to be coated with the resin. With the wrong amounts being added, this would either lead to thinner coating on the metal powders, with possible hydrogen generation being a later issue, or coating of metal powders with too much resin, which will result in possible poor performance, as it will take more energy to get through the coating to start a reaction with the metal. The first has later risk implications, the latter having more implications on quality, rather than risk.
- Too much mixing leads to increased work done on the product. However, unlike Premix 1, the metal powders will not undergo attrition. The most likely outcome for this is that heat may be generated, which causes the solvent to evaporate more quickly, leading to a possible uneven coating of resin on the metal, which then has the aforementioned risk associated with hydrogen generation.

For Premix 2, the predominant risk is independent of the failure mode and, as with Premix 1, allows a judgement to be made on the severity and hence risk of the failure occurring. In this case, as with Premix 1, detectability can be factored in, i.e. a product, metal powder that is not evenly coated with resin and is both a matter of experience and can be directly viewed through sampling. Direct observations (which is likely to back up operator experience) is easier to detect than for Premix 1, which lessens the risk to safety, because it can be monitored before any further process occurs. For Premix 2, it is most likely that this will be picked up initially via a Performance FMEA later in the lifecycle (tracers not functioning correctly), rather from than a safety aspect, but clearly the quality and hazard are interlinked.

4.5.3 Individual processes—final mix, blending premix 1 and premix 2

Blending the two premixes is normally carried out in a turbula (dry powder) mixer (table 4.8). Examining this blending (mixing) process, each of the failure modes can be analyzed:

- No mixing leads to the premixes remaining separate entities and, if filled, would give poor to no ignition. Premix 1 being 'just capable' of providing ignition whereas Premix 2 has no oxidant and would not be capable of combustion. The primary role of the metal is to increase the heat of combustion and aid conduction through the ignition transfer purposes.
- Too much mixing would lead to attrition and segregation of the two premixes due to the resulting differing particle sizes.
- Too little mixing is similar to no mixing but with a less distinction between function and non-function. A distribution of ignition times would vary from none to too fast; and potentially too bright a combustion when ignited.
- Intermittent mixing will not be properly mixed and the case is identical to that of 'Too Little Mixing'.
- Unintended mixing could create a variety of failure modes defendant, upon the nature of the unintended portion! Cases of the wrong metal or any contamination have produced sensitive compositions that have caused ignitions on pressing. A reemphasis of the need for rigorous stock control and clear labelling and segregation of ingredients. One metal powder, or white oxidant powder may look very similar to the untrained eye.

Thus, the Design, Functional-FMEA requires re-examining to investigate the functionality of the components of the mix so that an understanding can be made of how to control this one mixing process to obtain an acceptable product. This analysis may come up with check lists for operatives to carry out a process in the approved sequence and for the allotted times. Safety devices can be installed on the mixer so that only the appropriate speed can be selected. It is possible to remove some of the process out of the hands of the operatives, e.g. use automated dispensers (say a syringe pump) that will add the correct amount of binder to the composition at the push of a button or in some cases at a controlled, programmed rate during the mixing process.

How does this differ from the QFD approach? Effectively it doesn't, but is complementary. QFD 3 will identify the critical process, the critical means by which the critical component is made. The QFD delivers a statement of 'Watch the mixing

Table 4.8. Function against means table for final mix.

Final mix					
Function	Desired outcome	Means	Use	Hazards	Comment
Addition of Premix 1 and Premix 2 and final mixing	Even blending of the two premixes	Add Premix 2 to Premix 1 and mix	No	Dust generation, possible ignition hazard.	Premix 2 does not necessarily flow well, so more dust may be generated by turbulence when adding the mix, or it may not flow at all.
		Add Premix 1 to Premix 2 and mix	Yes	Dust generation, possible ignition hazard	Premix 1 is designed to be a free-flowing powder, less likely to generate dust (flammable hazard) and less likely to stick together and become stuck in the original vessel.

process' and from this information, use of 'What/Why' Tables and FMEA can establish the quality and safety controls effectively simultaneously, with some application and thought.

From the design perspective there is a requirement for the product to give a required delay in the ignition of the main tracer composition whilst remaining invisible to the naked eye. There is also a requirement for the composition to survive the journey up the barrel and the pressures and temperatures that entails.

In the subroutines, the functions needed to achieve these requirement goals have been listed and have included functionality such as 'bind' composition and 'provide fuel'. Via the design analysis it has been found that one of the composition ingredients is a key player in both these critical functions. A critical component has been found.

The next part of the QFD process is to examine the requirements on the manufacturing process and select means of achieving the manufacturing goals. Once these means have been selected, the critical design features can be cross matched with the process solutions and establish which process solutions deliver the key features.

In the analysis of the composition, proper execution of the mixing function delivers a product in which the binder and fuel are evenly distributed providing a composition which will perform well in the bullet and that can be manufactured safety. The binder not only provides structure in the design but also controls the risk of initiation during the manufacturing process.

A parameter diagram (P-diagram) for the process compares the items that are under control and the apparent noises on the system. This analysis tallies with the

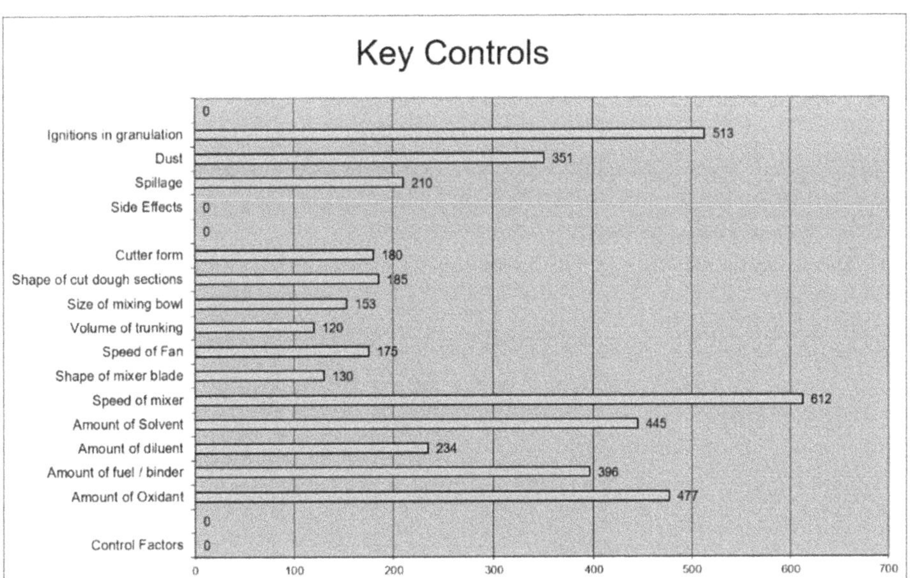

Figure 4.10. Controls diagram. Higher value controls are most influenced by identified process noises.

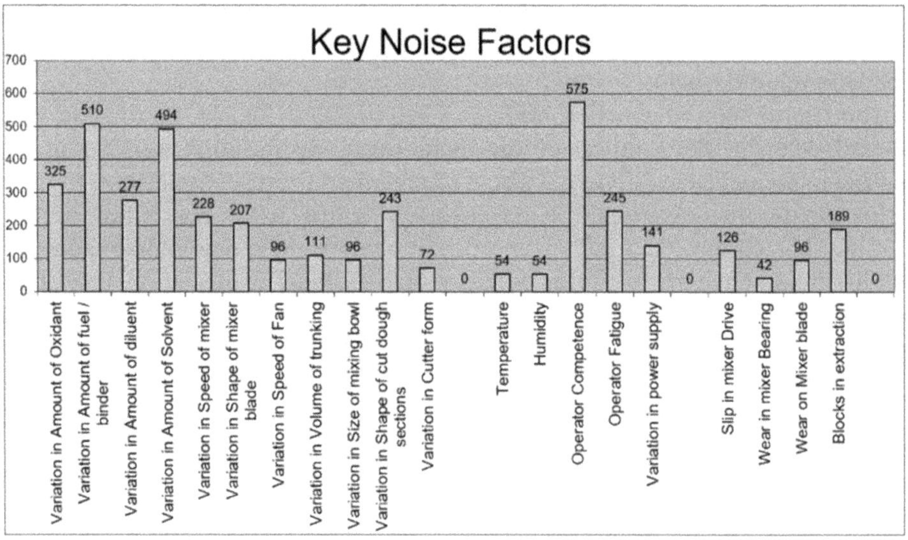

Figure 4.11. Noises diagram. Higher value noises impinge on critical process control factors.

P-FMEA in as much that operative error is the cause of most failures and operatives are the biggest noise.

In addition, examination of the 'What/Why' table helps to illustrate where control factors are susceptible to noise and which noises are providing the biggest influence on the controls that are trying to exerted (figures 4.10 and 4.11). The side effects for the process can also be examined to establish which are made more likely due to the influence of noise.

Significantly the human element and hence clear instruction and training figure very highly in the analysis. This is not desirable as humans are always prone to mistakes, so redesign and process aids to 'mistake proof' the process are required. As such, a bad system will beat a good person every time. This leads to the consideration that it might be preferable to redesign of the process and in light of the established knowledge of the controls and technologies that are now available, it might be best to redesign both.

Is there a better low temperature, organic fuel available today? Is there a better 'dark' oxidant? Is there a better binder? Can the process be re-examined and a different mixer used? Is a three-stage process required?

4.6 Conclusions

The chapter presents a Systems Thinking FMEA (Functional-FMEA) approach to the Risk Assessment of a complex system or process. A toolset for apply a Functional-FMEA based on a Quality function deployment (QFD) methodology is presented. The use of a QFD methodology is common place for assessing the quality of a process. The four QFD stage methodology of: (i) functional requirements, (ii) design solution requirements, (iii) system requirements and (iv) delivery requirements is applied to the safety of two worked examples of: model ship building and manufacture of pyrotechnic compositions.

The Systems Thinking FMEA toolset was applied to (1) a relatively low risk process of building a model wooden ship and (2) a higher risk process of formulating a pyrotechnic explosive for a small arms tracer round. In both examples the reader is led, step by step through the system thinking based toolset. These examples demonstrate that the application of a Functional-FMEA left shifts the traditional 'Plan, Do, Check, Act' Risk Assessment cycle, to focus on solving design and process issues early.

Whilst this sounds like a monotonous functional methodology, what is important to consider is that much of this already happens with a designed system and in the general thinking of the system and process. What this approach does is formalise this thinking into as logical a format as possible, allowing the trail of safety design and hence Risk Analysis to become clear and evident.

By stepping back and taking a functional view the design options can be examined and considered to select better, safer, and more rugged solutions by considering the F-FMEA before it ***becomes*** a D-FMEA or a P-FMEA. Once the designs or processes are cast in stone, they are expensive to fix. Solving the issues ***early*** is cheap. And it is probably time to automate!

References

[1] Ginn D M, Jones D V, Rahnejat H and Zairi M 1998 The 'QFD/FMEA interface *Eur. J. Innov. Manag.* **1** 7–20

[2] Ginn D M 1966 Effective application of quality function deployment: a flexible proposal for best practice within a changing quality improvement environment to ensure highest customer satisfaction *University of Bradford*

[3] DoD US 1980 *MIL-STD-1629A, Military Standard: Procedures for Performing a Failure Mode, Effects, and Criticality Analysis*

[4] MOD UK 2021 DE&S, Acquistion Safety and Environmental Management System (ASEMS) https://asems.mod.uk/toolkit

[5] Burge S 2018 *A Systems Approach to Failure Modes, Mechanisms, Effects and Causes* https://www.burgehugheswalsh.co.uk/

[6] Chan L-K and Wu M-L 2002 Quality function deployment: a literature review *Eur. J. Oper. Res.* **143** 463–97

[7] Moubachir Y and Bouami D 2015 A new approach for the transition between QFD PHASES *Procedia CIRP* **26** 82–6

[8] Burges S 2007 *A Functional Approach to Quality Function Deployement (Putting the Function Back into Quality Function Deployment)* Technical Paper 0001/sb

[9] Chan L-K and Wu M-L 2002 Quality function deployment: a comprehensive review of its concepts and methods *Qual. Eng.* **15** 23–35

[10] Fargnoli M, Lombardi M, Haber N and Guadagno F 2020 Hazard function deployment: a QFD-based tool for the assessment of working tasks—a practical study in the construction industry *Int. J. Occup. Saf. Ergon.* **26** 348–69

[11] Liu H-T and Tsai Y 2012 A fuzzy risk assessment approach for occupational hazards in the construction industry *Saf. Sci.* **50** 1067–78

[12] Li X, Wang Q, Chen Y, Li Y, He Z, Wang T and Wu X 2021 Safety analysis of energy storage station based on DFMEA *E3S Web Conf.* **236** 01006
[13] Krejci L, Schindlerova V, Bucko M, Hlavaty I and Mician M 2019 The application of PFMEA for roller bearings production *Manuf. Technol.* **19** 439–45
[14] Stavrou D I and Ventikos N P 2016 A novel approach in risk evaluation for ship-to-ship (STS) transfer of cargo using process failure mode and effects analysis (PFMEA) *J. Risk Res.* **19** 913–33
[15] Russomanno D J, Bonnell R D and Bowles J B 1993 Functional reasoning in a failure modes and effects analysis (FMEA) expert system *Annual Reliability and Maintainability Symp. Proc.* (Piscataway, NJ: IEEE) pp 339–47
[16] BS 4778-3.1: 1991 *Quality Vocabulary. Availability, Reliability and Maintainability Terms Guide to Concepts and Related Definitions* (London: BSI Standards)
[17] J1739_ 2009 *01 Potential Failure Mode and Effects Analysis in Design (Design FMEA), Potential Failure Mode and Effects Analysis in Manufacturing and Assembly Processes* (Process FMEA) (Warrendale, PA: SAE International)

IOP Publishing

Challenges in Risk Analysis for Science and Engineering
Development of a common language
Tracey J Temple and Melissa K Ladyman

Chapter 5

Influence of automation on human factor integration in Risk Assessment systems

B Thawani, R Critchley and R Hazael

Risk assessment processes have played an important role in the safe functioning of risk critical industries such as nuclear power plants. Over the years, the growth in industry created the need for development of Risk Assessment systems to account for more than just the technical factors. The inclusion of human factors in qualitative and quantitative Risk Assessment process allowed stakeholders in the industry to gauge the influence of human error on workplace safety. Subsequently, the development of automation and on-line process control systems added a level of complexity to industrial undertakings, resulting in further growth in Risk Assessment approaches. Case studies of the 3 Mile Island, Chernobyl, and Fukushima incidents have been discussed to provide context for the influence of human error, automation, and other non-technical factors in Risk Assessment at the time of the incidents.

5.1 Introduction

This chapter discusses the development of Risk Assessment systems in risk critical industries alongside the growth in industrial automation. It also aims to provide specific focus on the concept of human error and its inclusion in Risk Assessment systems. Accident case studies have been used to highlight the factors influencing human factors in Risk Assessment processes and the change in industrial approach to safety after the accident. The chapter is organised as follows: section 5.1 introduces Risk Assessment methods and their classification; section 5.2 discusses the accident case studies and the subsequent development of Risk Assessment processes; section 5.3 highlights future developments in the field of Risk Assessment; and section 5.4 marks the conclusion of the chapter.

5.2 Risk Assessment

Risk Assessment (RA) is the process of analysing an activity or undertaking to highlight potential hazards; estimate the subsequent risk in case of an accident; and

doi:10.1088/978-0-7503-3643-7ch5

© IOP Publishing Ltd 2022

suggest the possible steps to mitigate the risk and ensure maximum safety for the personnel involved [1]. Here, risk accounts for the personnel, financial, environment, and social aspects with regards to a given activity within an organisation [2]. For a given organisation, the factors considered while selecting a Risk Assessment methodology are (i) size of the organisation; (ii) complexity of the system; and (iii) time and resources available [3]. Additionally, Risk Assessment methodologies can be segregated into two groups, i.e., qualitative, and quantitative methods.

5.2.1 Qualitative Risk Assessment

Qualitative Risk Assessment methods function based on a known set of risks with a predefined rating scale to score events based on the likelihood of occurrence, or the impact to the project should the event occur [4]. These methods allow the organisation to define their own scale for RA based on their priorities and the predefined risk posed to various aspects of the project. As the name suggests, qualitative methods are more focused around highlighting the risks; their causes; and the subsequent effects rather than quantifying their corresponding probability [5]. Some examples of qualitative Risk Assessment methods are:

- Decision tree
- Bow tie analysis
- Risk matrix
- SWIFT analysis.

Decision tree analysis

A decision tree analysis is a method of Risk Assessment that allows the user to make project management decisions while accounting for uncertainties in the organisation [6]. This method can be used to assess a series of decisions taken by managers, their various outcomes, and the cost of these outcomes on the organisation. Here, the use of the word 'cost' accounts for the financial, environmental, personnel and social costs incurred to the organisation because of a series of decisions [6]. Figure 5.1 shows a sample structure of a decision tree.

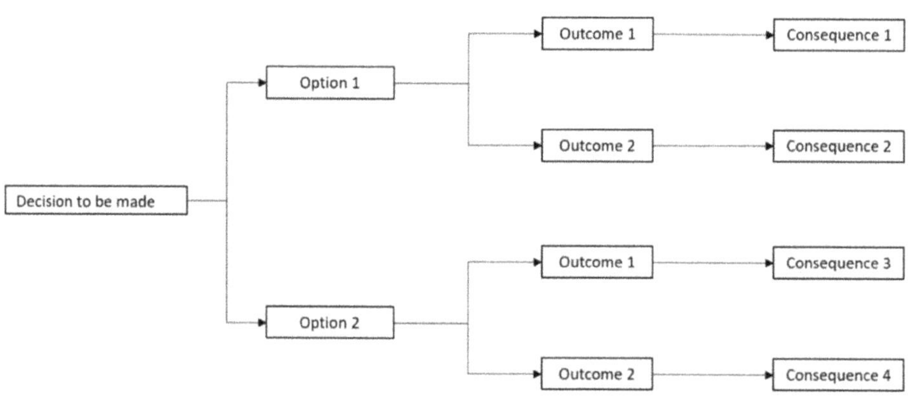

Figure 5.1. Example of qualitative Risk Assessment using decision tree analysis.

In this method, a set of probabilities are assigned to various outcomes and the subsequent effects of these outcomes are assessed to enable better decision making. After a decision tree is constructed, it is up to the manager to make a decision that balances the organisation's profits against the safety of the personnel involved.

Bow tie analysis

The bow tie analysis is a combination of the fault tree and event tree analysis. The fault tree is used to analyse the different series of events that can cause a particular accident and the various barriers in place to prevent it [7]. Subsequently, the event tree highlights the series of events that are bound to occur after the accident and the barriers in place to prevent them. Figure 5.2 shows the basic construction of a bow tie diagram. On the left of the accident is the fault tree diagram and the event tree on the right. The various barriers are displayed as blocks leading to and from the accident. Additionally, in some cases, there are situations where a plant condition escalates the hazard or the consequences. These are known as escalation factors and they require barriers too.

Risk matrix

The risk matrix is a straightforward means of assessing the risks associated with an undertaking and the various consequences of said risks [8]. Figure 5.3 shows the

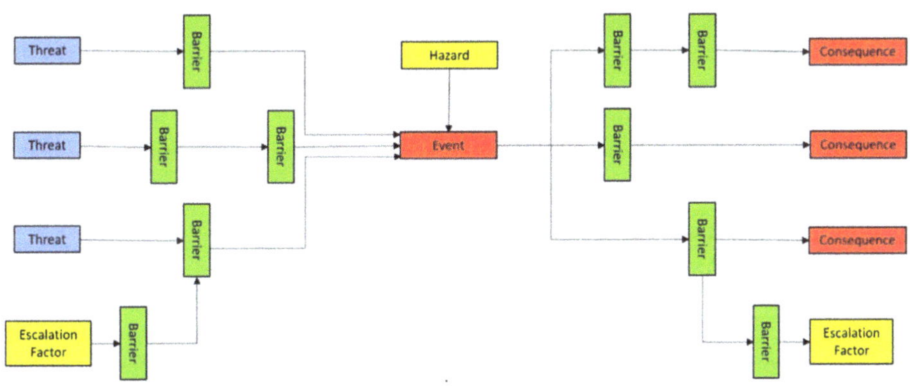

Figure 5.2. Example of qualitative Risk Assessment using bow tie analysis.

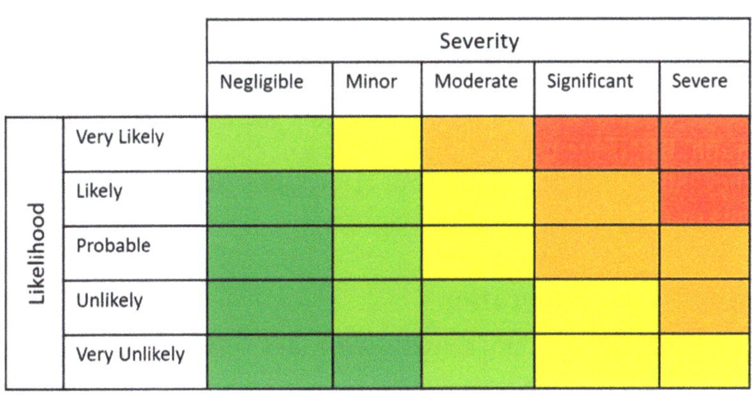

Figure 5.3. Example of qualitative Risk Assessment using risk matrix.

basic structure of a risk matrix. The rows show the probability of occurrence of an event, and the columns represent the severity of the consequences if the event were to occur. The probability of occurrence of an event is measured qualitatively by assigning labels ranging from 'low' to 'high' based on the discretion of the risk assessor. The various events are placed in the matrix based on the probability of their occurrence and the subsequent severity of the consequence. The risk matrix provides a visual representation of the various risks involved in a process but since the probability and consequences are measured qualitatively, it becomes tough for the user of the tool to make accurate decisions [4]. This is because the qualitative nature of assessment creates a level of uncertainty between multiple users due to a difference in judgement based on various factors [8].

SWIFT analysis

The SWIFT (Structured What If Technique) is a system engineering-based technique to identify risks using structured brainstorming. Structured brainstorming employs a mix of predefined guidewords and headings in combination with 'What if' prompts from the participants to allow for a careful examination of risks and hazards at various levels within the system [8]. Compared to conventionally detailed Risk Assessment methods such as HAZOP or FMEA, SWIFT analyses risk at higher level, which reduces the time and resource consumption for the assessment process [4]. The potential consequence of the SWIFT analysis is the risk of overlooking potential hazards because this method is not as detail oriented as HAZOP or FMEA. The simplest means ensure higher reliability of the process would be to combine the SWIFT analysis with another Risk Assessment method to reduce the scope of overlooking a potential risk [9].

Quantitative Risk Assessment

Quantitative Risk Assessment methods are a group of probabilistic Risk Assessment methods applied to the highest priority risks to quantify the possible outcomes of events; aid in decision-making processes within the organisation; and increase the accuracy of cost estimation and scheduling processes [9]. This process is conventionally carried out after a qualitative Risk Assessment because it requires a predefined set of risks and their subsequent prioritisation.

Bayesian belief networks

Bayesian belief networks (BBNs) are a kind of graphical model that use Bayesian inference for probability computation [9]. They aim to develop the conditional dependence between a series of events to model the causation with the help of a directed graph. BBNs are used for the modelling of cause consequence relationships because they allow for the expression of certainty in the form of probability distributions or conditional probability tables [9].

5.3 Human factor integration

Humans are at the heart of any activity being undertaken within an organisation. The involvement of humans in a process inherently creates the scope for human error. The act of having a human-centred approach towards engineering and design is known as Human Factor Integration (HFI) [10]. In the domain of RA, the aim of HFI is to actively account for the risk evolved with respect to human error and how the interaction of humans with a system will influence its efficiency [11].

5.3.1 Brief history of HFI

The core of HFI is based on systems engineering and the interaction of a human with the system. Human factors became prominent in the 1970s, with their core focus on the behavioural analysis of the operator. Subsequently, this was extended towards the effect of human error on the overall system while maintaining focus on the process of mitigation of hazards. Additionally, the advent of safety management systems has increased the value of an organisation's safety culture. Currently, the primary aim for HFI is to understand how the presence of a human influences the functioning of a system and how the active development of novel technologies needed to account for its interaction with humans [12]. The development of novel technologies coupled with the growth in large human–machine systems requires active consideration of human factors early in the design process to allow for better Risk Assessment and aversion. Furthermore, the focus on HFI shed light on the value of ergonomics in system design and the value of user experience on system efficiency [10].

5.3.2 HFI in RA systems

The application of HFI in RA allows key decision makers in an organisation to (i) estimate human error and the costs associated with it; and (ii) estimate the change in system efficiency after accounting for human factors [11]. There have been different approaches to execute the aforementioned processes due to various constraints such as, but not limited to, (i) size of the organisation; (ii) complexity of the process; (iii) cost of conducting RA; and (iv) the extent of human involvement in the process [3, 9, 12, 13].

Like machines, human performance is also affected by age, overuse (exhaustion) and damage (injury) [14, 15]. The aim of HFI is to account for these factors and estimate the change in probable risk for a given process. Based on the RA method and constraints, human factors are given a weightage that influences the final quantification of risk. Historically, human factors have been accounted for by using empirical data in Risk Assessments. Subsequently, growth in the field of ergonomics and an advent in man-machine collaboration in industry has resulted in an increased focus towards the inclusion of real time data for HFI in RA.

5.3.3 Human reliability assessment

With the advent of HFI and ergonomics, it became essential to provide specific focus on the influence of human factors on risk in an undertaking. Consequently, the inception of Human reliability assessment (HRA)—from its roots in Probabilistic Safety Assessment (PSA)—allowed for the use of methods and models to predict the occurrence and impact of human errors [16]. The primary roles of HRA are (i) Human Error Identification; (ii) Human Error Quantification; and (iii) Human Error Reduction [16, 17]. Given the role of the humans in a system, this approach to understanding human error allows for a better understanding of its overall vulnerability and reliability. Subsequently, results from the HRA play a part in the overall PSA, thus increasing the reliability of the PSA results.

5.3.4 Automation

The introduction of robots in industry was the first step towards automation in the manufacturing sector. This enabled a growth in productivity that subsequently resulted in the rise of profits [18]. Additionally, the use of robots allows for better control over product quality and ensure uniformity between product batches. Furthermore, the increased dependency on robots reduces the human intervention required within certain aspects of the production process, resulting in a reduced scope for human error [18]. A reduction in human intervention directly co-relates to the reduction in risk to personnel for a given industrial process. While the process of automation has proven its financial benefits, it has drawbacks worth considering.

The process of automation requires the installation of a control system to ensure smooth functioning and constant monitoring of the robots involved. The addition of a control system adds a level of complexity the industrial process [12]. Since a human operator is required to monitor the control system, the increase in complexity creates a scope for human error, consequently increasing the risk involved in carrying out a given process [13]. There are two different approaches that have been suggested towards reducing the human error with respect to automated processes—Machine learning and Human Robot Collaboration (HRC).

5.3.5 Machine learning

The first approach is that of machine learning, which is aimed at improving the decision-making capability of the robots/machines involved by simulating the human mind. In essence, this process uses data to train the machine to make accurate, statistics-based decisions each time [19, 20]. As the machine is exposed to more data, the decision-making process becomes more accurate. The primary goal of using machine learning is to reduce error in the decision-making process within an organisation. The use of machines to make statistics-based decisions reduces the mental pressure on the human, allowing them to make more rational decisions [18–20]. Furthermore, the focus on machine learning reduces the influence of social and organisational factors for decision-making processes within the industry. Additionally, the use of machine learning for Risk Assessment allows the organisation to respond to threats faster, thus allowing better control in an accident scenario. The challenge with this approach is that it is still in its early stages and requires human intervention for processes that are relatively more dynamic.

5.3.6 Human robot collaboration

The human robot collaboration (HRC) approach is a relatively new approach that, as the name suggests, relies on the collaboration between robots and humans. Here, the robots are expected to carry out the repetitive and relatively mundane tasks whereas the humans are expected to participate in the dynamic aspect of the industrial process pertaining to decision making and creative thinking [21].

The aim of this approach is to combine the strength of robots with human cognitive and tactile abilities, which is expected to increase productivity and enable an ergonomic work environment [21]. Additionally, HRC is aimed at reducing the exhaustion and fatigue of the workers by sharing their workload [22]. Furthermore, HRC is beneficial for the robots as well because it increases the flexibility and range of tasks that a robot can do under the supervision of a human [22, 23]. Compared to the status quo, HRC has the potential to increase productivity of the workers and robots, resulting in a positive growth for the industry [21–23].

While this approach has potential, it brings about its own challenges. The primary concern is safety in the new workspace. Conventionally, humans and robots work in separate spaces independent of each other but in the case of HRC, the two share the same space. The shared space requires different processes to ensure safety of both, the robot, and the human [23]. The secondary concern is the interface for HRC. Given the fact that the human and robot are working in a close knit environment, it is important to recognise that the interface for HRC has to be intuitive to ensure proper execution of tasks [21, 22]. To ensure an intuitive interface, the workers need to be trained to work around the robots, while the robots must be taught to make smarter decisions with minimal processing time to keep up with the dynamic work environment [22, 23].

5.4 Scope

The aim of this text is to shed light on how human error is accounted for in high risk industries; and how that has changed alongside the development in technology; Risk Assessment processes; and organisational approach to safety. This has been done by understanding the safety culture and Risk Assessment practices in a risk critical industry. The extent of HFI in the Risk Assessment processes can be correlated to the measurement of human error in the industrial processes. Furthermore, this chapter aims to understand how major incidents in the industry caused a change in the way regulatory bodies and organisations approached the idea of Risk Assessment and safety. To understand Risk Assessment processes in high risk industries, case studies of nuclear power plants are being considered. This industry has been considered because it is a risk critical industry that is dependent on both, automation, and human intervention, to have reliable functioning.

5.5 Case studies and the development of Risk Assessment systems

Risk Assessment processes in high risk industries play an important role in ensuring smooth running of processes while preventing personnel and financial losses. Furthermore, RA processes allow the supervisors to gauge the extent to which an industrial process can be pushed before posing a threat to the workers and equipment. While the value of RA has been widely recognised in recent years, it is important to understand how our understanding of it has changed historically and some of the important events that caused this growth. This section will look at case studies from the nuclear and chemical industries, with focus on the Risk Assessment and the inclusion of human factors in these processes. The nuclear power industry has been chosen because it is a risk critical industry requiring a balance between human and robot intervention. The case studies have been selected based on their recognition (based on a literature review) as events that led to a reform in safety management processes in their industry. Nuclear power plants are risk critical industries largely because of the use of radioactive materials and the risk they pose in case of an accident of any kind. The release of nuclear waste of any kind poses a long-term risk to humans and the environment alike. Due to this, the safety of the plant, its workers and the surrounding area must be ensured in case of a mishap.

5.5.1 Mile Island

The incident on 3 Mile Island was a loss-of-coolant accident and did not result in any casualties [24]. A partial meltdown occurred, and radioactive material was released into the environment. This incident was significant because it highlighted the glaring errors in the RA and safety culture that existed in risk critical industries at the time [24, 25]. The incident on 3 Mile Island was triggered with the tripping of a steam turbine. The system responded with an emergency shutdown (SCRAM) and control rods were inserted into the core to arrest the chain reaction. Decay heat from the reactor was generating steam in the reactor but it was not being used by the turbine. Conventionally, auxiliary pumps would be used to pump water into the reactor, but these pumps were undergoing maintenance at the time. This was one of the key failures in this accident scenario. Subsequently, the rise in temperature in the reactor increased its pressure, resulting in the relief valve being opened. Ordinarily, a relief valve closes automatically once the pressure has been normalised but, in this case, a mechanical fault in the valve left it open. On the control front, indicators showed the plant workers that the valve was shut, resulting in another error. This depressurisation resulted in the loss of coolant from the reactor core and filling up of the containment building sump [25, 26]. Excessive pressure in the relief pressuriser tank, caused a rupture resulting in coolant leakage in the main building. The consistent rise in pressure and temperature finally resulted in an exposed reactor core which caused a fire due to its reaction with air [24–26]. Consequently, radiation alarms were sounded due to coolant leakage and an exposed reactor core.

5.5.2 Human factor considerations in the 3 Mile Island accident

Post-accident investigations showed that this incident was caused by both human and technical errors. Consequently, this shed light on the organisational factors that enabled these errors, resulting in the emergency scenario [27, 28]. The primary human error in this case was the act of shutting down the reactor without ensuring the proper functioning of the auxiliary pumps. Improper communication and supervisor absence exacerbated the human error, causing the chain of events leading to the loss of coolant [27]. Subsequently, the technical error in the relief valve also played a significant part in the accident scenario. This faulty valve and sensor misrepresented the actual situation of the tank, and since there was no other means to confirm valve function, the sensor indication was trusted without confirmation of any kind [24, 27]. The lack of means to accurately verify sensor information in a situation like this highlighted the flaws in safety culture at the time. It also shed light on the organisational attitude towards safety and the risks associated with it. Furthermore, the accident scenario was worsened by design flaws in the control room, which hindered the workers' ability to identify the source of the alarms and decide an adequate course of action [28]. Additionally, the absence of the supervisor at the time of the accident slowed down communication, resulting in a slower response to the threat. In this case study, these organisational factors led to human and technical errors resulting in an emergency. But this also shed light on the importance of improving RA processes in risk critical industries as well as considering the value of human factors in industrial safety processes.

5.5.3 Influence of Risk Assessment and automation on process safety

During the 1970s, RA processes focused largely on the technical safety and the concept of human factors was still in its nascent stages (figure 5.4) [28, 29]. The focus on technical safety resulted in a lack of focus towards understanding human errors and their effect on system stability. Consequently, the Risk Assessment systems at the time did not adequately account for human–machine interaction, resulting in the formation of an inaccurate sense of process safety [30]. Given that the field of human factors was under-developed, it was inherently difficult for risk assessors and regulators to predict and account for technical issues due to human error [25, 27, 30]. Furthermore, the lack of awareness about HFI allowed significant design flaws (such as the non-ergonomic design of the control room) in the nuclear power plant (NPP) to go unchecked. Since these design flaws were not accounted for, the risk posed by them could not be accounted for either [27]. In addition to this, the plant workers' attitude towards Risk Assessment was very negligent, resulting in regular disregard for work regulations and the creation of an unsafe work environment. In this case, the use of the SCRAM function while the auxiliary pumps were under maintenance was a clear example of worker negligence of regulator guidelines [27, 30]. This action also highlighted the lack of communication within the workplace and how it worsened the emergency.

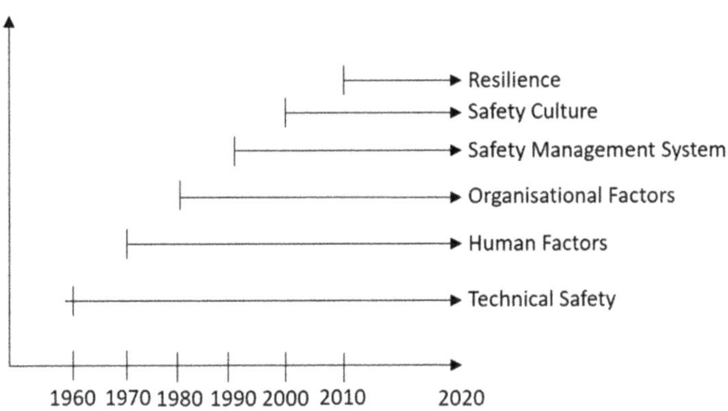

Figure 5.4. Timeline of progression of Risk Assessment processes [28, 29].

At the time of the accident, an on-line process control system was not present in the NPP. An on-line process control system is a computer software that aids in actively monitoring the plant, ensuring desired output and process safety [31]. The absence of such a system increases the decision making load on the workers, creating the possibility of an error [18]. Furthermore, the process control system allows the workers to take the best possible decisions in an emergency to reduce losses on any kind. The decision-making process in a process control system is based on a quantitative Risk Analysis that highlights the outcomes of various actions and their subsequent probability [30, 31]. In this case study, the lack of automation increased the risk for the NPP and the workers. Furthermore, it can be inferred that the presence of a process control system could have potentially prevented this accident by creating a safety barrier between the worker and the emergency shutdown process [28].

5.6 Chernobyl

Chernobyl Reactor Unit 4 suffered a catastrophic accident during a safety test in 1986 resulting in an exposed reactor core. The environmental pollution due to radioactive materials is significant to this day and its effects will be observed in the long-term future as well [32, 33]. This accident can be looked at from different lenses to understand the various factors, technical and non-technical, that contributed to the accident and its subsequent consequences.

Reactor unit 4 used an RBMK reactor to create steam for the turbine. The notable features about an RBMK reactor are its positive power coefficient and the positive void coefficient. A positive power coefficient means that an increase in power input will increase the power of the reactor while a negative power coefficient means that an increase in power input will reduce the power of the reactor [32]. Similarly, a positive void coefficient means that a higher fraction of steam in the reactor will increase the reactor power while a negative void coefficient reduces the power of the generator for the same condition [34]. Additionally, this type of reactor uses graphite (surrounding the fuel channels) as a moderator instead of water. These three aspects setup the base for the technical failure of the NPP in this scenario. Another factor that added to the technical failure was the design flaw in the graphite tipped boron control rods. The graphite tips caused a spike in reactor power each time they were lowered in [34]. The reactor conditions at the time of the accident showed high temperature and pressure which was suggestive of a large void coefficient [32–34]. When the control rods were lowered into the reactor with the emergency SCRAM button, the graphite tips of the control rods caused an increase in power which was further exacerbated by the positive void coefficient. This finally caused a large increase in reactor power, resulting in a high-pressure explosion in the reactor core. From this situation, it is easy to point out the design flaws in that process, but it must be noted that multiple non-technical factors caused these design flaws to be overlooked [32–37].

5.6.1 Human factor considerations in the Chernobyl accident

Admittedly, there was safety system in place to notify reactor engineers that the safety of the reactor was compromised based on the number of control rods in the reactor. On the night of the accident, these warnings were ignored due to the nature of the type of safety test being conducted [35, 36]. This act of ignorance and the errors in the process of the safety test can be chalked up to human error which resulted in this man-made disaster. To put the human error into context, it is important to understand the social and organisational factors that resulted in the erroneous behaviour.

The USSR in 1986 had a prominent culture of secrecy that was driven by the social focus on hierarchy and bureaucracy. This factor played a large role in the workers' understanding of how the RBMK reactor functions because only limited information was provided to them, resulting in a belief of 'complete safety' [36]. Additionally, the rapid growth in the nuclear industry at the time caused a shortage of experienced and trained operators. Beyond the scope of plant personnel, malpractice by the state committee played a large role in this accident scenario [35, 36]. The safety test being conducted on the night of the accident should have been conducted before the reactor unit was cleared for use. Furthermore, there was a last minute change in the safety test programme, which was not planned for, thus adding to the lack of safety considerations [35]. These decisions by the state committee created an unpredictable risk for the plant workers, surrounding communities and the environment. Based on how the events played out before and after the Chernobyl accident, it is safe to conclude that plant designers had not considered all possible outcomes due to human activity at the plant [37, 38]. Furthermore, the design of the control room was not ergonomic and only experienced plant engineers could operate the instrumentation smoothly [35]. An outcome of all these non-technical factors was the inadequate human–machine interaction which added to the chain of events resulting in the accident.

5.6.2 Influence of Risk Assessment and automation on process safety

Around the time of the Chernobyl accident, RA processes had developed and became relatively more inclusive of human factors. This was done by the introduction of HRA methods in conjunction with conventional RA processes in the industry (figure 5.4) [28, 29]. This in turn allowed plant supervisors and industrial regulatory bodies to account for human–machine interaction better, thus improving the level of safety in the workplace [28]. Furthermore, HRA models allowed risk assessors to better understand human behaviour in the workplace by quantifying factors such as, but not limited to, worker experience, stress levels, task complexity and ergonomics [9]. Quantification of these relatively empirical factors allowed regulators to provide an estimation of their influence on worker and process safety. While these aspects of RA processes had developed since the 3 Mile Island accident, their application was not visible in NPPs in Chernobyl [34, 35]. When investigations were carried out after the Chernobyl accident, the effect of organisational factors on process safety was observed. As mentioned earlier, factors such as the secrecy culture and organisational corruption played a key role in the workers' awareness about safety processes and the functioning of the NPP. Consequently, these organisational factors led to the accident and the subsequent mismanagement of the fallout as well.

The Chernobyl accident shed light on the importance of organisational factors on safety in the workplace and how the organisation's safety culture plays a large role in the development of the workers' attitude towards safety [35–37].

In addition to the organisational factors, the lack of automation in the NPP also enabled the chain of events to play out as they did. The only process control system present in the NPP was the PRIZMA program used to estimate the Operational Reactivity Margin (ORM), which is a measure of the equivalent number of control rods present in the core to ensure stable functioning [34]. This system only provided the workers with information regarding the safety of the reactor core and did not have the capacity to override manual control. Furthermore, the location of the console running the PRIZMA program was not ergonomic and usually out of view of the workers. Consequently, it had little influence on the decisions made by plant workers, thus resulting in the negligence of risk and safety [34] . In this case study, the lack of an on-line process control system allowed the plant workers and engineers to carry out a risky test without a proper safety system resulting in a catastrophic accident.

5.7 Fukushima

In 2011, the Fukushima Daiichi NPP suffered multiple explosions due to a reactor failure after the NPP was hit with a tsunami. While this catastrophe was not unprecedented, it shed light on serious design flaws and active ignorance of the safety management systems in the plant [39, 40]. Furthermore, this accident reinforced the value of human; organisational; and cultural factors in RA processes in a risk critical industry. The Fukushima accident is largely considered a man-made disaster because Tokyo Electric Power Company (TEPCO) could have avoided a catastrophe of this scale if the appropriate safety principles and RA processes had been applied [39, 41, 42]. This RA case study can be broken down into the technical and non-technical factors to better understand the direct and indirect causes of the accident.

The Fukushima Daiichi NPP was undergoing maintenance when an earthquake of magnitude 9.0 caused a loss of offsite power, resulting in the activation of emergency on-site AC power to continue the cold shutdown process in the plant. Following the Earthquake, 16m high tsunami waves flooded the plant causing the failure of the emergency on-site power. The tsunami caused the failure of backup DC power resulting in the loss of instrumentation, sensors and alarms that monitored the reactor conditions. The loss of power caused the failure of all reactor cooling and monitoring systems. Consequently, cooling water in the reactor started the evaporate, resulting in a pressure increase in the reactor vessel. Furthermore, the isolation condenser had been shut down before the tsunami and its availability was unknown after the loss of power. High-pressure conditions in the reactor ruled out alternatives like low pressure water injection to control the reactor conditions [39, 43]. Subsequently, unreliable readings on the water level monitors showed that the core was submerged but that was not the case as plant workers detected radiation outside the reactor building [43]. By the end of day one after the tsunami, the reactor pressure had surpassed the design pressure, but nothing was done to start the ventilation process [43]. By the time the ventilation was done, and emergency power restored, an explosion occurred due to a hydrogen-air reaction. The design flaw in the seawall started the chain of events leading to this accident. A seawall of height 5.5m was not nearly enough to protect the NPP and was a sign of inadequate organisational RA processes [41, 43]. Additionally, a notable design flaw in the plant was the underground placement of DC power sources, which resulted in the loss of vital reactor condition data during the loss of AC power [41]. Consequently, the loss of DC power resulted in the failure of hydrogen sensors to detect hydrogen release due to the high-pressure reaction between water and zirconium in the reactor core. These technical factors combined with social and organisational failures created the chain of events resulting in this disaster [37, 41, 44].

5.7.1 Human factor considerations in the Fukushima accident

TEPCO, being the primary stakeholder in the Fukushima NPP had actively put organisational interest above the safety of the plant, subsequently resulting in the catastrophic failure of the safety systems [Ref]. Furthermore, it was later revealed that TEPCO had a history of concealing minor incidents/near misses, causing stunted organisational learning for safety management [41, 42, 44, 45]. While 21st century Japan is not USSR, unpunished concealment of key information does imply that the government colluded with TEPCO. Additionally, Japan does not have an independent regulatory body for nuclear safety, which creates a conflict of interest as the organisation responsible for promoting nuclear energy also must regulate it [43, 46]. Both these factors resulted in the creation of a work environment that was not conducive for the development of a healthy safety culture. TEPCO's collusion with the government combined with the lack of a regulatory body led to a work culture where open discussion was not promoted, resulting in the workers' inability to speak up against safety violations and inadequate Risk Assessment [43, 44, 46]. This organisational failure was further worsened by social factors. The workers' inability to engage in an open discussion or share their concerns prevents the formation of an open safety culture. The lack of an open safety culture is suggestive of an unjust blame culture that can compromise the psychological wellbeing of the employees [37, 43, 44]. Additionally, this threat to the workers' psychological wellbeing is worsened by a culture where the obedience to authority is given primary importance, often resulting in wilful blindness towards serious levels of risk [43]. All these aspects of the Japanese work culture have created a sub-culture of groupism where workers are too scared to step out of the shadow of a group and speak up against malpractice or point out any safety lapses. These social factors created a work culture where no employee could point out organisational failures, creating a false sense of safety and stunting organisational growth [41, 43, 44, 46]. Consequently, when the NPP was in an accident scenario, the workers were inadequately trained to handle the crisis and the organisational failures exacerbated the technical failures, resulting in a man-made disaster that could have been avoided.

5.7.2 Influence of Risk Assessment and automation on process safety

Between the Chernobyl accident and the Fukushima accident, RA processes grew with the development of concepts like safety management systems; safety culture; and safety resilience (figure 5.4) [28, 29]. These concepts allowed for a more structured approach to assessing risk and accurately quantifying the empirical sources of risk. With respect to human factors, concepts like safety management systems and safety culture helped workers become more mindful of risk in the workplace [9]. Furthermore, these concepts were more inclusive of human–machine interaction, thus allowing for the inception of HRC. With respect to the Fukushima accident, the primary causes can be chalked up to organisational and technical factors [37, 43]. Conventional RA processes were regularly carried out in the NPP but the social structure in the organisation impaired its ability to actively include worker feedback regarding the safety of the plant [13]. Furthermore, the lack of a formal regulatory body most likely resulted in the ignorance of recommendations from the RA processes. In addition to this, design flaws in safety and emergency systems hampered the system response to the emergency. The primary example of the organisation's ignorance of risk was the construction of a short seawall. The short seawall was the primary barrier of defence to ensure plant safety in case of a tsunami. When the seawall failed, the plant was left vulnerable to a chain of hazardous events [39, 42].

While the plant had a process control system that was operational, design flaws and sensor malfunctions rendered it useless at the time of the emergency. When the primary source of power was lost, the emergency on-site power was started but that failed due to flooding. The design flaw of putting the emergency power at ground level was evidence of improper Risk Assessment [39, 43, 47]. Additionally, the failure of the emergency systems removed power from the various sensors in the reactor and turbine, resulting in inaccurate information being supplied to plant workers. The lack of sensor data, left the release of hydrogen unchecked, resulting in multiple fires.

5.8 Human factors and Risk Assessment: the way forward

The case studies show how the considerations for RA processes changed after the introduction of human factors. Furthermore, these case studies also showed the change in industrial attitudes towards RA and the influence it has in ensuring a safe work environment. It is also important to note how organisational factors have played a large role in HFI in the workplace and the workers' attitude towards maintaining a safe workplace. The common trend noted from all three case studies was the need to quantify the intangible aspects of human behaviour to accurately measure risk in a workplace undertaking [5, 9, 48]. This requirement stands true even now because multiple studies have shown the importance of understanding the physical and mental health of workers while conducting Risk Assessment for any industrial process [14, 47, 49]. Additionally, the need for a dynamic RA system has arisen to aid plant workers in making accurate decisions while balancing organisational and legislative factors [49]. Here, organisational factors refer to the profitability and efficiency goals of the organisation; whereas legislative factors refer to the required industrial safety standards as suggested by the regulatory body. To make a more dynamic RA system, existing RA processes should be bolstered with the application of concepts such as machine learning and BBNs. Both these concepts are compatible with quantitative RA processes because they are probabilistic and do not require additional data collection [5, 48]. Furthermore, the use of machine learning can help simulate emergency conditions to allow workers to understand the decision-making process to optimise emergency response. Additionally, machine learning data can be coupled with the process control system in the industry to improve its decision-making process in an emergency, thus reducing the stress on the worker [18, 49]. In essence, machine learning can provide a larger and constant stream of industry data to ensure higher reliability of the decision-making process [4, 5, 9]. Subsequently, the use of BBNs allows the user to understand the cause and effect of an accident better by showing the probabilistic correlation between multiple events. Essentially, the use of BBNs will enhance the use of processes like Bow Tie analysis, fault trees and event trees [5]. Furthermore, BBNs have the potential for representing the influence of human error (and its effect on a chain of events) in a graphical manner, thus improving the reliability of methods like the Bow Tie analysis.

While machine learning and BBNs have shown potential to improve the reliability of RA processes, it is important to note that qualitative data about the workers' wellbeing is still the base over which these quantitative processes have been developed [4, 50]. Regular checks of the workers' wellbeing can play a large part in identifying sources of risk in the workplace as well as highlighting potential stressors in the safety culture [12–14, 50]. Subsequently, ensuring worker wellbeing reduces their mental stress, allowing them to make more informed decisions. Consequently, this reduces the probability of human error, allowing for a safer work environment. Additionally, the influence of cultural and social factors must be accounted for in the Risk Assessment process to increase the reliability of the data collected. Cultural and social factors play a large part in the workers' attitude toward safety as well as their personal wellbeing [27, 30, 35, 43]. Accounting for these can allow for an additional layer of personal safety for the workers, resulting in an additional layer of safety for the industrial process, especially in the case of risk critical industries.

5.9 Conclusion

Risk Assessment in the Nuclear Industry has exponentially grown in the last 50 years. Major accidents like the 3 Mile Island, Chernobyl and Fukushima accidents played a large role in highlighting the industrial gaps in RA processes and attitudes towards safety. Additionally, these accidents shed light on the importance of understanding the human, social and cultural factors that play a role in accident causation, along with the technical factors. Based on observations from all three case studies, it was noted that quantitative Risk Assessments alone are not enough to ensure an adequate safety culture. The safety culture in risk critical industries requires the development of dynamic RA processes that are also inclusive of conventionally qualitative aspects pertaining to human factors and ergonomics. In addition to this, the increase in automation and development of on-line process control systems adds a level of complexity to plant control while reducing the mental load on the plant workers. Subsequently, the combination of a dynamic RA system with an on-line process control system should allow workers to have real time risk data, allowing them to better respond to plant conditions in case of an emergency. Machine learning and BBNs have been suggested as potential methods to aid in the development of dynamic RA processes due to their ability simplify the inclusion of non-technical factors while estimating risk. Lastly, it must be noted that a sense of 'complete safety' can be effectively worked towards if all the stakeholders (the organisation; workers; and regulatory bodies) in the industry work in balance towards the same goal.

References

[1] *Risk Assessment: OSH Answers* https://ccohs.ca/oshanswers/hsprograms/risk_assessment.html

[2] Ornai D, Elkabets S M, Kivity Y, Ben-Dor G, Chadad L, Gal E, Tavron B, Gilad E, Levy R and Shohet I M 2020 A methodology of Risk Assessment, management, and coping actions for nuclear power plant (NPP) hit by high-explosive warheads *Adv. Eng. Inf.* **46** 101192

[3] Barlow J and Pillay M 2017 Case study in best practice incident investigation *Advances in Safety Management and Human Performance* ed P Arezes (Berlin: Springer)

[4] Cox L A, Babayev D and Huber W 2005 Some limitations of qualitative risk rating systems *Risk Anal.* **25** 651–62

[5] Kwag S and Oh J 2019 Development of network-based probabilistic safety assessment: a tool for risk analyst for nuclear facilities *Prog. Nucl. Energy* **110** 178–90

[6] Hulett D T 2006 Decision tree analysis for the risk averse organization https://pmi.org/learning/library/decision-tree-analysis-expected-utility-8214

[7] CGE Risk Management Solutions 2019 *The Bowtie Method—CGE Barrier Based Risk Management Knowledge Base* p 1 https://cgerisk.com/knowledgebase/The_bowtie_method

[8] Shuttleworth M 2017 *5 Useful Qualitative Risk Analysis Techniques—Project Risk Manager* p 1 https://project-risk-manager.com/blog/qualitative-risk-techniques/

[9] Steijn W M P, Van Kampen J N, Van der Beek D, Groeneweg J and Van Gelder P H A J M 2020 . An integration of human factors into quantitative risk analysis using Bayesian belief networks towards developing a 'QRA+' *Saf. Sci.* **122** 104514

[10] Sandom C and Harvey R S 2004 *Human Factors for Engineers* (London: The Institute of Electrical Engineers) p 361

[11] Green B 2017 When is a human factors review appropriate? Development of a human factors screening tool for nuclear regulatory commission project managers *Advances in Human Factors in Energy: Oil, Gas, Nuclear and Electric Power Industries* ed P Fechtelkotter (Cham: Springer)

[12] Waefler T, Fischer K, Kunz C and Saric N 2017 Managing tacit knowledge in safety critical systems *Advances in Safety Management and Human Performance* ed P Arezes (Cham: Springer) p 1204

[13] Pillay M 2017 Advancing organisational health and safety management: are we learning the right lessons *Advances in Safety Management and Human Performance* ed P Arezes (Cham: Springer)

[14] Wagner V, Grossi N R and Kallus K W 2017 Age-related impairments like reduced hearing capacity—a safety issue for the working world? *Advances in Safety Management and Human Performance* ed P Arezes (Cham: Springer)
[15] Deng M and Wu F T 2017 Analysis of Chinese coal miners' vision ability *Advances in Human Factors in Energy: Oil, Gas, Nuclear and Electric Power Industries* ed P Fechtelkotter (Los Angeles, CA: Springer)
[16] Hollnagel E 2005 Human reliability assessment in context *Nucl. Eng. Technol* **37** 159–66
[17] Kirwan B 2008 Human reliability assessment *Encyclopedia of Quantitative Risk Analysis and Assessment* (Chichester: Wiley)
[18] Ogle R A, Morrison D T and Carpenter A R 2008 The relationship between automation complexity and operator error *J. Hazard. Mater.* **159** 135–41
[19] Hegde J and Rokseth B 2020 Applications of machine learning methods for engineering Risk Assessment—a review *Saf. Sci.* **122** 104492
[20] Zoppis I, Mauri G and Dondi R 2018 Kernel machines: introduction *Encyclopedia of Bioinformatics and Computational Biology: ABC of Bioinformatics* (Milan: Elsevier) pp 495–502
[21] Poot L, Johansen K and Gopinath V 2018 Supporting Risk Assessment of human–robot collaborative production layouts: a proposed design automation framework *Procedia Manuf.* **25** 543–8
[22] Villani V, Pini F, Leali F and Secchi C 2018 Survey on human–robot collaboration in industrial settings: safety, intuitive interfaces and applications *Mechatronics* **55** 248–66
[23] Ajoudani A, Zanchettin A M, Ivaldi S, Albu-Schäffer A, Kosuge K and Khatib O 2018 Progress and prospects of the human–robot collaboration *Auton. Robots* **42** 957–75
[24] Sorensen J 2014 Three Mile Island *Encyclopedia of Toxicology* 3rd edn (Amsterdam: Elsevier) pp 562–3
[25] US NRC 2016 Three Mile Island Accident of 1979 https://nrc.gov/reading-rm/doc-collections/nuregs/knowledge/km0001/r1/index.html#pub-info
[26] Wahba A B and Steinhoff F 1989 Analysis of the Three Mile Island Unit 2 accident *Nucl. Technol.* **87** 1086–96
[27] Le Bot P 2004 Human reliability data, human error and accident models—illustration through the Three Mile Island accident analysis *Reliab. Eng. Syst. Saf.* **83** 153–67
[28] Swuste P, van Gulijk C, Zwaard W, Lemkowitz S, Oostendorp Y and Groeneweg J 2016 Developments in the safety science domain, in the fields of general and safety management between 1970 and 1979, the year of the near disaster on Three Mile Island, a literature review *Saf. Sci.* **86** 10–26
[29] Jain P, Pasman H J, Waldram S P, Rogers W J and Mannan M S 2017 Did we learn about risk control since Seveso? Yes, we surely did, but is it enough? An historical brief and problem analysis *J. Loss Prev. Process Ind.* **49** 5–17
[30] Swuste P, Groeneweg J, van Gulijk C, Zwaard W and Lemkowitz S 2018 Safety management systems from Three Mile Island to Piper Alpha, a review in English and Dutch literature for the period 1979 to 1988 *Saf. Sci.* **107** 224–44
[31] Choi E, Ha J, Hahm D and Kim M 2020 A review of multihazard Risk Assessment: progress, potential, and challenges in the application to nuclear power plants *Int. J. Disaster Risk Reduct.* **53** 101933
[32] Chesser R K and Rodgers B E 2014 *Chernobyl. Encyclopedia of Toxicology* 3rd edn (New York: Elsevier) pp 822–9

[33] Kortov V and Ustyantsev Y 2013 Chernobyl accident: causes, consequences and problems of radiation measurements *Radiat. Meas.* **55** 12–6
[34] Mullner N 2019 *Three Decades after Chernobyl: Technical Or Human Causes?* (Berlin: Springer) pp 323–40
[35] Munipov V M 1992 Chernobyl operators: criminals or victims *Appl. Ergon.* **23** 337–42
[36] Britkov V and Sergeev G 1998 Risk management: role of social factors in major industrial accidents *Saf. Sci.* **30** 173–81
[37] Ohba T *et al* 2021 The SHAMISEN Project: challenging historical recommendations for preparedness, response and surveillance of health and well-being in case of nuclear accidents: lessons learnt from Chernobyl and Fukushima *Environ. Int.* **146** 106200
[38] Hamer R, Waterson P and Jun G T August 2021 Human factors and nuclear safety since 1970—a critical review of the past, present and future *Saf. Sci.* **133** 105021
[39] Shimura T, Yamaguchi I and Kunugita N 2019 Fukushima nuclear disaster: emergency response to the disaster *Encyclopedia of Environmental Health* 2nd edn (Amsterdam: Elsevier) pp 96–106
[40] Sezen H, Hur J, Smith C, Aldemir T and Denning R 2019 A computational Risk Assessment approach to the integration of seismic and flooding hazards with internal hazards *Nucl. Eng. Des.* **355** 110341
[41] Amorim A P A, Figueiroa Filho C L S and Assis E M d 2021 An INRS cause tree method assessment for complex accidents analysis: application to the Fukushima Nuclear Power Plants accident *Eng. Fail. Anal.* **120** 105062
[42] Yamaguchi I and Kunugita N 2019 Fukushima nuclear disaster-monitoring and Risk Assessment *Encyclopedia of Environmental Health* 2nd edn (Amsterdam: Elsevier) pp 107–20
[43] Murata A 2021 Cultural aspects as a root cause of organizational failure in risk and crisis management in the Fukushima Daiichi disaster *Saf. Sci.* **135** 105091
[44] Berglund J 2020 After Fukushima: safety culture and fostering critical thinking *Saf. Sci.* **124** 104613
[45] Evangeliou N, Balkanski Y, Cozic A and Møller A P 2014 How 'lucky' we are that the Fukushima disaster occurred in early spring. Predictions on the contamination levels from various fission products released from the accident and updates on the Risk Assessment for solid and thyroid cancers *Sci. Total Environ.* **500–501** 155–72
[46] Bansal S and Selvik J T 2021 Investigating the implementation of the safety-diagnosability principle to support defence-in-depth in the nuclear industry: a Fukushima Daiichi accident case study *Eng. Fail. Anal.* **123** 105315
[47] Murata A and Moriwaka M 2017 Anomaly in safety management: is it constantly possible to make safety compatible with economy? *Advances in Safety Management and Human Performance* ed P Arezes (Cham: Springer)
[48] Cai Y and Golay M W 2020 Formulation of a Risk Assessment framework capable of analyzing nuclear power multiunit accident scenarios *Reliab. Eng. Syst. Saf.* **202** 107040
[49] Villa V, Paltrinieri N, Khan F and Cozzani V 2016 A short overview of risk analysis background and recent developments *Dynamic Risk Analysis in the Chemical and Petroleum Industry: Evolution and Interaction with Parallel Disciplines in the Perspective of Industrial Application* (Amsterdam: Elsevier) pp 3–12
[50] Clougherty J E and Rider C V 2020 Integration of psychosocial and chemical stressors in Risk Assessment *Curr. Opin. Toxicol.* **22** 25–9

Chapter 6

Minimising the risks to decision-making by selecting representative experimental methods for environmental science research

E Gutierrez Carazo, F Persico, M Ladyman, T Temple and F Coulon

Representative experimental methods for environmental science are essential to ensure that environmentally relevant decisions such as scale of remediation efforts, mitigation plans and management strategies are supported by accurate and reliable data. Risks associated with these decisions can determine the success or failure of the project, and thus, experimental work needs to be representative of the real environment under study. Therefore, the aim of this chapter is to highlight the importance of incorporating representative methods into environmental science research in order to minimise the risks associated with poor decision-making.

On one side, laboratory or indoor experiments allow researchers to study what happens in the real environment by controlling environmental conditions and isolating specific processes. This affects representativeness and therefore, researchers must engage in working with both concepts and reach a balance between them. On the other side, the use of computational methods is growing due to the cost and time efficiency for achieving results. They require the use of experimental datasets to be used as inputs in the models and to validate the simulations. Although computational methods can support decision-making, it is the researcher who makes the final decision.

Poor decisions can lead to a variety of impacts varying in significance from mostly insignificant to severe economic, legislative, environmental and social consequences. Neither laboratory experiments nor computational methods can be utilised as standalone processes as they are not suitable for defining a representative body of data. Therefore, current research suggests that a hybrid approach (e.g. observational/experimental and computational) is needed to ensure that the research is properly carried out and the risks associated with decision-making are minimal.

doi:10.1088/978-0-7503-3643-7ch6 © IOP Publishing Ltd 2022

6.1 Introduction

Environmental science research mostly focuses on the evaluation, collection and determination of qualitative and quantitative data, to diminish the adverse impact of a specific threat (e.g. contaminants, climate change). Consequently, the risks associated with the decision-making process involved in developing a representative experimental method to define consequences and possible remediation of the threats, are considerably high. Being representative has, therefore, been identified to be a focal point when considering environmental science research. Consequently, reflecting and reproducing environmental changes, which are known to be spatially and temporally heterogeneously distributed, requires attentive planning before attempting to recreate and reproduce experimental studies (Warren *et al* 2005). This reflects on poor decision-making when developing and completing Environmental Risk Assessment (ERA).

ERA involves the determination of the main risks that can cause damage to the environment and the evaluation of their likelihood and severity and this is informed by qualitative and quantitative data preventing small and catastrophic environmental accidents. A poor data collection will reflect in even worse decision-making, increasing the risks of not identifying what the real problems are and how to avoid catastrophic events.

Widely and commonly accepted is that a representative environment can be reproduced only if good planning for the collection of a representative sample is identified (Hadley and Bruce 2014). Nonetheless, the word 'representative', is shaped around the objectives of the research, which are answered following comprehensive research and data collection (Ramsey and Hewitt 2005). The scientific approach involves experimental methods to observe how a change of one or more variables are influencing an entire system, helping the researcher during data collection; although, due to the enormous complexity of the environment, is unlikely that a single experimental method can be used to inform final decision-making. To overcome this problem, a step further has seen the development of data-based computational modelling, where more variables can be included in the system and a more complex environment can be represented. Informing computational modelling requires a wide body of data, and therefore, cannot be representative if used without being supported from experimental methods.

Therefore, the aim of this chapter is to highlight the importance of incorporating representative methods into environmental science research to ensure effective decision-making based on the collection of a comprehensive body of data. This will be achieved by discussing the main approaches used to obtain reliable and representative data from empirical and experimental work going from a comprehensive comparison for the use of indoor facilities to computational modelling reflecting on the differences in replicating a real scenario.

6.2 Incorporating representativeness into experimental methods

Representativeness is mostly associated with qualitative data gathering, where information is collected through surveys relying on people to be representative of a bigger population (Schouten *et al* 2009). The same concept can be applied during acquiring and collecting quantitative data, which should be representative of the studied environment (Sparrow *et al* 2020). Even though different techniques have been developed for field and laboratory work, to overcome these problems, when it comes to representativeness, the main source of error is a consequence of a poorly planned management system for the sampling plan procedure (Webster and Sell 2014).

For this chapter, a sample is defined as a small part of a bigger population, which can comprise of living (e.g. people) or material (e.g. soil, water, air) systems.

The importance of having a well-established sampling plan originated from the need of having well-planned activities to ensure that the outcome of the research is representative of the studied environment. Although different are the problems that who oversees the planning stage (e.g. the researcher) has to face, and several are the questions that should be addressed:

1. How many samples should be collected?
2. What should the size of the samples be?
3. Where should the samples be collected?
4. How should the samples be collected?
5. When can the study area be defined as assessed?

In addition, representativeness of samples is a function of sampling error. If no sampling error exists, the samples would be by definition representative. Thus, determination of sampling error is a critical component of assessing sample representativeness. There is, however, always some degree of sampling error unless the entire population or area is analysed. Analysis of the entire population of interest is typically cost-prohibitive or impractical. Designing a sampling plan for fieldwork requires a high degree of studying to be comprehensive of the studied environment and minimise sampling error.

The population representative of the studied environment would be equal to the volume of the matrix (e.g. soil, water, etc) collected, to underline the need to make a sampling plan before the collection is completed (ITRC 2012, Hadley and Bruce 2014). However, the accuracy of the group of samples collected should be site-specific and representative of the aim of the analysis (EPA 1995) where, inhomogeneities are influenced by various phenomena (e.g. microbiological activities, physical properties) (De França *et al* 2009). The instruments used to collect the sample can be influential if the sample is correctly collected although this is dependent on the objectives (Pepper *et al* 2000). Representativeness is though dependent on the research and the sampling objectives, thinning the line for a universally representative sample (Hadley and Bruce 2014). Therefore, a well-established sampling plan is based on well-determined objectives.

'*The objectives of a sampling plan should describe what you hope to accomplish by implementing a sampling program*' strictly related to the aim of the study (Plans 2006). Once the objectives are properly determined a sampling plan can be developed to ensure that the aim of the research is properly met. Different are the decision that must be made depending on the objectives previously defined, for example: for soil depth related analysis a core sampler must be preferred to a shovel; or because of water mobility an intermittent water sample collection must be preferred to a single collection; or the impingement technique must be preferred for collection of aerosols to analyse airborne microorganisms and so on (Pepper and Gerba 2015).

Following the analyses and description of the objectives, fieldwork, therefore sampling collection, has a huge influential contribution due to the possibility of continuous errors that can happen during on-site research. Jackson (1987), for the first time defines 'useful fieldwork', which described the act of collecting valid information following data gathering. When it comes to environmental analyses, there is an infinite number of variabilities that can compromise the possible results rendering difficult the planning of a representative collection plan. Therefore, several groups of researchers have attempted to develop sampling methodologies to ensure that the sampling collection was as much as representative of the analysed field (ITRC 2012, Hadley and Bruce 2014). Multi-increment sampling has been thus far, the most representative methodology for the collection of a representative sample, although representativeness cannot universally be achieved because of the infinite variability that the environment offers (ITRC 2012).

Storage of the samples after collection can also be significantly influential on possible variability in the integrity of the samples influencing representativeness. Depending on the type of sample and on the type of chemical expected, different storage conditions can have different influences on the samples. Therefore, a well-established plan should contain a section strictly related to how and where the sample will be stored together with the motivations associated with the decision-making process. Following the storage of the samples, indoor facilities are used to detect, analyse and define results from the collected samples.

6.3 The use of indoor facilities: data representativeness uncontrolled versus controlled conditions

Over the years, the use of laboratories or indoor facilities has enabled scientists to control environmental conditions when performing experiments attempting to reproduce outdoor settings. Some scenarios can have a high environmental complexity or can be poorly understood, leading researchers to minimise external conditions by conducting experiments with isolated processes. To that end, researchers commonly control conditions for a better understanding of how the individual mechanisms work and to perform better assessments of the changes that occurred in a system. However, when researchers isolate processes and control variables in indoor facilities, they tend to lose representativeness of outdoor conditions, leading the study to inaccurate data that might not be useful or misguide decision-making. Outdoor environments are usually governed by multiple variables that cannot be simultaneously controlled and that hamper the study of complex scenarios. This raises the challenges of to what extent indoor experiments are representative of outdoor conditions and how the data obtained can be used for decision-making, therefore a balance between controlled conditions and representativeness is needed (figure 6.1).

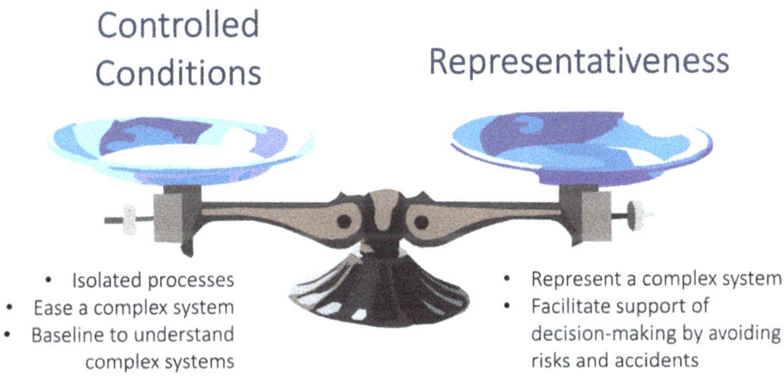

Figure 6.1. Objectives of controlled conditions and representativeness using indoor facilities and laboratories to understand more complex environmental systems.

Assessing sample representativeness is a critical component of any environmental investigation and should be performed before any conclusion is reached. Although, because a sample cannot be entirely representative of a real-life scenario, the main goal of having laboratory experiments is to collect reliable data and ensure that the experimental methods are replicable and reproducible by the entire research community (Anderson and Strich 2016). The advantage of this process is to thoroughly understand different behaviours and relationships between the change in variables. Thus, the optimal laboratory experiments must be kept simple but representative.

For example, the environmental fate and transport of a contaminant can be influenced by an infinite number of environmental variables (e.g. temperature, rainfall, physiochemical properties of both the soil and the contaminant). Organic matter % is one of the most influential parameters involving major changes in the contaminant fate and transport, therefore a low organic matrix (e.g. sand) is chosen for the first experiments. Consequently, once the data from the previous experiments are collected, other variables can be changed and explored, for instance: changing the pH, which can entail the transformation of the contaminant in more toxic products or immobilisation of the contaminant, can define a bioaccumulation process and therefore possible contamination of the flora and fauna; changing the amount of rainfall can determine a faster transportation of the contaminant. Subsequently, a change in two variables can entail numerous changes which cannot be studied during the same experimental design. The complexity can increase if involved in the process are mixture instead of unique compounds.

There are plenty of different experiments that can be performed to obtain representative data. The number of combinations is infinite and thus, it develops the creativity and research skills of the researchers when they require the design of an experimental setup. Therefore, researchers must carefully choose what material/equipment/tools are the most suitable for their indoor experiments that will allow an efficient reproduction of environmental conditions at the same time they offer adequate and representative data. These decisions can influence the data collection and can be different if the design changes (table 6.1).

Table 6.1. Examples of environmental science research studies comparing controlled conditions to experimental representativeness.

Example	Soil (Temple *et al* 2018)	Air (Galante *et al* 2020)	Water (Fawcett-Hirst *et al* 2020)
Main research conducted	Fate and transport processes of two explosive formulations	Evaluation of small-scale combustion of an insensitive munition	Adsorption behaviour of explosive compounds on commercial activated carbons
Controlled conditions	– Sieved and dried soil – Room temperature – Rainfall volume – Use of distilled water	– Use of controlled heat source – Contemplate potential external sources of contamination (material of septum)	Initial chemical concentration – Use of pre-approved extraction methodology Sample shaking time and speed
Representativeness	Triplicates of soil samples (heterogeneity) – Natural soil (sand and sandy loam)	– 25 replicates per sample – Methodology for capturing released gases.	– Use of chemicals-activated carbon as a possible future solution – Use of different chemical concentrations
Outcome	– Representative data – Gravity column might create preferential flow paths – Reliable degradation and transport rates of IHE over time	– Butane as a heat source might not be representative enough – Features such as wind or humidity were not evaluated – Overview of environmental impact posed by insensitive munition	– Baseline dataset of IHE isotherm models and adsorption to commercial carbons – Efficiency of biochar to remove IHE from wastewater

Temple *et al* (2018) studied the mobility and degradation of explosive compounds in two types of soil (sand and loam) simulating what would happen to an explosive residue after detonation. The aim was to determine the fate and transport of explosive compounds under static experiments (degradation) and soil columns (mobility) in the laboratory. Sieving and drying the soil ensured homogeneity of the particle size, avoiding possible problems related to the heterogeneity of the soil and the contaminants transport. The controlled irrigation system ensured that the amount of water trespassing the soil columns was equivalent to the UK rainfall average, to ensure representativeness of a real-life scenario. However, the use of distilled water might have altered the dissolution and immobilisation/mobility of these Insensitive Munition compounds due to the lack in dissolved ions.

Temperature and pH also play an important role and although pH was constant over time, temperature was not recorded creating an opportunity for future work to fill that gap. Although, as the experiments were set in an indoor facility, the temperature supposedly remained constant over time, could be slightly higher than outdoors. This might have potentially affected the processes that the chemicals underwent, leading to a possible higher degradation rate and an early transformation into products. Therefore, the study provides an overview of how these chemical compounds behave in soil systems, leaving gaps to be fulfilled during future studies, defining a starting point for future research.

Galante *et al* (2020), identified the combusted signature of vapour chemical compounds produced during combustion of insensitive munitions and described a small-scale lab method to burn, catch and analyse combustion gases. The experimental setup developed, based on previous work, contained a small vial with 20–30 mg of explosive formulation placed above a heat source and inside of an aluminium block. The gases were collected by an aluminium chromatograph seal. To ensure representativeness, four different septa were used to prove that none of them interfered with the gas analysis, although one of them showed evidence of mechanical deformation due to heat. In addition, 25 replicates were performed per sample, which enabled a more reliable set of data. Temperature gradient was also recorded, although the gas used (butane) might not be enough representative of the temperature that explosive formulations can reach during open burning detonation. Further to this, the study showed a methodology representative enough to study the vapour signature in which two out of three chemical compounds tested in the formulation showed reliable results, providing adequate data that could support reliable decision-making. However, there was another chemical compound whose oxygen balance was lower, leaving some residues in the experiment that probably will not appear outdoors as the concentrations of available oxygen in the air are limitless.

Fawcett-Hirst *et al* (2020) determined sorption of insensitive high explosives formulations on commercial activated carbons in water media as a possible solution of removal of explosives from wastewater. The work studies the interactions of explosive residues with different activated carbons to decontaminate wastewater from manufacturing sites. Main controlled conditions required knowing the explosive initial concentration and the carbon mass to calculate the amount of explosive removed from the water. To determine maximum adsorption, and fewer interactions with the explosive material, distilled water was used. Samples were shaken for 7 days on a horizontal electric shaker at constant speed. However, in wastewaters, insensitive munition chemicals are usually in combination and often contain other salts/ions coming from their synthesis, which could compete with the target chemicals for the activated carbon surface, reducing the maximum sorption capacity. The addition of salts can suppose a change in water pH, which would affect equilibrium of chemical isomers and probably activated carbon sorption. In conclusion, the study provides insight on possible solutions for IHE removal and removal and quality results on how insensitive high explosive chemicals behave compounds under a chemical point of view.

Therefore, all the considered studies showed that entire representativeness of a real-life scenario cannot be achieved, leaving different studies that in the future can be merged to determine a clearer picture. Although, previous work (Temple *et al* 2018, Fawcett-Hirst *et al* 2020, Galante *et al* 2020) showed that replicates are clearly a good indicator as it allows the inclusion of a wider variability of a sample or population. Typically, the addition of positive and negative controls (the contaminant with no media and the media with no contaminant) is added to ensure that the recorded changes are specifically related to the interaction between the two elements (contaminant and matrix).

Even though the use of indoor facilities is widely utilised and experimental settings are replicable and consequently easy to be reproduced from different research groups to ensure continuity of the research, alone can be time-consuming and economically disadvantageous. Hence, recently, to overcome the difficulties in having more research explaining changes of different variables, computational modelling has been introduced, where, using a more limited body of data, is possible to analyse possible changes to better inform Risk Assessments during the decision-making stage.

6.4 Computational modelling and the art of assessing representativeness

In order to deal with an increasingly complicated world and complex problems difficult to resolve, more sophisticated approaches are required for assessing the risk in Environmental Science and Engineering. To that, mathematical or computational models offer a different approach to studying the environment and making decisions by simulating in a model what could potentially happen under specific conditions. A model is defined as any representation or concept that helps the understanding of the world whenever common sense or direct observations are not sufficiently accurate (Egeghy *et al* 2016). Their use has notoriously increased over the years due to the main advantages posed against laboratory and field works, as they provide insight into common issues by using algorithms or mathematical formulas and establishing certain assumptions. The outcome enables, therefore, the creation of a narrative about why an event occurs, what might happen and how it evolves, which can then be used to inform about a system and to aid decision-making. Nevertheless, computational models cannot work independently to laboratory experiments, as they need data for their simulations and they need to be validated, two activities that require empirical experiments. Thus, it raises the challenge of how much to rely on computational modelling and how to assess their representative of real environments (UK Government Office for Science 2018).

Over the years, the investigation of contaminants in the environment for environmental risk purposes has gained awareness due to the toxicity problems that have been associated with uncontrolled release of pollutants, spillages and chemical accidents (Ozkara *et al* 2016). To that end, mathematical models are continuously employed to represent, understand, simulate and manage environmental systems as they provide a computational representation of the Earth's processes at a wide variety of spatial scales across land, water and air. Computational modelling is commonly used to test hypothesis, validate theories and ultimately improve the understanding of natural processes. Plus, models also help predict future system conditions, which is useful for assessing potential environmental impacts and negative risks by modelling simulation and what-if analysis. This allows model users to investigate these mechanisms and forecast their responses to both natural and human-induced change (Calder *et al* 2018). Understanding processes, regulating parameters, trends and differences between spatially or temporally differing data is very often the ultimate goal in science and models serve as tools towards that end (Alewell *et al* 2019). Therefore, computational modelling is a powerful tool to study the surrounding environment and help decision-making with a different approach when it is representatively constructed. In terms of Risk Assessment, models were created and are expected to inform and support decision-making processes from early warning systems to the sustainable planning and management of natural resources (figure 6.2) (Amigioni and Schiaffonati 2014).

Overall, computational models are less time-consuming than laboratory and field experiments. It takes time and effort to develop good models, but once achieved they can repay this investment many times over. In addition, they require less analytical equipment and laboratory material and tools, decreasing the costs of the research and disposal treatment. The use of computational models is frequently considered environmentally friendly, as there is no need to contaminate or alter the environment, there is no manipulation of toxic or hazardous compounds and there are no toxic or hazardous residues or solvents to dispose of after. This advantage reduces the risk of workers being injured, broken or replaced equipment and the need to deal with Risk Assessment and control of substances hazardous to health paperwork. Furthermore, laboratory experiments are quite often not reproducible and cannot be extrapolated to other locations as features of the medium (water, air, soil) vary in

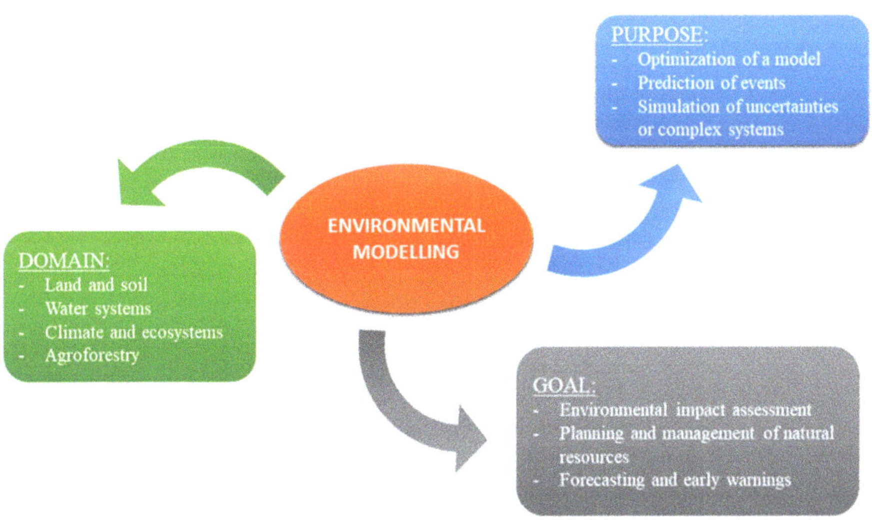

Figure 6.2. Application domain, purpose and main goals of the use of environmental modelling in Science and Engineering fields.

each geological region and tend to change over time. Often, researchers tend to extrapolate data from one region to another, but this carries but the risk of not being representative of their own system. Finally, computational models allow the study and simulation of large environments, when they are difficult to reproduce in small facilities, allowing a better understanding of the mechanisms involved.

An aspect that must be considered when assessing representativeness is the purpose of the model, that is, the question the model requires to address (Grimm *et al* 2020). The questions must be specific in order to obtain more representative outcomes of modelling. For instance, they can be used to predict or forecast what might happen in a specific environment and anticipate unknown data, helping users make decisions. They can also be used to explain processes by utilising plausible mechanisms to match outcome data in a known defined matter. This purpose is important for managing complex systems where processes are poorly fathomed. Another purpose is to understand or clarify theories by establishing the consequences of determined mechanisms, that is, to explore the results of some mechanisms where analytic mathematics is infeasible. This one is commonly used to provide evidence or refute hypotheses. Furthermore, models can be created to provide a new way to think about a determined system or environment under study by mapping what is has been already modelled in a flexible way and providing new insights. However, not all purposes can be framed as the answer to a question and all of them carry out some risks when they are used (table 6.2).

The second question needed to be asked when using the outputs of computational mode is how the mode is organised. Models have many technical aspects, such as data, mathematical expressions and algorithms, yet these are not sufficient for a model to be useful or representative of a specific environment. Conversely, asking a model to answer more and more detailed questions can be counterproductive,

Table 6.2. Different model purposes, examples of questions that can address and particular risks for not being representative (adapted from UK Government Office for Science 2018).

Purpose	Examples of question addressed	Risks
Prediction	*What will happen if there is a contaminant spillage? Where do contaminants end?*	Conditions of application unclear
Explanation	*What mechanisms govern the transport of contaminants in soil? Which ones affect pollutants the most?*	Minor changes in the setup might result in a non-representative fit to explained data
Understanding theory	*Which water pathways do contaminants follow to reach receptors? How does temperature affect its solubility and transport in water environments? Does contaminant spread change the addition of physical barriers?*	Mistakes in the model specification and inadequate coverage of possibilities
Analogy	*Is there an alternative way of considering contaminant pollution? Are there new insights not included yet?*	Model gives no support to empirical claims

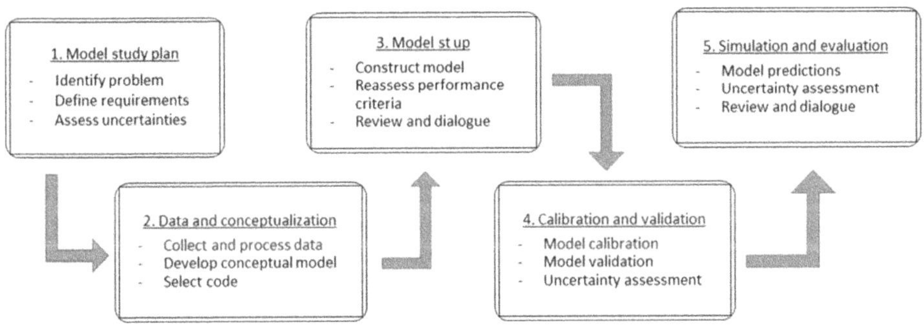

Figure 6.3. Main steps followed in the modelling process. (Reprinted from Refsgaard *et al* 2007, Copyright (2007), with permission from Elsevier.)

because it would require even more features of the real system to be included in the model (Calder *et al* 2018). In order to ensure representative results, the project model requires a model study plan and experimental data to be used as inputs (figure 6.3). Once the model is set up, it needs to be validated by comparing experimental results to the outputs of the simulation. The last step involves an evaluation of the outputs, and it is here when the users decide whether the model works. Unfortunately, there are no available standardised methods to decide whether a model is representative enough and whether it can substitute experimental designs (Refsgaard *et al* 2007).

The third question modellers need to ask is related to the way the existent evidence that the model works (e.g. previous use of the model), how it does (e.g. reproduced patterns), how the model makes predictions (e.g. environmental conditions used, uncertainty addressed) (Grimm *et al* 2020). The use of not representative models for assessing environmental risks can limit the conclusions modellers draw, just as bad experiments can limit the ability to test hypotheses (Donkin *et al* 2014). Real systems are defined by features that change over time, but they can be quantified (e.g. Dissolution rates). Computational models are developed to reproduce those features; however, they can be idiosyncratic, which often hampers the work done by experts when they define the importance of the characteristics of a system. As a result, the optimal way to deal with this uncertainty is by being explicit about the defined features and inputted data and why they are considered essential or crucial (Grimm *et al* 2020). Nonetheless, computational models require representative simulations of the system to be able to answer the questions posed by modellers and thus, should be able to demonstrate that incorrect assumptions lead to poor replication of main features.

Furthermore, computational modelling has always come across certain limitations. Quite often, the algorithms used cannot explain the behaviour of the systems in a representative manner, misleading the simulation to fatal decision-making. Some studies have reported limitations in their model's capacity related to bigger scales, in which key processes might vary and change the system (Beven 2002, Klamt 2017, Alewell *et al* 2019). This fact can, then, lead modellers to wrong or non-representative simulations in large systems, which might carry negative consequences to stakeholders and organisations. Another limitation of the use of computational modelling in Environmental

Science is the wrong extrapolation of data (both inputs or outputs) to other environments with at first sight, similar conditions. This is important, especially when studying dynamic systems like water or soil, as particle movement and environmental conditions are continuously changing over time. Last, the inclusion of uncertainties commonly enables the user to obtain more representative environments, however, the misuse of data or the lack of knowledge can ruin the model if inputs are not well selected and outputs are not well interpreted (Beven 2002).

The use of computational modelling to support decision-making is growing, as models tend to look realistic enough to answer specific questions posed by researchers and stakeholders. In Environmental Science, they have been used as a tool to understand how the system would work and might fail in order to mitigate effects and apply satisfactory corrections. However, it is still important to remark that (a) models need inputs and hence cannot work without laboratory and field experiments or previous datasets and (b) models can only support decisions, which means that the responsibility for relying on and using model outputs lies with decision-makers. Nevertheless, the use of computational models requires a continuous review as decisions made are based on a weight-of-evidence approach (Weed 2005). Thus, decisions are required to be updated when new data and models are available.

6.5 The consequences of choice: the final step in decision-making overall

Assessing the representativeness of a scientific study raises the challenge of selecting the most accurate approach to obtain the most suitable results to help decision-making. Currently, there is no standardised procedure in Environmental Science that can lead towards representative results and work efficiently every time it is needed. The importance of having a comprehensive decision-making process lays in avoiding catastrophes and spreading of misleading information. These risks can go from small incidents to fatal accidents and usually cover a wide range of economic, legal, environmental and reputational consequences (figure 6.4) (Ladyman *et al* 2019).

This chapter explored different experimental methodologies that are currently used in environmental sciences to gather a body of data informing decision-making during ERA procedure. Although, none of these methodologies can be use as stand-alone as not suitable for defining a representative body of data. Therefore, the current research suggests that a hybrid approach (e.g. observational/experimental and computational) is needed to ensure that the research is properly carried out. The data gathered from these experimental methodologies are the baseline for weighted choices during decision-making processes. Often, depending on the number of events that can take place if the risk is not correctly evaluated, the grade of uncertainty in decision-making can be greater than expected and the risks of not obtaining or using representative data can guide research into misleading information and poor decision-making. The levels of uncertainty are different and often risks are difficult to identify; although, nowadays, risks can be avoided with a high grade of certainty if laws, previously informed by research, are in place (Reis and Spencer 2019).

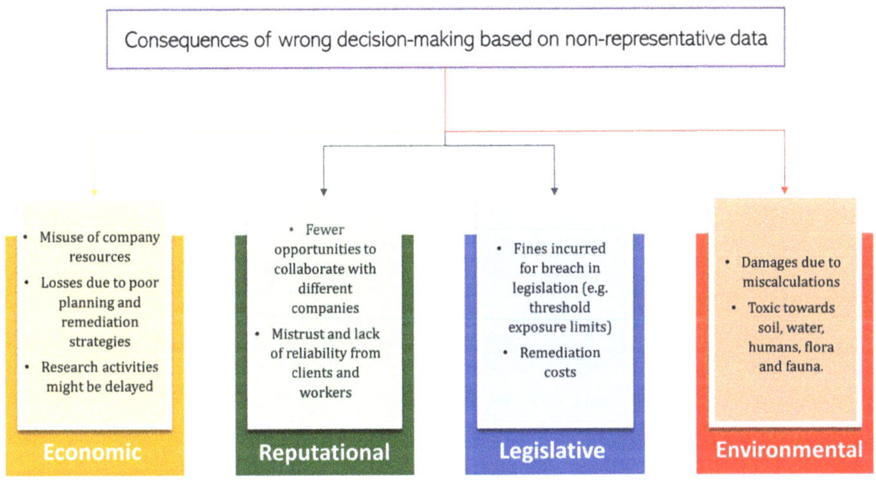

Figure 6.4. Main risks and some consequences of wrong decision-making based on non-representative dataset.

6.6 Conclusions and recommendations

- There are some existing risks when we make decisions in Environmental Science and this chapter aimed to highlight and mitigate them;
- To minimise the risks it is necessary to have a good selection of the data used to inform those decisions, including the heterogeneity of the samples and the evaluation of their representativeness of an entire system;
- Indoor facilities allow researchers to control environmental conditions, although it might reduce data representativeness. Therefore, researchers must find an equilibrium between these two concepts in order to comply with their objectives and make the optimal decision;
- Computational models are cost-effective and low-time consuming, however there are some questions that need to be answered before selecting and using a model;
- Models are created to support decision-making but the last word can only be made by researchers/modellers based on their evidences/ experiments;
- To answer the research questions in a simple but efficient way, more than one approach should be utilised to achieve a comprehensive dataset to better support decision-making.

References

Amigioni F and Schiaffonati V 2014 *Methods and Experimental Techniques In Computer Engineering* (Cham: Springer)

Anderson D and Strich J 2016 Goal clarity, task significance, and performance: evidence from a laboratory experiment *J. Public Adm. Res. Theory* **26** 211–25

Alewell C, Borrelli P, Meusburger K and Panagos P 2019 Using the USLE: chances, challenges and limitations of soil erosion modelling *Int. Soil Water Conser. Res.* **7** 203–25

Beven K 2002 Towards a coherent philosophy for modelling the environment *Proc. Royal Soc.* **458** 2465–85

Calder M *et al* 2018 Computational modelling for decision-making: where, why, what, who and how *Royal Soc. Open Sci.* **5** 172096

De França E, De Nadai Fernandes E, Bacchi M, Bode P and van Soldt R 2009 Improving sample representativeness in environmental studies: a major component for the uncertainty budget *J. Radioanal. Nucl. Chem.* **282** 125–32

Donkin C C, Tran S and Nosofsky R 2014 Landscaping analyses for the ROC predictions of discrete-slots and signal-detection models of visual working memory *Atten. Percept. Psychophys.* **76** 2103–16

Egeghy P P, Sheldon L S, Isaacs K K, Ozkaynak H, Goldsmith M R, Wambaugh J F, Judson R S and Buckley T J 2016 Computational exposure science: an emerging discipline to support 21st-century Risk Assessment *Environ. Health Persp.* **124** 697–702

Environmental Protection Agency (EPA) 1995 Sampling Design Guidelines https://epa.nsw.gov.au/~/media/EPA/Corporate%20Site/resources/clm/95059sampgdlne.ashx

Fawcett-Hirst W, Temple T, Ladyman M and Coulon F 2020 Adsorption behaviour of 1,3,5-trinitroperhydro-1,3,5-triazine, 2,4-dinitroanisole and 3-nitro-1,2,4-triazol-5-one on commercial activated carbons *Chemosphere* **255** 126848

Galante E, Mai N, Ladyman M, Gill P, Temple T and Coulon F 2020 Evaluation of small-scale combustion of an insensitive high explosive formulation containing 3-nitro-1,2,4-triazol-5-one (NTO), 2,4-dinitroanisole (DNAN), and 1,3,5-trinitroperhydro-1,3,5-triazine (RDX) *J. Energ. Mater.* **39** 85–99

Grimm V, Johnston A S, Thulke H H, Forbes V E and Thorbek P 2020 Three questions to ask before using model outputs for decision support *Nat. Commun.* **11** 10–2

Hadley P W and Bruce M L 2014 On representativeness *Environ. Forensics* **15** 1–3

Jackson B 1987 *Fieldwork* (Champaign, IL: University of Illinois Press)

ITRC I 2012 Incremental sampling methodology *Interstate Technology and Regulatory Council, ISMT. ISM-1* (Washington, DC: Interstate Technology and Regulatory Council, Incremental Sampling Methodology Team)

Klamt A 2017 The COSMO and COSMO-RS solvation models *Wiley Interdiscip. Rev.: Comput. Mol. Sci.* **8** e1338

Ladyman M, Temple T, Piperakis M, Fawcett-Hirst W, Gutierrez-Carazo E and Coulon F 2019 Decision framework for the environmental management of explosive contaminated land *Sci. Total Environ.* **690** 730–38

Özkara A, Akyıl D and Konuk M 2016 Pesticides, environmental pollution, and health *Environmental Health Risk-hazardous Factors to Living Species* (Rijeka: IntechOpen)

Pepper I L and Gerba C P 2015 Aeromicrobiology *Environmental Microbiology* (New York: Academic) pp 89–110

Pepper I L, Gerba C P, Gentry T and Maier R M 2000 Environmental sample collection and processing *Environmental Microbiology* 3rd edn ed I L Pepper, C P Gerba and T J Gentry (San Diego, CA: Academic) ch 8 pp 157–75

Plans B S 2006 *The Wastewater Treatment Plant Operators Guide to Biosolids Sampling Plans* New England Interstate Water Pollution Control Commission (Available at: www.neiwpcc.org

Ramsey C and Hewitt A 2005 A methodology for assessing sample representativeness *Environ. Forensics* **6** 71–5

Refsgaard J C, van der Sluijs J P, Hojberg A L and Vanrollenghem P A 2007 Uncertainty in the environmental modelling process—a framework and guidance *Environ. Modell. Softw.* **22** 1542–56

Reis J and Spencer P S 2019 Decision-making under uncertainty in environmental health policy: new approaches *Environ. Health Prev. Med.* **24** 1–8

Schouten B, Cobben F and Bethlehem J G 2009 Indicators for the representativeness of survey response *Surv. Meth.* **35** 101–13

Sparrow B D, Foulkes J N, Wardle G M, Leitch E J, Caddy-Retalic S, van Leeuwen S J and Lowe A J 2020 A vegetation and soil survey method for surveillance monitoring of rangeland environments *Front. Ecol. Evol.* **8** 157

Temple T, Ladyman M, Mai N, Galante E, Ricamora M, Shirazi R and Coulon F 2018 Investigation into the environmental fate of the combined insensitive high explosive constituents 2,4-dinitroanisole (DNAN), 1-nitroguanidine (NQ) and nitrotriazolone (NTO) in soil *Sci. Total Environ.* **625** 1264–71

UK Government Office for Science 2018 *Computational Modelling: Technological Futures*

Warren C, Mackay D, Whelan M and Fox K 2005 Mass balance modelling of contaminants in river basins: a flexible matrix approach *Chemosphere* **61** 1458–67

Weed D L 2005 Weight of evidence: a review of concept and methods *Risk Anal.: Int. J.* **25** 1545–57

Webster M Jr and Sell J 2014 Why do experiments? *Laboratory Experiments in the Social Sciences* (New York: Academic) pp 5–21

Chapter 7

The importance of the conceptual site model in the assessment of land contamination

P Burden, A-M Deloughry and J Morgan

The conceptual site model (CSM) lies at the heart of the assessment of land contamination in the UK. It was central to the earliest UK land contamination guidance and remains the primary method to both represent and understand the site relationships between potential sources of contamination and the receptors who may be affected by them. There are a wide range of potential sources of chemical contamination to be considered; receptors tend to fall into either humans or the water environments (surface waters or groundwater).

This chapter will set out the legislative background to UK guidance, how this guidance works in practice when assessing potential land contamination and finishes with a case study that demonstrates the approach on a real-life site.

7.1 Introduction

The CSM lies at the heart of the assessment of land contamination in the UK. The primary guidance for England is Land Contamination: Risk Management (LCRM), which is published on the gov.uk website. The LCRM [1] defines the CSM as 'a representation of the characteristics of the site. It shows the possible relationships between contaminants, pathways and receptors. It will form the basis of your initial assessment and all future decisions as you progress through LCRM.' This chapter describes how the Ministry of Defence (MoD) uses the CSM to assess and manage potential land contamination hazards.

7.2 Contaminated land legal/regulatory background

The UK has had a variety of legal controls on the assessment and management of land contamination for at least 40 years. There are broadly two approaches, through Part 2A of the Environmental Protection Act (EPA) 1990 [2] (enacted in England

doi:10.1088/978-0-7503-3643-7ch7 © IOP Publishing Ltd 2022

through the Contaminated Land Regulations 2006) and through the Planning Regime, whereby restrictions or requirements to ensure land is suitable for use are placed on developers when changing the use of land (e.g. constructing new houses or converting existing buildings for alternative uses).

This section will not attempt to cover the wide breadth of legislation and guidance related to this, nor the regional variations between England and the other nations of the UK. It will provide an overview of the current primary legislation in England, a short history of where this legislation has come from and give an indication of other key areas which may be of interest to the reader for further research.

7.2.1 A brief history

In 1987 the second edition of ICRCL Guidance Note 59/83 [3] was published, with a cover price of £2.25; this quickly became a key reference for assessing land contamination through the late 1980s until it was withdrawn in 2002. The ICRCL guidance published 'trigger' and 'intervention' values for a number of common contaminants, including metals and gasworks related compounds such as phenols and coal tar.

Although the guidance may appear simplistic compared with modern assessment methods, the key principles of source–pathway–receptor, desk study research, site investigation and use of upper and lower assessment criteria are all established. These principles/aspects form the basis of current guidance; namely a risk-based approach with the site summarised/encapsulated by the Conceptual Site Model (CSM).

7.2.2 Current contaminated land legislation and guidance

Part 2A of the Environmental Protection Act (EPA) 1990 [2] introduced a statutory regime for the identification and remediation of 'Contaminated Land'. It introduced, for the first time in the UK, a statutory definition of Contaminated Land based on significant harm or the likelihood of significant harm (including to human health) or the pollution or likely pollution of controlled waters (e.g. groundwater, inland waters and estuaries).

Local authorities are the primary regulator under the Part 2A regime with a duty to determine whether land in their area is Contaminated Land, with the Environment Agency responsible where sites are classed as 'special sites' as defined in the Contaminated Land (England) Regulations 2006 [4]; these are sites with particular or unusual potential hazards, which include MoD sites.

UK Government objectives with respect to land contamination are set out in the Department for Environment, Food and Rural Affairs (DEFRA) Contaminated Land Statutory Guidance 2012 [5] as:

- to identify and remove unacceptable risks to human health and the environment;
- to seek to ensure that contaminated land is made suitable for its current use; and

- to ensure that the burdens faced by individuals, companies and society as a whole are proportionate, manageable and compatible with the principles of sustainable development.

A precautionary approach to the risks raised by contamination is recommended, while considering the benefits and costs of taking action to remove/reduce such risks. Increasingly, sustainability is a factor which needs to be considered as part of the overall assessment.

Primary guidance for assessing and managing land contamination is presented in the UK Government guidance Land Contamination: Risk Management (LCRM) [1]. This provides a technical framework for identifying and managing land contamination through the application of a risk-based approach. LCRM also includes guidance on assessment of remediation options.

7.2.3 Planning and construction policies and guidelines

The National Planning Policy Framework (NPPF) [6] states local planning policies and decisions should ensure that 'the site is suitable for its proposed use taking account of ground conditions, contamination and land instability risks, including from natural hazards and former activities, pollution arising from previous uses and any proposals for land remediation and after remediation, as a minimum,

- land should not be capable of being determined as Contaminated Land under Part 2A; and
- where a site is affected by contamination or land stability issues, responsibility for securing safe development rests with the developer and/or landowner.'

Furthermore, construction activity is subject to Building Control under the Building Regulations 2010 [7] and subsequent Building (Amendment) Regulations 2013 [8] highlight the precautions that shall be taken to avoid danger to health and safety caused by substances found on or in the ground of building sites.

Approved Document C [9] identifies that site investigations should be carried out to determine the extent of contamination and that ground covered by the building must be reasonably free from material that might damage the building or affect its stability. Precautions must be taken against ground gases such as landfill gases or radon, which pose an asphyxiation or explosive risk.

7.2.4 Other consideration and further guidance

There are a number of related legislation and guidance documents that need to be considered alongside those described earlier in this chapter. These cover aspects such as groundwater protection, environmental protection, pollution prevention, health and safety, sustainability and waste management. LCRM includes links to many of these aspects as well as more detailed guidance relating to the Risk Assessment process, remediation, cost benefit analysis and specific contaminants such as asbestos.

7.3 Developing the conceptual site model (CSM)

The CSM clearly presents the source–pathway–receptor relationships as a pictorial, schematic, cross section and/or a table and is supported by an appropriate written justification. The level of detail of the CSM depends on the site conditions such as the complexity of the geology and nature and likely extent of the contaminants. To keep the CSM manageable and easy to understand, it can be presented with contaminants and pathways, be they potential or otherwise, grouped together rather than treated individually.

Detailed guidance is provided in the UK Environment Agency's 'Guide to Good Practice for the Development of Conceptual Models and the Selection and Application of Mathematical Models of Contaminant Transport Processes in the Subsurface' (2001) (National Groundwater & Contaminated Land Centre report NC/99/38/2) [10].

Figure 7.1 shows an example of a simple CSM for a leaking fuel tank close to a river. The source, pathway and receptors are shown on the indicative cross section

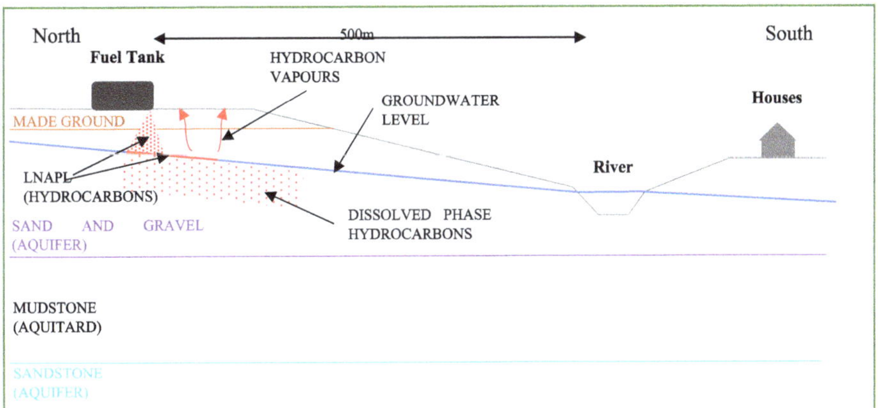

Source: Leaking Fuel Tank leading to light non-aqueous phase liquid (LNAPL) (hydrocarbons) in ground beneath the tank.

Pathways:
Migration of LNAPL through the ground to the aquifer.
Migration of dissolved phase hydrocarbons in groundwater towards the river.
Migration of hydrocarbon vapours from LNAPL in ground back towards the surface

Receptors:
Groundwater of the aquifer.
Surface water in the river.
Humans.
Ecology of the river.

Assumptions: Groundwater flow is north to south, towards the river.
Geological strata are horizontally bedded.
Groundwater is in continuity with the river.
The mudstone is impermeable and protects the underlying Sandstone aquifer.

Uncertainties: Permeability and porosity of the sand and gravel deposits.
Rate of migration of dissolved phase hydrocarbons towards the river.
Whether the river is used for abstraction.
Composition, variability and extent of the made ground.
Presence of any preferential pathways in the made ground (e.g. land drains which flow to the river).
Volume of fuel that has been lost to the ground.

Figure 7.1. A simple CSM for a leaking fuel tank close to a river.

and described in the text box below. The text box also describes some of the key assumptions and uncertainties to be confirmed and/or investigated to refine the CSM.

By identifying the potential sources, pathways and receptors as part of the CSM, the likely pollutant linkages are identified. The risk is then estimated i.e. what degree of harm or pollution might result and how likely it is to occur? Each pollutant linkage will then be refined through the various tiers of Risk Assessment.

For the UK MoD, the CSM and Risk Assessment is presented in a report called a Land Quality Assessment. The Land Quality Assessment (LQA) proceeds through a number of phases, as more information is gathered about the site. The objective of a LQA is to quantify the contaminated land risks and the associated liability (health, environmental, legal and financial) in a logical and rational manner achieving both economy in the expenditure of resources and confidence in the end result, such that the LQA provides the basis for defensible and auditable decision making.

UK MoD is committed to undertaking a programme of Land Quality Assessments (LQAs) to assess the land quality across the defence estate in order to provide a proper knowledge of the condition of the estate and ensure that it is 'suitable for use' and not causing harm to the environment. Where it is identified that unacceptable risks are posed by the presence of contamination, action must be taken to reduce and control the risks to an acceptable level.

The first phase of LQA (also known as a desk study) develops a Preliminary CSM and Preliminary (Qualitative) Risk Assessment (PRA) based on information and data gathered about the site from reference and site-specific documents and a site walkover. The CSM must also state the assumptions and uncertainties as these are areas that will need to be proven or investigated as the CSM is developed. It also helps the reader to understand the level of certainty with which the CSM should be considered.

To develop the CSM the following process is completed to understand the source–pathway–receptor relationships.

7.3.1 Source identification

The definition of a contaminant under part 2A of the Environmental Protection Act is—*a substance that is in, on or under the land and has the potential to cause harm.*

As a starting point, the potential sources of contamination need to be identified by examining the site history, previous uses and current operations on the site.

Historical features

The following are examples of features that could indicate areas of potential concern from historical uses/activities on site, often using historical maps and satellite imagery:

- Building footprints that appear to have changed over the years—A change in layout of buildings could indicate potentially demolished buildings which could be an indication of asbestos in soils from building materials where historically the demolition material was used to level the ground or build bunds in the area of, or close, to the building.

- Areas on the site marked as 'infill, spoil, or disused quarries'. This could indicate historic landfill areas. These areas could contain contaminants from discarded materials. For example, metals, asbestos, inert material, waste, radium dials.
- Structures labelled as 'tanks'—MoD sites have fuel tanks for heating, Military vehicles and aircraft. The storage tank sizes range from small domestic tanks for heating to large bulk fuel installations (BFI's) for fuelling military vehicles (e.g. aircraft, ships). Due to the age of the majority of MoD sites, there is potential for spills and leaks from the tanks and/or associated pipework.
- Other 'uses' prior to MoD ownership. Some sites may have had a previous use before the MoD occupied the site. If the site was used prior to this the historic maps can help understand if the site may have contamination from an historic source pre-dating the MoD ownership. This could be indicated by structures labelled as 'Works' on this historic map.

There are some activities that are potentially contaminative which are specific to MoD. These activities/features can be identified using historic mapping, site plans, satellite imagery and/or during the site walkover and discussions with site staff. Some examples of these activities are:

- Explosive Storage Areas. These are areas on site where explosives are stored or have been in the past. There is the potential for residues relating to explosives to be present in these areas.
- Burning Areas. After WWII there were significant aircraft surplus to requirement. In these cases, the aircraft were dismantled on MoD sites. A common practice when aircraft were dismantled was for aircraft to be initially burnt to remove combustible material, considered to be low value, leaving the metal structure to be more easily recycled. The luminous dials in the aircraft which contained radium 226 were either fully or partially combusted, however radium 226, the radioactive element used to make the dials luminous, was often left behind in the ash, which was often buried or left *in situ* on sites at the location of the burning area. Satellite imagery can identify areas where it shows aircraft 'parked' in areas on the site awaiting dismantling, aircraft are partly dismantled or burning areas are identified. Burning areas can also be associated with destruction of explosives, which may leave behind explosive residues.
- Other areas to look out for which may have contamination associated with the use/activities are workshops, Motor Transport (MT) sections, Petroleum, Oil Lubricant (POL) points, cross base pipelines, firing ranges, fire training areas and runways.
- During site walkovers it is important to look for visual evidence of contamination present on site. For example, vegetation dieback in areas around fuel installations may indicate a spill.

7.3.2 Receptors

A receptor is defined in the LCRM as:

'Something that could be adversely affected by a contaminant, for example a person, controlled waters, an organism, an ecosystem, or Part 2A receptors such as buildings, crops or animals.'

Once the potential sources of contamination have been identified the next step is to identify the potential receptors. Expanding on the definition above, the receptors can be described as:

Humans. These could be workers on the site or residents of houses either on an MoD site or on adjacent land. Sometimes we need to consider future users of the site, for example if the site use is likely to change due to selling it to a housing developer.

Controlled Waters. For groundwater, is the site over a sensitive aquifer designation, are there abstractions on or close to the site, or is the site in a Source Protection Zone (SPZ). For surface waters, what are the closest surface water features to the site? Are they sensitive?

Other receptors. Is land used for grazing animals or growing crops? Is the site in a sensitive ecological area? Are there any environmental designations, such as Sites of Special Scientific Interest? Are there buildings or structures which would be impacted by potential contamination?

7.3.3 Pathways

Once the potential receptors have been identified the next step is to understand how the potential contaminants can move through the environment to impact the receptors.

These will vary according the both the contaminant and the receptor; some examples of potential pathways are given below.

The migration routes for humans to be affected by contaminants are ingestion, inhalation and dermal contact. For controlled waters, infiltration and contaminant migration through permeable strata in the unsaturated zone will lead to potential groundwater contamination. Groundwater can also be a pathway for surface water receptors. Surface water can also be impacted by surface water drainage or direct runoff.

Ground gases and vapours such as landfill gases (e.g. carbon dioxide, methane) or volatile hydrocarbons can migrate into buildings through the building structure and collect in spaces such as basements or confined cupboard spaces. For wider environmental receptors, pathways include ingestion by animals and direct contact and uptake by plants.

7.4 Assessing the risk

To fully understand the risks using the data that has been collected in the CSM there needs to be consideration of:

- the probability or likelihood of the linkage occurring;
- the severity/significance of the potential consequence, taking into account the nature of the pollutant linkage and the context of the current and/or envisaged land use.

If there is an absence of a pollutant linkage this means, there is no risk. A view will need to be taken on whether there is sufficient data to provide the necessary confidence to discount a source at the Phase 1 LQA stage.

There are a number of ways to calculate the risk in the preliminary Risk Assessment of the Phase 1 LQA, but the most common is to assign numerical scores for likelihood and severity for each pollutant link and multiply them together to calculate a risk value.

The evaluation of the risk for each source–pathway–receptor linkage is then be used to decide whether a risk is unacceptable. If the risk(s) are unacceptable then the site assessment moves on to the next tier, the Phase 2 LQA.

7.4.1 Phase two LQA—the site investigation

In a Phase 2 LQA the potentially unacceptable risks that have been identified in the preliminary CSM are estimated and further evaluated by testing and refining the Preliminary CSM using quantitative, site-specific data, such as laboratory derived concentrations of contaminants in the soil.

The aim of a phase 2 site investigation is to provide a reliable assessment of land quality and in doing so confirm the presence and quantify the nature and extent of contamination, set out the level of uncertainty (inherent in the sampling, sampling preparation and analysis), assess the significance in terms of the risks associated with environment and health and make recommendations on how to manage the risks and liabilities in a cost-effective manner. The locations of the exploratory holes should be designed to maximise the data and reduce the uncertainties as much as possible. An approach which is often taken is to target areas of potential concern (e.g. the ground around a buried fuel tank) while also including more general locations across the site to get an understanding of the site as a whole and try to identify areas of potential contamination not identified as part of the phase 1 LQA desk study.

At this tier of Risk Assessment an exploratory investigation is usually a limited investigation. The aim is to:

- reduce uncertainty
- update the initial conceptual site model
- provide information to support any future detailed investigation

7.4.2 Site investigation methods

Exploratory holes can be advanced using various methods, with the most common being:

- **Trial pits.** Hand or machine excavated pits which can be extended up to 4 metres depth using a 'standard' JCB backhoe excavator. Trial pits cannot be extended as deep as boreholes, but have the advantage of exposing a much larger area of subsurface ground allowing the site engineer to see the structure of the near surface layers of soil and any man-made structures or obstructions which may be of interest (e.g. buried foundations or abandoned tanks);
- **Boreholes.** Boreholes can be advanced to greater depths than trial pits, with the subsurface soils and rock being brought to the surface for inspection and sampling. The advantage of boreholes over trial pits is that they can be extended though both loose soils and solid rock and can be extended into the

underlying groundwater which may be a receptor or pathway within the CSM. Another advantage of boreholes is that permanent ground and groundwater monitoring equipment can be installed, whether standpipes to allow subsequent groundwater sampling or more sophisticated equipment (e.g. for 'real time' measuring of temperatures in landfill material);
- **Probe holes** using a window sampler (or similar). This is relatively quick technique to make a small diameter exploratory hole, to depths greater than for trial pits if in suitable soil conditions. The sampler has a 'window' cut out of it along one side which allows the soils sampled to be observed as a relatively undisturbed soil column. This method allows a relatively large number of locations to be investigated quickly without creating too much disturbance of the area. Its benefits are a cross between a borehole and trial pit; and
- **Collection and analysis** of soil, groundwater, ground gases, surface water and other media where required.

7.4.3 Revised CSM

The Preliminary CSM must be revised once all the chemical analyses and site-specific data has been derived to confirm or exclude the potential pollutant linkages identified by the Phase 1 LQA.

Only then should the process of risk estimation by means of Quantitative Risk Assessment (QRA) should be undertaken.

Quantitative Risk Assessment—tier 2 and 3

There are two types of QRA: Tier 2—Generic QRA (GQRA) and Tier 3—Detailed QRA (DQRA). Ordinarily a GQRA is carried out first, whereby the determined contaminant concentrations are compared to appropriate and justifiable generic evaluation criteria before potentially moving to Tier 3 DQRA. Examples of evaluation criteria are given below.

Examples of evaluation criteria include:

- category 4 screening levels (C4SLs)
- human health toxicological assessment of contaminants in soil (SR2) tolerable daily intakes and index doses
- site-specific assessment criteria derived using remedial targets methodology (RTM) and a regulatory compliance point
- drinking water standards (DWSs)
- environmental quality standards (EQSs)
- ecosystem end-points which take into account the ecological value of a site—for example, used under Part 2A for assessing significant harm to ecosystems
- other sources of generic assessment criteria developed by industry

The following text explains how generic evaluation criteria should be used and is taken from the LCRM [1].

As evaluation criteria take into account a degree of uncertainty they are inherently conservative and as such an exceedance should be considered an indication that further works may be required. Where measured concentrations of contaminants fall below them it can be concluded that there is no significant risk providing the guidelines have been applied correctly. It is essential to understand the limitations of the application of the various evaluation criteria, in particular the exposure scenarios to which they relate.

The DQRA is where site-specific parameters are applied to the model in order to take account of the conditions at the site and the proposed use. Some of the conservatism that is built in to the evaluation criteria can be removed by applying site-specific knowledge, whether related to the site use or the nature of chemicals in the ground, such as using bioavailable values rather than total values where ingestion of soil and dust is being considered as a pathway.

Following the GQRA/DQRA the CSM can be updated and the risk classifications updated accordingly.

If there are still unacceptable risks either further intrusive works or remediation can be carried out. These works can be used to further refine and update the conceptual model, delineate contaminant areas and plumes, understand remediation requirements, identify suitability for different types of remedial options and provide information on post-remediation monitoring. Remediation can include source removal, *in situ* or ex situ soil treatment or containment to remove the pathway between the source and the receptor.

The following case study puts into practise the phased approach of assessing land quality outlined in the previous sections. It highlights the importance of gathering information both on current and historical activities undertaken on site and the environmental setting, developing a Preliminary CSM through Source–Pathway–Receptor relationships and then testing those linkages through the gathering of site data to determine risk to receptors.

7.5 Case study: Caesar's Camp, Area B6, Aldershot [12]

Area B6, Aldershot is a 270-hectare site comprising open grassland and woodland located to the west of Aldershot town as shown in figure 7.2. The site is primarily used for dry training (using blank or 'dummy' ammunition) including basic infantry tactics and driver training. There is public access onto the area, and it is frequently used for dog walking and cycling. The site is within a designated Site of Special Scientific Interest (SSSI) due to the acid heathland.

Between 2007 and 2011, the Aldershot training area heathlands were grazed by cattle managed by the Hampshire and Isle of Wight Wildlife Trust (HWT). The cattle were not used for meat or milk production. During this time there were two incidents of lead poisoning of the cattle, the first between 2007 and 2008 when eleven cattle died, the second in 2011 when four cattle died. In the first incident cattle grazing on the south-eastern area were affected; a post-mortem determined likely cause of death was lead poisoning. The deaths were linked to an area of dumped lead and linseed-based paints and putty in the south-eastern part of the site. An investigation carried out in this area (Apple Environmental, Environmental Report Intrusive Contaminated Land Investigation Caesar's Camp Odiham Road

Figure 7.2. Site location plan. Contains OS data © Crown copyright (UK MoD 2018).

Farnham, 2008) took a total of six soil samples and two sediment samples from the base of a pond in the area. These showed levels of lead of 5–360 mg kg^{-1} in soil and 19 and 140 mg kg^{-1} in sediment. The linseed-based putty was shown to have lead concentrations of 27 000 mg/kg. The south-eastern area was subsequently visually inspected, and any further potential sources of lead removed.

In the second incident, lead poisoning was confirmed as the cause of death of four cattle with metallic lead being found to have been ingested by the cattle (Apple Environmental, Environmental Report Lead Contamination Investigation Caesar's Camp Bourley Road Aldershot, 2012) [11]. A search of the area by HWT and Aldershot range staff did not identify any significant or obvious sources of lead or other pollutants. As a precaution all the remaining cattle were removed from the site.

At this point the issue was passed onto the Defence Infrastructure Organisation's Technical Services department, to a team specialising in land contamination. A Phase One Land Quality Assessment was commissioned which included interviews with range staff, a site walkover, information relating to geology, hydrology and hydrogeology, review of historical maps and publicly available information to identify potential sources of contamination and develop a preliminary conceptual site model.

7.5.1 Site history

Area B6 has several former firing ranges located on it, including a 600-yard gallery range called Caesar's Camp in the north-east, see figures 7.3 and 7.4. The area was developed as part of the wider Aldershot training area in the 1850s and Caesar's Camp was used from the 1870s through to the 1970s when anecdotal evidence

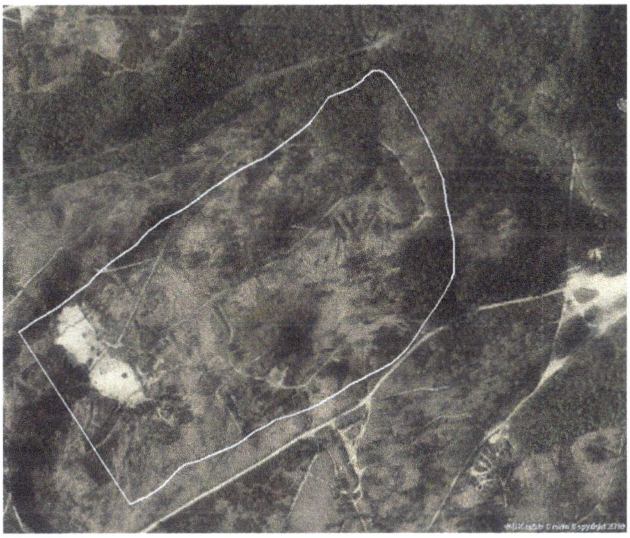

Figure 7.3. Aerial image of Caesar's Camp 600 yd range. The former stop butt is denoted by bare sand in the south-west corner [12].

Figure 7.4. Looking north east from above the stop butt. © UK MoD Crown Copyright 2021 [12].

suggests that the stop butt material was pushed over and left. During the site walkover clay pigeon skeets and shot gun cartridges were identified; range staff reported that clay pigeon shooting had historically also taken place in the area.

7.5.2 Environmental setting

Geology

The site is underlain by sand and gravel formations, these are subsequently underlain by London Clay up to 110 m thick which protects the Chalk Aquifer.

Hydrogeology

The sand and gravel formations are all classified as Secondary A aquifers[1]. The area is not within a groundwater protection zone nor are there any licensed abstractions on the site.

Hydrology

There are several streams across the site, some of which are ephemeral. The Claycart Stream links to a network of drains present in Caesar's Camp and also collects water from a southern stream whose source is a spring in the south-west of the range. The flow of Claycart Stream is from the base of Caesar's Camp north-eastwards towards Bourley Road. From here, the stream flows northwards and eventually discharges to the Basingstoke Canal which is a Site of Special Scientific Interest (SSSI) located 1.8 km from the site boundary. A hydraulic connection between the stream and the canal was confirmed as part of these investigations.

Ecology

Area B6 is part of the Bourley and Long Valley SSSI which was designated in 1993[2] and a Special Protection Area (SPA)[3] due to the acid health grassland present at the site. The site is defined as 'unfavourable recovering' indicating the SSSI units are not yet fully conserved but all the necessary management measures are in place. Natural England, in their views on management of the SSSI, indicate that low intensity cattle grazing is a suitable means of managing areas of dry heath.

7.5.3 Preliminary conceptual site model

Using all the collated information, several areas were identified as significant potential sources of lead and other heavy metal contamination due to the presence of bullets, and clay pigeon skeets (indicating historic use of lead shot). These areas were part of or located near Caesar's Camp 600 yd range (figure 7.5). In addition to cattle other potential receptors identified included people using the site, surface water of the Claycart Stream which flows into the Basingstoke Canal about 1.8km from the site and groundwater in the sand and gravel aquifer at the site.

The Preliminary CSM for these eight areas is presented in table 7.1.

7.5.4 Summary of site investigations

The focus of the Phase Two Land Quality Assessments was on the Caesar's Camp range; there were four phases of investigation between 2012 and 2016. The first phase was carried out in August 2012 to assess potential lead sources identified during the Phase 1 desk study; a second phase was carried out in September and October 2013 to address specific uncertainties identified in the first investigation and

[1] **Secondary A:** Aquifer-permeable layers capable of supporting water supplies at a local rather than strategic scale, and in some cases forming an important source of baseflow to rivers.

[2] https://designatedsites.naturalengland.org.uk/PDFsForWeb/Citation/1006761.pdf

[3] https://designatedsites.naturalengland.org.uk/SiteGeneralDetail.aspx?SiteCode=UK9012141&SiteName=Bourley&countyCode=19&responsiblePerson=&unitId=&SeaArea=&IFCAArea

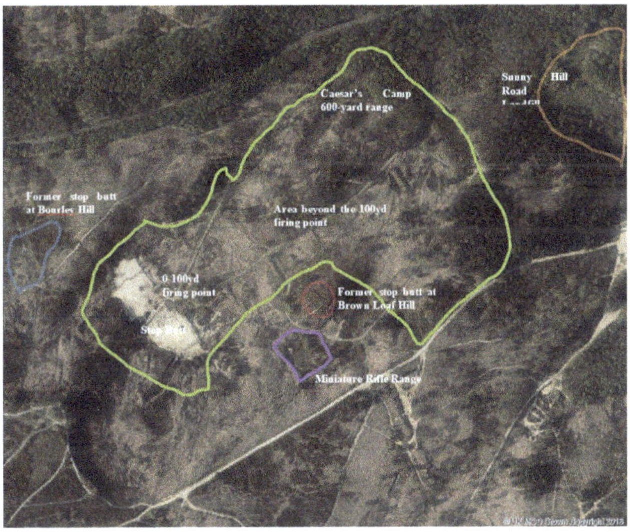

Figure 7.5. Areas identified as potentially significant sources of lead and heavy metal contamination [12].

Table 7.1. Preliminary CSM [12].

Source	Pathway	Receptor
Firing Range Caesar's Camp (Site walkover identified bullets present throughout firing area. Lead confirmed elevated in soil vegetation and groundwater, sourced from previous report).	• Ingestion of shallow soil • Ingestion of soil and sediment in streams • Ingestion of water in streams • Inhalation of shallow soil dust • Consumption of vegetation • Direct ingestion of lead shot and lead bullets	Cattle
	• Accidental ingestion of soil • Inhalation of dust • Consumption of surface water (by military personnel)	Military personnel, dog walkers,cyclists, runners and vehicle users
	• Vertical migration • Leaching from sand and sediment • Leaching from bullets and casings	Groundwater (Secondary A aquifer)
	• Runoff of sediment • Leaching from sand and sediment • Leaching from bullets and casings • Baseflow of groundwater to streams	Claycart Stream

Miniature Rifle Range at base of Brown Loaf Hill. (Site walkover identified bullets. Some stress on vegetation away from sand)	• Ingestion of shallow soil • Ingestion of soil and sediment in streams • Ingestion of water in streams • Inhalation of shallow soil dust • Consumption of vegetation • Direct ingestion of lead shot and lead bullets	Cattle
	• Accidental ingestion of soil • Inhalation of dust	Military personnel, dog walkers,cyclists, runners and vehicle users
	• Vertical migration • Leaching from sand and sediment • Leaching from bullets and casings	Groundwater (Secondary A aquifer)
	• Migration via groundwater	Claycart Stream
Sunny Hill Road Landfill (Former revolver range)	• Ingestion of shallow soil • ingestion of soil and sediment in streams • Ingestion of water in streams • Inhalation of shallow soil dust • Consumption of vegetation	Cattle
	• Accidental ingestion of soil • Inhalation of dust	Military personnel, dog walkers,cyclists, runners
	• Vertical migration	Groundwater (Secondary A aquifer)
	• Migration via groundwater	Claycart Stream
Two other firing range stop butts: Location of former stop butt at Brown Loaf Hill; and Location of former stop butt on Bourley Hill.	• Ingestion of shallow soil • ingestion of soil and sediment in streams • Inhalation of shallow soil dust • Consumption of vegetation	Cattle
	Accidental ingestion of soil. Inhalation of dust. Consumption of surface water (by military personnel)	Military personnel, dog walkers, cyclists, runners and vehicle users
	Vertical migration	Groundwater (Secondary A aquifer)
	Migration via groundwater. Potential for runoff to pond from the former stop butt near the pond although limited by vegetation.	Claycart Stream and pond

to investigate the impacts on controlled waters and human health; a third phase of investigation was carried out in December 2014 to address further uncertainties related to groundwater quality down gradient, the flow volume in the streams and further delineate lead in shallow soils; the final phase undertaken in 2016 was surface water sampling at monthly intervals along Claycart Steam and Basingstoke Canal in order to help better understand the risks to surface water from heavy metals.

A detailed description of the four phases is presented in table 7.2. The techniques selected were the most appropriate to meet the objectives of each phase of work and mindful of the ground conditions, and ecological sensitivities of the site.

Table 7.2. Phase two site investigations.

Site investigation techniques	Objective
First phase of investigation—August 2012	
164 shallow hand dug pits	To allow X-ray fluorescence (XRF) screening of concentrations of lead in the field and sample soil and vegetation across the source areas.
Four window sampling locations	To assess the depth of stop butt sand and the variation of lead with depth in the vicinity of the stop butts at the Caesar's Camp range.
Sampling of surface water at 15 locations	To measure surface water quality across the Caesar's Camp range.
Topographic survey	To survey elevation of surface water and soil sampling locations.
Laboratory analysis	To analyse soil samples for heavy metals and pH; and lead in plant material (divided into leaves and roots) and soil in root zone at the same location; and selected samples for leachability of heavy metals and for particle size distribution and/or bioaccessibility.
Second phase of investigation—September and October 2013	
Trial pitting at 14 locations using excavator	To assess the depth profile of lead using XRF within the stop butts at Caesar's Camp, at the Miniature Rifle Range and in the vicinity of the spring upstream of the Caesar's Camp range.
Hand excavation at 12 locations	To allow further investigation and sampling of shallow soil in less accessible areas and surface screening using XRF.
Seven boreholes (Cable percussion)	To establish groundwater sampling locations across Caesar's Camp and up gradient.
Surface water sampling on and off site	To confirm the variability of 2012 data and to assess the stream quality off site and background levels.
Invertebrate sampling	To sample Claycart Stream and compare with other streams representing background in the area to support the assessment of surface water quality in the streams.
Topographical survey	To survey elevation of trial pits and boreholes.
Catchment modelling	To assess the relative size of the catchments of the two streams in Caesar's Camp range.

Laboratory analysis	To analyse soil, surface water and groundwater samples.
Third phase of investigation in December 2014 to January 2015	
Three additional boreholes and monitoring of groundwater on two occasions	To confirm the groundwater flow direction and assess the groundwater quality down hydraulic gradient of the Caesar's Camp Areas 0 and 1.
Permeability testing of groundwater	To assess the hydraulic conductivity and usage of the groundwater in the vicinity of the site.
Flow monitoring and sampling of surface water on two occasions	To assess the relative impact of each stream on the off-site migration of metals in surface water.
Sediment sampling on one occasion	To assess the thickness of the sediment along both streams (Claycart and Southern) and provide data to inform any remediation of the sediment in the streams.
Surface screening of soilusing an XRF	To further delineate shallow lead concentrations in the soil through Area 2.
Topographical survey	To survey elevation of boreholes and confirm location of surface samples.
Laboratory analysis	To analyse soil, surface water and groundwater samples.
Fourth phase of investigation from February 2016 to November 2016	
Monthly surface water sampling on and off site	To better understand the risks of heavy metal soil contamination at Caesar's Camp to surface water by gathering a set of statistically significant results so that concentrations in the surface water receptors can be compared to annual average Environmental Quality Standards (EQS). A further objective was to establish whether the Claycart Stream and the Basingstoke Canal are in direct hydraulic continuity. It was important to determine this as elevated concentrations of heavy metals could be discharging into Basingstoke Canal, a SSSI.

7.5.5 Assessment of soil and controlled waters

To assess risks to cattle a detailed quantitative Risk Assessment was used to derive a site-specific threshold of 2500 mg kg^{-1} for lead for soil and plant material based on risks of chronic lead poisoning to cattle.

To assess risks to human health from lead a detailed quantitative Risk Assessment was carried out to determine site-specific assessment criteria of 3500 mg kg^{-1} for risks to current site users. The risks from other metals were assessed based on generic assessment criteria for risks to people conservatively based on thresholds in a residential garden.

Groundwater and surface water results were screened against Environmental Quality Standards and Drinking Water Standards. The invertebrate sampling was used to determine if exceedance of the Environmental Quality Standards was being realised in an effect on the invertebrate ecology.

7.5.6 Summary of findings

Within the Caesar's Camp range and the Miniature Rifle Range the following was concluded:

- Lead bullets are present in the stop butt areas and in the 0–100 yard firing point. Lead is also present in the range itself beyond the 100 yard firing point, particularly south of an area of skeets indicating it may derive from clay pigeon shot. Lead bullets and shot decompose to form lead salts which preferentially bind to finer particles of soil. This is confirmed by the high concentrations in the finer fraction of the soil. Bullets are also a potential source of copper and nickel from the bullet jackets whereas lead shot does not have such jackets.
- The presence of lead bullets, lead shot and lead in the finer fraction of the soil are above the site-specific threshold for cattle grazing.
- The lead appears to be present in the shallow soil in the main range probably from lead shot but extends to greater depths in the stop butt area and in the 0–100 yard firing point up to 3m below ground level based on the depth of sand.
- Copper concentrations were also recorded in the stop butt area and in the 0–100 yard firing point and may be associated with the presence of bullet casings.
- Lead concentrations in plant roots were higher than in the leaves. The concentrations of lead in leaves were generally below the thresholds for cattle. Those for plant roots exceeded this threshold assuming the cattle's diet was solely from these plants.
- The key pathways for lead exposure to people relate to accidental ingestion of soil and to inhalation of soil dust from around vehicles.
- It has been established that the Claycart Stream and the Basingstoke Canal are in direct hydraulic continuity.
- The Southern Stream is fed by water from a source area in the west of the area whereas Claycart Stream in the north receives water from the stop butt area.
- Above the confluence of the stream, the Claycart Stream shows high copper, lead and nickel consistent with cupronickel jacketed bullets from the stop butt and 100 yard firing point. The relatively low copper but high lead concentrations in the Southern Stream are consistent with lead shot source as found the area beyond the 100 yard firing point.
- The monitoring data demonstrate that dissolved and bioavailable concentrations of lead, copper, nickel and zinc decrease along the stream with distance from Area B6.
- The 2016 annual average bioavailable concentration data shows that copper and lead in the on-site stream are above the EQSs; that copper in the off-site stream is marginally above the EQS; and that zinc and nickel are below their EQSs within the surface water bodies.
- The annual average bioavailable concentrations for lead and copper within the canal were below the EQSs. This indicates that the Basingstoke Canal is not being adversely impacted.

- The groundwater flow is to the north of the site rather than in the same direction as the Claycart Stream. Groundwater monitoring down hydraulic gradient of the stop butt area indicates that the heavy metals have not migrated far and hence there is a no evidence of a significant plume of groundwater contamination.
- In the streams, the loose sediment is fairly thin and while sediment and suspended solids could be acting as a secondary source in the streams their extent is limited.

For other former stop butt and landfill areas it was concluded:

- In the areas of the former stop butts that were present in the 1870s (Bourley Hill and Brown Loaf Hill), there is evidence of higher levels of lead in the soil than are found elsewhere on site away from the ranges. These elevated concentrations were particularly on the Bourley Hill former range (Area 6) where one reading exceeded the thresholds for cattle. No evidence of bullets was identified however the investigation in these areas was limited.
- Lead concentrations in the surface soil at the landfill to the east of Sunny Hill Road were below the site-specific thresholds.

Table 7.3. Revised CSM and environmental Risk Assessment [12].

Source	Pathway	Receptor	Severity of associated hazard	Likelihood of Occurrence	Risk
Bullets and lead shot in Caesar's Camp and Miniature Rifle Range. Elevated concentrations of lead identified.	Ingestion of bullets and lead shot.	Cattle	Medium	High	High – if cattle allowed to graze there in future.
Bullets and lead shot in Caesar's Camp and Miniature Rifle Range. Elevated concentrations of lead in soil and plant roots associated with degrading bullets and shot.	Ingestion of soil and plant roots, surface water and sediment.	Cattle	Medium	High	High – if cattle allowed to graze there in future.
	Direct contact, ingestion, inhalation including dust generated from vehicles on soil, ingestion of stream water (military only).	Military personnel, dog walkers, cyclists, runners and vehicle users	Medium	Likely	Moderate
	Leaching through permeable strata	Groundwater (Secondary A aquifer)	Mild	Likely – Acidic conditions at Aldershot likely to aid leaching of lead out of the soil and infiltrating through soil into the groundwater.	Moderate/Low
	Surface water runoff.	Surface water – Claycart & Southern Streams	Medium	Likely	Moderate
	Migration in surface water runoff from Claycart Stream.	Surface water – Basingstoke Canal	Medium	Low	Moderate/Low
Bullets and lead shot in Caesar's Camp and Miniature Rifle Range. Elevated concentrations of copper, nickel and antinomy.	Leaching through permeable strata.	Groundwater (Secondary A aquifer)	Mild	Likely	Moderate/Low
	Surface water runoff.	Surface water – Claycart & Southern Streams	Medium	Likely	Moderate
	Migration within surface water.	Surface water – Basingstoke Canal	Medium	Low	Moderate/Low
Elevated lead in the other three stop butt areas.	Ingestion of bullets, soil, plant roots, surface water and sediment.	Cattle	Medium	Low	Moderate/Low
	Direct contact, ingestions, inhalation.	Military personnel, dog walkers, cyclists, runners and vehicle users	Medium	Unlikely	Low
	Leaching through permeable strata.	Groundwater	Mild	Low	Low
	Surface water runoff.	Surface water – Claycart & Southern Streams	Medium	Likely	Moderate
	Migration within surface water.	Surface water – Basingstoke Canal	Medium	Low	Moderate/Low

Probability (likelihood)	Consequence: Severe	Medium	Mild	Minor
High Likelihood	Very High Risk	High Risk	Moderate Risk	Moderate/Low Risk
Likely	High Risk	Moderate Risk	Moderate/Low Risk	Low Risk
Low likelihood	Moderate Risk	Moderate/Low Risk	Low Risk	Negligible Risk
Unlikely	Moderate/Low Risk	Low Risk	Negligible Risk	Negligible Risk

Classification	Definition
Very High Risk	Severe harm to a receptor may already be occurring OR a high likelihood that severe harm will arise to a receptor, unless immediate remedial works / mitigation measures are undertaken. Realisation of that risk is likely to present a substantial liability to MOD
High Risk	Harm is likely to arise to a receptor, and is likely to be severe, unless appropriate remedial actions / mitigation measures are undertaken. Remedial works may be required in the short term, but likely to be required over the long term. Realisation of that risk is likely to present a substantial liability to MOD
Moderate Risk	Possible that harm could arise to a receptor, but low likelihood that such harm would be severe. Harm is likely to be mild. Some remedial works may be required in the long term. Realisation of that risk is unlikely to present a substantial liability to MOD, but further work may be required to determine whether this is the case
Moderate/Low Risk	Possible that harm could arise to a receptor, but where a combination of likelihood and consequence results in a risk that is above low, but is not of sufficient concern to be classified as mild. It can be driven by cases where there is an acute risk which carries a severe consequence, but where the exposure is unlikely. Such harm would at worse normally be mild. Unlikely to present a substantial liability to MOD. Limited further investigation may be required to clarify the risk and liability. If necessary remediation works likely to be limited in extent.
Low Risk	Possible that harm could arise to a receptor. Such harm would at worse normally be mild
Negligible Risk	Low likelihood that harm could arise to a receptor. Such harm unlikely to be any worse than mild. No liability.

Figure 7.6. Risk matrix and category classification definitions.

7.5.7 Revised conceptual site model and environmental Risk Assessment

The approach taken to assess the risks of land contamination in this case used both Tier 2, interpretation of site investigation results against generic assessment criteria, and Tier 3, interpretation of site investigation results against site-specific criteria. This has also allowed the Source–Pathway–Receptor linkages to be revised from the initial Conceptual Site Model and a risk assigned to each linkage, see table 7.3.

The Land Quality Management Practitioner Guide [13] was used for the risk matrix and definitions of the risks as presented in figure 7.6.

7.5.8 Conclusion

There is a high risk of lead poisoning of cattle in the Caesar's Camp range and in the adjacent Miniature Rifle Range from lead bullets, lead shot and lead salts in the finer fraction of soil and in plant roots. To mitigate against this risk the Caesar's Camp range and miniature rifle range have been fenced off; this has allowed cattle to return to the wider Area B6 to help maintain the condition of the SSSI. The installation of the fence has also acted to manage the risk to people using that area by preventing them from entering.

There are ongoing 'moderate' risks to the Claycart and Southern Streams, and potential moderate to low risks to the Basingstoke Canal. There are also moderate to low risks to groundwater posed by lead, nickel and antimony in the Caesar's Camp range area. These higher heavy metal concentrations appear to be localised to the areas containing bullets: Area 0—Former stop butt, Area 1–0–100 yards firing point and Area 3—Former miniature rifle range. The groundwater is a Secondary A aquifer which is not abstracted locally.

The next phase of works for Caesar's Camp and the miniature rifle range is to develop a remedial strategy that will remove the source and break the Source–Pathway–Receptor Linkages and greatly reduce and/or remove the risks.

References

[1] Environment Agency 2021 www.gov.uk/government/publications/land-contamination-risk-management-lcrm

[2] 1990 *Part 2A of the Environmental Protection Act (EPA)*

[3] 1990 *ICRCL 59/83 Guidance on the Assessment and Redevelopment of Contaminated Land*

[4] 2006 *Contaminated Land (England) Regulations*

[5] Department for Environment, Food and Rural Affairs (DEFRA) 2012 *Contaminated Land Statutory Guidance*

[6] National Planning Policy Framework (NPPF), Ministry of Housing, Communities and Local Government 2021

[7] 2010 *Building Regulations*

[8] 2013 *Building (Amendment) Regulations*

[9] 2013 *Site Preparation and Resistance to Contaminates and Moisture: Approved Document C*

[10] 2001 *Contaminant Transport Processes in the Subsurface, National Groundwater & Contaminated Land Centre report NC/99/38/2*

[11] 2012 *Environmental Report Lead Contamination Investigation Caesar's Camp Bourley Road Aldershot, Apple Environmental*

[12] 2017 *Assessment of Lead at Area B6 in Aldershot, Targeted Phase 2: Land Quality Assessment, Jacobs*

[13] 2017 *Land Quality Management Practitioner Guide 2017/01*

Chapter 8

Land Release: a Risk Management approach for mine action

R Evans

Land Release is a Risk Management model developed for Humanitarian Mine Action (HMA). This chapter explains how Land Release came into being, how it works, and how it can be improved.

There are finite resources to clear land contaminated by not only Anti-Personnel and Anti-Vehicle Mines but also Explosive Remnants of War. Targeting those resources as effectively as possible by means of Land Release means such contamination can be cleared more efficiently.

At its core Land Release involves risk assessing land based on evidence. The key is to actively collect that evidence, which can then be used as criteria on which risk decisions can be taken. This evidence gathering occurs through the three stages of Land Release; Non-Technical Survey, Technical Survey and Clearance. Evidence, often the actual explosive ordnance, can be found right up until the last day of clearance, and decisions concerning what land to clear can be influenced accordingly.

While the Land Release model has seen some limited success, it can at best be seen as a work in progress. Like many Risk Management systems it relies on good quality data. Often this is not collected. The model also relies on criteria for designating land as either suspected or confirmed hazardous areas, but such criteria are rarely agreed. The model also relies on concepts of 'confidence' and 'all reasonable' effort that are yet to be sufficiently defined and are open to subjective interpretation. Nevertheless, while Land Release might be deemed an underdeveloped Risk Management model, for HMA it is the only viable option. The alternative would be to clear everything where there is a suspicion of explosive ordnance contaminating land, and accepting the opportunity cost, and more exactly the likely human cost, such inefficiency would entail. With further development, Land Release can be the right Risk Management model for deciding what land HMA should clear.

doi:10.1088/978-0-7503-3643-7ch8 © IOP Publishing Ltd 2022

8.1 Introduction

The explosive ordnance left behind after conflict presents significant human security, economic and environmental risks. Removing this contamination draws on finite resources. How those resources are allocated to reduce the ongoing risk to civilians in the most efficient and effective way practicable remains an ongoing challenge for the HMA sector.

Removing this explosive ordnance is not a straightforward process since it is invariably not always clear exactly where the contamination is located. This is especially so if that contamination was concealed when it was emplaced, as is often the case with minefields. Time spent clearing land that isn't contaminated is time not spent clearing land which is, and implicitly carries the risk of more civilian casualties that may not have happened if that land had been cleared earlier [1]. In short, there is an opportunity cost in not accurately identifying the contaminated land prior to clearance.

How then to approach the problem of efficiently allocating resources to clear the contamination when the location of the explosive ordnance itself is often hard to confirm until after clearance has taken place? Or put another way, how best to manage the risk that, with limited initial information, the wrong priority for clearance is chosen, even with the best of intentions? The Mine Action sector's attempt to answer this ongoing dilemma is a Risk Management process called Land Release.

8.2 The basis of mine action

Mine Action is based on international treaties. Foremost amongst these is the Convention on the Prohibition of the Use, Stockpiling, Production and Transfer of Anti-Personnel Mines and on their Destruction, known as the 1997 Mine Ban Treaty. Article 5 of this treaty requires that '*each State Party shall make every effort to identify all areas under its jurisdiction or control in which antipersonnel mines are known or suspected to be emplaced and shall ensure as soon as possible that all antipersonnel mines in mined areas under its jurisdiction or control are perimeter-marked, monitored and protected by fencing or other means, to ensure the effective exclusion of civilians, until all anti-personnel mines contained therein have been destroyed*' [2]. The key words in this statement, at least in regard to efficient and effective removal of explosive contamination, are 'every effort'. For a long time, 'every effort' was the basis on which any location where there was a suspicion of anti-personnel mine contamination was not only identified, but also cleared. The same stipulation was found in the 2008 Convention on Cluster Munitions, Article 4 of which requires that states make '*every effort to identify all cluster munition contaminated areas under its jurisdiction or control*' and to '*clear and destroy all cluster munition remnants located in areas under its jurisdiction or control*' [3].

Given the humanitarian cost of landmines [4–6] and submunitions [7], the treaties were undoubtedly justified in requiring complete survey and clearance of both anti-personnel mines and cluster munition remnants. However, states and operators soon required more guidance on how this might be achieved in the most efficient and effective way in a context of limited resources. A 2004 study by the Geneva International Centre for Humanitarian Demining (GICHD) of fifteen mine affected countries, showed that of 290 000 000 m^2 that was originally claimed to be contaminated, on average only 2.09% was recorded as contaminated [8]. In 2007 it was openly acknowledged that '*demining programs continue to allocate scarce resources to conduct operations on land which is then discovered not to be contaminated.*' [9] Therefore, the concept of Land Release, discussed since the mid-1990s [10], was developed from the mid-2000s [11, 12], when it was increasingly accepted that the best way to apply finite resources was to design a system that would encourage those resources to be targeted where evidence of contamination was actively identified [13]. While retaining the treaty requirement that the '*...aim of manual mine clearance is the clearance of every mine, and every item of unexploded ordnance (UXO) or abandoned explosive ordnance (AXO), from a given area,*' and acknowledging that '*there are many situations in which such a level of clearance is necessary or is demanded*' it was for the first time recognised in 2005 that '*the full clearance standard is not necessarily universally applicable, nor is it by any means universally achieved. For this reason, the mine action community and its stakeholders need to urgently consider moving to a more explicit Risk Management approach*' [11, 14]. In September 2006 the term 'Land Release' started to appear in official treaty documentation [15]. In October 2008, Norway, as Coordinator of the Resource Utilization Contact Group, achieved the first real recognition of Land Release by States Parties to the Mine Ban Treaty [16]. In 2009 Land Release took shape as a recognisable Risk Management approach in International Mine Action Standards (IMAS) [17]. In the same year the Land Mine Monitor agreed that '*any land that is not contaminated but is physically cleared represents inefficiency and a potentially huge waste of resources for a national demining program*' [18].

8.3 What is Land Release?

Land Release is a means to '*ensure that assets are deployed to achieve as much as possible, for the minimum cost in the shortest time. Application of the Land Release process consists of establishing and improving the definition of where mines/Explosive Remnants of War (ERW) are to be found (and where they are not) through the application of all reasonable effort, until it can be shown with justifiable confidence that mines/ERW are either not present in an area or, if they were found to be present, have all been destroyed or removed from that area*' [19]. Practically attaining the application of 'all reasonable effort' is achieved by promoting the primacy of evidence in decision-making process [20]. The problem might be put in its most basic form by asking the simple question; 'where are the mines?' Prioritising evidence in all activities allows this question to be answered as quickly as practicable, ideally with the minimum cost. One of the fundamental challenges to achieving this process is the human cost of getting it wrong.

Land Release normally involves three basic stages, Non-Technical Survey, Technical Survey and Clearance [21]. All of these stages should prioritise the collection of evidence to best target the expensive resources required to ultimately release an area. If, for example, there is an area which is believed to be a minefield, it will first be surveyed in a way where no physical assets such as manual deminers, machines or Mine Detection Dogs (MDDs) will enter that hazardous area. This involves gathering all information available on the minefield. Should reliable minefield records be available, this is a hugely significant advantage and will often be the key factor that can speed up clearance [22]. This may then be backed up by looking at accident data geolocated to the area in question. The types of injuries can give a good indication as to the type of explosive contamination present. For example, lower limb injuries can be a good indicator of anti-personnel blast mines [23–25]. Added to this might be injuries to animals such as grazing livestock. Other relevant evidence will also be collected. For example, it is sometimes possible to observe small craters from the air where mines have already detonated or observe trench systems that, in a given context, are known to be typically protected by anti-personnel mines. The significance of a given piece of evidence changes from country to country. A trench might be very significant in one place and have little significance in another. This initial process is termed Non-Technical Survey, although this is not necessarily the best label since assessment of relevant evidence requires a lot of technical knowledge and experience to be done competently. A better title would be non-intrusive survey since it involves finding relevant information without taking physical risk by entering into a potentially hazardous area. Once all evidence has been collected, a decision will made as to whether the evidence justifies the deployment of significantly more expensive clearance resources to intrusively survey (technically survey) and potentially subsequently clear the area.

A degree of prioritisation is possible at this stage dependent on the strength of evidence available for a given area. If the evidence indicates that there is a reasonable suspicion of contamination, the area will be designated a Suspected Hazardous Area (SHA). The exact type of evidence that would serve as basic criteria for such a designation is left to a National Mine Action Authority (NMAA) to specify in their National Mine Action Standards [26]. However, IMAS, do identify likely examples of what is termed Indirect Evidence. These examples include:

- Potentially productive land not in use
- Verbal reports from local population/former combatants
- Mine/ERW records, where the reliability of such records remains open to doubt or has not been assessed
- Analysis of other known contamination areas, tactics and historical sources
- Former combatant zones
- Evidence from previous surveys, not supported by direct evidence of the presence of contamination
- Mine/ERW accidents or incidents where the location of the event cannot be accurately determined [27].

If the evidence is strong enough that the nature of the hazard can be confirmed, it will be termed Direct Evidence and a Confirmed Hazardous Area (CHA) will result. As with indirect evidence it is up to the NMAA to specify what constitutes direct evidence, although IMAS also lists likely examples:

- Mine/ERW records, where the reliability of such records has been confirmed during previous operations
- Visual observation of mines/ERW, mine/ERW parts, fragmentation or craters
- Detonations during fires or by animals
- Mine signs, fencing, ancillary equipment (boxes, canisters) etc associated with contamination
- Mine/ERW accidents or incidents where the location of the event can be accurately determined [26].

Unless items such as mines are confirmed through sight, it is often necessary to breach into a hazardous area using techniques usually associated with clearance in order to confirm the presence of mines. The process is termed Technical Survey [28], although it might more accurately be described as intrusive survey, since it involves physically entering into the hazardous area to gather evidence. Should ordnance be found, this could be the basis on which the area is designated a Confirmed Hazardous Area. Typically, since the assets required to conduct this form of intrusive survey are the same as required for actual clearance, the clearance stage invariably follows in short order if explosive ordnance is found. Indeed, Technical Survey of pattern minefields is often indistinguishable from the first stages of clearance. '*The term Clearance in the context of mine action, refers to tasks or actions to ensure the removal and/or the destruction of all mine and ERW hazards from a specified area to a specified depth*' [29]. Clearance may involve assets that assist processing land, such mechanical tillers or flails. However, with the exception of machines excavating large areas of topsoil, clearance will usually require a deminer to physically clear the contaminated ground in a labour-intensive way. Depending on the technique used, for an Anti-Personnel (AP) mine hazardous area, deminers can typically clear 5–20 square metres per day [30]. For this reason alone, the importance of making sure such manual demining efforts are as targeted as practical, as enabled by the Land Release approach's emphasis on evidence, becomes apparent.

8.4 The importance of operational data

One of the key indicators as to the presence of explosive ordnance is the presence of other explosive ordnance. This is especially the case with contamination such as pattern minefields and cluster munition strikes. Finding one mine or one submunition can often be the best indicator of the presence of others. To make Land Release decisions, deminers need to know exactly what was found and where. In addition, they must know exactly where has been processed to a given standard up to clearance and where has not. For the Land Release process to work well, the contaminated areas therefore need to be correctly and accurately identified and this requires the accurate location recording of certain individual explosive items and where has been cleared, and where has not. On a typical site of a few hectares in area, this entails the recording of thousands of location points that can then be recorded and represented using GIS software. The data collected is not limited to location points. A key-criteria of any clearance task is a clearance depth. This depth can only be decided based on depth data of hazards found in the past. Therefore, depth of all items must be recorded as well.

The system for capturing this information, while increasingly dependent on good information technology, also depends on simple procedures that are more prosaic. Hugely significant is the design of the forms that capture information and feed the databases, whether they are electronic, or paper based. A technologically impressive information management system is of limited use, and can mask a host of failings, if the forms that feed it are not as good as they could be. Risk management depends on accurate, operational data, and well-designed data collection forms are the means by which that information is obtained. A poorly designed form that does not capture the right data impairs effective Risk Management. Land Release, as a Risk Management process, depends on good quality data, collected by means of well-designed forms.

Mine Action currently incorporates several information management systems. On a national level, the Information Management System for Mine Action (IMSMA), is used. First developed in 1999, IMSMA aims to enable states to report satisfactorily under Article 7 of the Mine Ban Treaty [31] and Convention on Cluster Munitions [32] respectively, where they are required to report how many square metres have been cleared of what contamination. IMSMA at a national level is usually fed by the actual field operators who typically use their own systems to collect data. Some of the systems commonly used include Fulcrum [33] and MS Infopath/Sharepoint [34], enabling the partial digitalisation of reporting in the sector over the last decade. The programme databases that this technology supports tend to be more detailed than the national databases, with operators using their discretion about what extra information they wish to collect beyond their obligations for national reporting. In any case, information is often inputted into unique systems more than once, increasingly the risk of poor-quality data entry. Scripts that link databases can overcome this frailty, but such data links are still relatively rare.

Collection of information is not standardised in mine action. Different operators collect differing levels of information, enabling different degrees of practical Risk Management. At a national level an item of ordnance may only be recorded as an item of Unexploded ordnance (UXO) with no further information, such as the ordnance sub-category or model, logged. Improvised Explosive Devices (IED) present another challenge. Counter-IED doctrine lists five basic elements of an IED, the main charge, the power source, the container, the means of initiation and the switch. The need to record just these five basic data points in order to build even the most basic IED threat picture has led to a re-examination of the level of detail about all explosive hazards that operators and NMAAs should gather from the field. Ideally the full potential of mobile information management on handheld devices will be exploited so that greater detail of exactly what is found, and where, and in what condition, is available for operational Risk Management (threat assessment) at all levels. Historically, large scale collection of such information for mines has been shown to be possible with just simple paper forms [35–37]. It is hoped that further development of modern data collection and processing will enable such information to be routinely available in the future.

Location data is also especially important for Land Release Risk Management, particularly for clearance of minefields. For a pattern minefield, it is now rare for 100% of an area designated as a CHA to be fully cleared. Typically, a clearance organisation will only clear where they have evidence to show where mines will be. Mines laid in a pattern that can be demonstrated not to have moved, enable reduced clearance of the area designated as hazardous in order to find all the mines. The final clearance pattern resembles a form of messy grid that must be mapped accurately. This requires Differential GPS (D-GPS), a more precise means of geolocation than is routinely used. D-GPS uses differential correction from a base station in order to calculate any potential error in a roving GPS receiver [38]. The accuracy this system allows is essential for the most advanced applications of Land Release Risk Management. Exact recording of what was found and where constitutes the key evidence on which risk decisions are taken. The diagram represented in figure 8.1 shows why exact position reporting is essential for efficient clearance of pattern minefields. In this example the lanes represent an area of 37 400 m^2 (15.9%) out of an initial SHA of 235 716 m^2. Within the 37 400 m^2 cleared, 2168 mines were found. Both the location of the mines, and each turning point demarcating where is clear and where is not, had to be exactly geolocated using D-GPS.

D-GPS is not necessarily required for all contamination found. If there is good reason for an entire area to be cleared, only the turning points around the polygon may

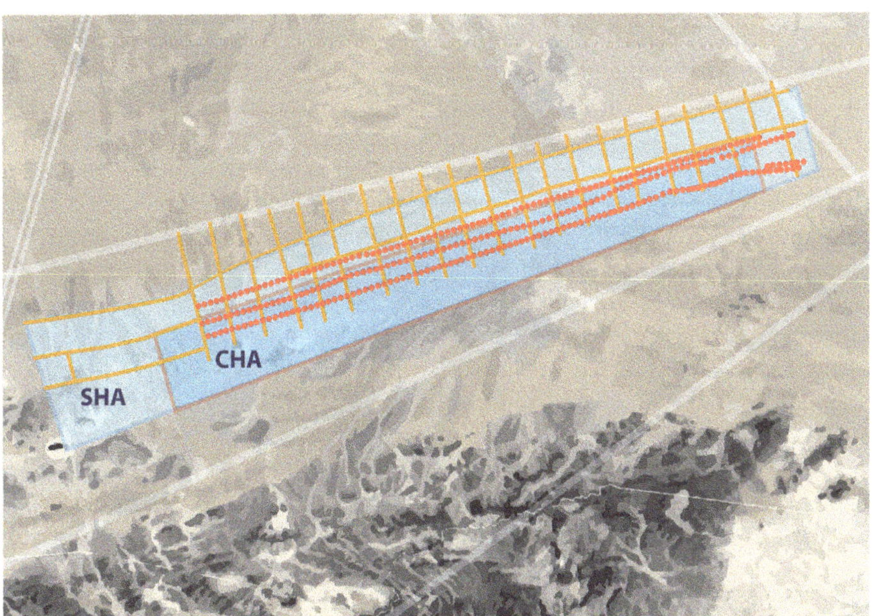

Figure 8.1. The lanes represent an area of 37 400 m^2 (15.9%) out of an initial SHA of 235 716 m^2. Within the 37 400 m^2 cleared, 2168 mines were found.

need to be recorded. Typically, this does not require the accuracy associated with a D-GPS. The same is true for individual items of explosive ordnance that are not categorised into a SHA or CHA. Exact accuracy to geolocate these items is not necessarily required. The risk such items present can be managed if a reasonably accurate geolocation associated with standard GPS is used. Managing risk within the Land Release framework is not uniform but dependent on the type of contamination. For mine contamination found in minefields where the cost of expensive clearance must be factored almost down to the metre squared, the management of risk requires detailed and accurate data. For the mortar round left beside a road on the surface, where no actual ground clearance will necessarily take place, a different level of detail will enable effective Risk Management.

8.5 The problem of low-density contamination

Low-density minefields, traditionally referred to as nuisance minefields, contain contamination where there is no or a limited pattern to the positioning of mines. In the cold logic of emplacing a minefield, a handful of mines can potentially have the same impact on an opposing combatant as a more deliberate pattern minefield since the suspicion an area is mined is often enough for individuals not to enter and to effectively deny that area. What is true for soldiers during a war is also true for civilians during peace. Just the suspicion of the presence of mines will often be enough for civilians not to use an area and for the land to be denied to those who invariably are in precarious economic circumstances.

Low-density minefields present issues of opportunity cost and clearance prioritisation all the more acutely. Put simply, is it hypothetically worth clearing 10 000 m^2 to find one mine? Is the time spent clearing the 9999 m^2 where no mine could be found, time where more heavily contaminated ground could be cleared instead? Then again, the treaties demand clearance of such areas [39, 40], and in most circumstances the only time field operators can be sure there is only one mine is at the end of the clearance process, when the resources have already been spent.

How does the Land Release process deal with such dilemmas? The short answer is, partially. The International Standards Organisation (ISO) defines risk as the effect of uncertainty on objectives [41]. Anything less than full clearance of the hypothetical low-density minefield of 10 000 m^2 will entail a less than certain state about the safety of that land. In short it will entail risk. Given there will never be enough resources in mine action to clear all the low-density contamination, despite the treaty commitments, the essential premise of Land Release is that a certain degree of residual risk is inherent in the process. Practitioners will protest that even in low-density minefield there is usually a reason why a mine has been placed where it has, perhaps to cover a path or deny a piece of dead ground, and if such likely places are cleared, the residual risk of mines being found in unlikely places is low. The acid test is how many mines result in accidents in areas released that are not subject to full clearance. If there are no accidents over a given timeframe of intensive land use (perhaps 5 years), that could also be perceived as 'full certainty'. Unfortunately, while there are examples of this [42], data is not as clear as policy makers would wish, and this is just one area where better data would enable more confidence in Land Release as a Risk Management model. There is also the consideration that mines might still be present but may have aged beyond the point where they can function as designed [43]. Such mines, if not recorded in an original minefield record, might never be discovered since they don't cause accidents. Such mines might not present an explosive hazard as originally intended but they remain a mine that must be cleared in order to comply with the Mine Ban Treaty [44].

8.6 The problem of routes

Routes are perhaps the areas where low-density mine contamination can have the highest impact. One or two anti-tank mines or IEDs, on a route of 100 km in length, can deny that route. This is all the more unfortunate since routes are essential for the stabilisation and reconstruction efforts that necessarily follow a conflict [45, 46]. One or two mines can have a disproportionate impact on the recovery of a given area. Full clearance of routes is a slow and difficult process since metal contamination, leading to false positive indications on metal detecting equipment, tends to be higher near areas of human activity, such as routes. Furthermore, minimum metal mines, used increasingly since the 1940s, can make the use of many metal detectors redundant, requiring selection of slow full excavation techniques. In many ways minimum metal mines used on routes exemplify the low-density minefield risk dilemma for operators. In a hypothetical example of looking for one anti-tank mine along even a narrow route of 6 m in width (including verges), for a distance of just one kilometre, would entail searching 6000 m^2 of invariably metal contaminated ground, for one item. The best metaphor for trying to find mines on routes might be the cliché of finding a needle in a haystack.

The solution again lies in a sensible Risk Management approach. The term 'route clearance' is used less and less in mine action since it is recognised that full clearance as defined by most standards on every square metre of the route surface, and a set distance either side, is simply not achievable, let alone practicable. Only areas where it is likely that a mine will have been laid will undergo full clearance. Other areas might undergo some search technique that does not amount to full clearance. For example, a single pass with a MDD might be deemed to provide sufficient confidence that no mines are present, especially if combined with a dataset that shows there is no instance of mines being found in an equivalent location in the past. Machines are often used to process routes to a level below that normally accepted as clearance. For example, mine rollers might be used to increase confidence about a route without providing certainty that there are no explosive hazards that the rollers did not initiate, either due to uneven ground, the fuzing of the mines no longer functioning or the mine changing in orientation so that the pressure of the rollers did not cause initiation. It is worth noting that the treaties do not just require mines that are functional to be removed from an area, but all mines including those that are not functional as well. Means of processing land that require the mine to be functional in order to initiate it and by these means detect it, have obvious limitations and increase the risk of explosive hazards being left behind.

8.7 The problem of criteria

The IMAS that detail what Land Release is, mention the need to develop criteria for the various stages and outputs of Land Release. Current IMAS state that such criteria should be developed for a given national environment by a NMAA in their National Mine Action Standards. Those detailed in IMAS are only illustrative [26]. However, to date no NMAC detail criteria that differs from those illustrative examples in IMAS.

The criteria required are not just lists of the type of evidence that may justify the creation of a SHA or CHA. They should also include criteria for cancelling land that has mistakenly already been designated hazardous, and criteria for reducing land by means of technical survey. For example, if it is observed that no evidence of AP mines is present and that land has been tilled by local farmers for a given number of years, these might be deemed reasonable criteria to cancel land previously designated contaminated. Logically there should also be criteria for what constitutes clearance in a given operating environment. Such criteria might include clearance depth, or what combinations of mechanical and MDD processing, combined with manual clearance, constitute clearance. Until criteria applicable for the range of operating environments existing in mine action are developed, Land Release will not be fully realised or implemented.

8.8 The issue of explosive remnants of war

Land Release was a process primarily developed to find and clear minefields. It was subsequently adapted to Cluster Munition contamination, especially after the signing of the Convention on Cluster Munitions in 2008. Land Release soon became applicable to all types of explosive ordnance [47] contamination. However much of the contamination found in former combat zones is not congregated so as it contaminates an area, but is found on its own, as a spot. In such circumstances it is typically dealt with as an individual spot task, typically by an Explosive Ordnance Disposal (EOD) team. If there is a significant concentration of spots, the area will most likely be dealt with as Battle Area Clearance (BAC) task, involving a systematic clearance of ground to a specified depth in a way that can be like demining but often with essential differences.

Even in the most heavily mine affected countries, a significant majority of the explosive ordnance found is not anti-personnel or anti-tank mines. For example, in Cambodia during the period March 1992–June 2002, only 18.6% of ordnance destroyed were AP mines, 0.04% were AV mines and 81% were ERW [48]. In Afghanistan in 2012 what is reported as UXO was deemed to account for two-thirds of civilian casualties, with landmines only accounting for a third [49].

Protocol V of the Convention on Certain Conventional Weapons (CCW), defines ERW. ERW is an umbrella term for explosive ordnance that does not include mines, including Cluster Munition Remnants (CMR). ERW is defined as being either UXO or Abandoned Ordnance (AXO) [50]. Both AXO and UXO are often found as spots rather than areas.

How does Land Release contribute to the Risk Management of clearing contamination spots of UXO or AXO? The Land Release methods focus on areas and spots of contamination not covered in any of the respective documents that detail Land Release. Nevertheless, such contamination represents risks associated with all explosive ordnance, albeit risks not necessarily consistent with a victim operated device. Land Release does have its limitations in relevant scope then, and whether it can be adapted to meaningfully enable Risk Management of all ERW contamination remains to be seen. There are already efforts to try to integrate ERW spots into the Land Release process [51]. That Land Release is not an all-encompassing model, at least not yet, should not detract from how appropriate it is from area-based contamination. Risk Management approaches continue to evolve, and Land Release, as a form of Risk Management for particular types of explosive contamination, is no different.

8.9 The influence of ISO

Much of mine action terminology, and Land Release guidance in particular, is influenced by ISO. Mine action is guided by its industry standards, the international version of which is known as IMAS. IMAS started in 1997 and from early in their development ISO language was often adopted, including language concerning risk. The ISO definitions of 'shall', 'should' and 'may' are the basis of IMAS, where 'shall' indicates requirements that must be applied, 'should' indicated preferred requirements and 'may' indicates a possible method or course of action. IMAS uses the ISO definition of risk. Risk is 'the effect of uncertainty on objectives' [52]. In IMAS 'confidence' is deemed the opposite of uncertainty, and levels of confidence are the means by which Land Release decisions can be made. '*Before land can be cancelled, reduced or accepted as cleared, it should be established, with a sufficiently high level of confidence, that there is no longer any evidence that the area contains mine/ERW contamination*' [53]. How do decision makers develop the requisite confidence? '*Confidence can only be gained after all reasonable efforts have been made to investigate whether mine/ERW contamination is present and, when contamination is found to be present, to remove it*' [53]. '*All reasonable effort*' in mine action represents the effort that it is reasonable to expect should be applied in order to achieve the desired level of confidence that cancelled, reduced and cleared land is free of mine/ERW contamination within specified limits' [53].

One issue for Land Release is that neither '*confidence*' or '*all reasonable effort*' have ever been fully defined. Confidence is a subjective term that can differs between individuals and organisations. It is notable that the IMAS never actually defines or gives a clear example of what confidence should be practically or what level of confidence is required to make a decision about clearing land. Does a field operator need to be very confident, or just confident enough? No further definition is found in any National Mine Action Standard (NMAS) or in any Standard Operating Procedure. Detailed answers as to what confidence actually requires in practice remain elusive in mine action and tend to result in individual judgements that could be deemed subjective. The evidence criteria against which an area can be designated a suspected or confirmed hazardous area also remain ill-defined with nothing developed at a national level beyond the suggested illustrative list in IMAS. Whether it is confidence, defining exactly what all reasonable effort is in a given context or the criteria against which key land designations are made, there is a sense that Land Release is a house half built and requires further development as an underpinning doctrine.

Confidence need not necessarily be subjective when making Land Release decisions. In order to be objective it needs to be strictly aligned with verifiable data that provides an evidence base. A clearance operator may be confident that there are no mines in an area of land beneath 30 cm depth. If that confidence is based on the recorded depth of every mine ever found within that context, it might be deemed objective. If it is not based on such evidence, it might be little more than intuition. This key link between confidence and evidence is mentioned in the IMAS that detail Land Release [54] but it is not a strict requirement mandated by a 'shall' statement. Until such a link becomes a strict requirement the sense that Land Release is not yet a mature Risk Management model will remain.

8.10 Land release success stories

Given that demining is often a slow process, the issue of land mine contamination can easily be seen as an intractable problem without an end. However, since Land Release has been progressively adopted since 2009, there are clear examples of countries effectively removing significant contamination by applying Land Release. Mozambique was heavily contaminated with anti-personnel landmines as a result of the 30 year civil war that ended in 1992 [55]. However, in December 2015. Mozambique was able to declare formal compliance with Article V of the Anti-Personnel Mine Ban Treaty, having cleared the last known contamination [56]. From 2008–2014 a National Action Plan was successfully implemented where 86392 AP mines and 136 anti-tank mines were found and destroyed. This resulted from the survey and clearance of over 55 km^2 [57]. Expensive clearance resources were progressively better targeted over time to areas where there was reasonable evidence of contamination. The risk of clearing areas without contamination, with the associated opportunity cost of not clearing areas with, was continuously reduced with the eventual success of the country being declared mine free [58]. The application of Land Release in Mozambique was not without difficulties, but it was ultimately successful.

8.11 Future developments of land release

While the ostensible purpose of Land Release is to manage risk on the basis of evidence, it could also be seen as seeking to improve the performance of mine action. In its most simple terms, the more squared metres cleared that are actually contaminated, not only the quicker reduction of risk from explosive hazards, but also the better performance of mine action operators. However, there is scope for further improvement on the issue of performance, not least since it is only very recently defined within IMAS [59]. Performance has been defined as '*how well a person, machine, etc does a piece of work or an activity*' [60]. ISO define performance as referring to '*the measurable results that activities, processes, products, services, systems and organizations are able to achieve. Whenever they perform well it means that acceptable results are being achieved and whenever they perform poorly, unacceptable results are achieved*' [61]. However, Mine Action is still to develop agreed indicators for performance. Suggested Key Performance Indicators were developed in 2019–2020 by the GICHD. However, these are published in a Technical Note for Mine Action (TNMA). Technical Notes are neither legal documents nor IMAS. They are purely advisory and are designed solely to supplement technical knowledge or to provide further guidance on the application of IMAS' [62]. Therefore, if it is accepted that managing risk and managing performance are inextricably linked, consistent use of performance indicators are a further step required to realise a more comprehensive Risk Management system within mine action in general, and Land Release in particular.

8.12 Conclusion

It would be wrong to present Land Release as a complete answer to the problem of best targeting resources to clear explosive hazards. It remains a work in progress rather than a polished Risk Management model. Nevertheless, the basic precepts of the Land Release approach, chief amongst which is an emphasis on evidence to inform risk-based decision making, remain relevant and appropriate. The approach will need to be continually refined, with an ever-increased emphasis on operational data and better definition of previously ambiguous terminology. At its best the Land Release framework is the only practicable approach available to a mine action sector that will invariably be limited in the resources it can apply to the problem. Land Release will continue to evolve as a workable Risk Management approach in the future, but its fundamental precepts will remain.

Land release is a Risk Management framework and not a comprehensive solution in itself. It does not amount to a silver bullet. As is the case with any framework, its success is subject to how well it is applied. Land release remains to be fully implemented in many countries, not least because despite the use of common ISO definitions, the concepts of risk and Risk Management vary across the globe. Applying a common global understanding of Risk Management is possible. Aviation shows how this can be achieved [63, 64]. The level of universally accepted Risk Management inherent in that industry shows how much further Land Release still has to go. Land Release is also hamstrung by the limited relevant data that operators and National Authorities gather from the field. The evidence base on which decisions are made remains limited. Nevertheless, in little over a decade this Risk Management framework for mine action can fairly claim significant and tangible success in better targeting scarce resources to mitigate the threat of explosive ordnance to civilians worldwide. The challenge for the next decade is to implement practical Land Release more widely, based on much better data, and to inculcate Risk Management thinking at all levels of the sector.

References

[1] Geneva International Centre for humanitarian Demining 2005 *A Study of Manual Mine Clearance* p 2 https://gichd.org/en/resources/publications/detail/publication/a-study-of-manual-mine-clearance/

[2] Convention on the Prohibition of the Use, Stockpiling, Production and Transfer of Anti-Personnel Mines and on their Destruction 18 1997 https://apminebanconvention.org/fileadmin/APMBC/text_status/Ottawa_Convention_English.pdf

[3] Convention on Cluster Munitions 2008 https://clusterconvention.org/files/2011/01/Convention-ENG.pdf#page=1

[4] Doswald-Beck L, Herby P and Dorais-Slakmon L 1995 Basic facts: the human cost of landmines. International Committee of the Red Cross, 1 https://icrc.org/en/doc/resources/documents/misc/57jmcy.htm

[5] Boutros-Ghali B 1994 The land mine crisis: a humanitarian disaster *Foreign Affairs* **73** 8–13

[6] Krug E G and Gjini A 1999 Number of land mine victims in Kosovo is high *Br. Med. J.* **319** 450

[7] Riordan K 2008 Introductory note. Convention on cluster munitions *Dublin* Audiovisual Library of International Law https://legal.un.org/avl/ha/ccm/ccm.html

[8] GICHD 2004 A Study of Mechanical Application in Demining p 57 https://gichd.org/en/resources/publications/detail/publication/a-study-of-mechanical-application-in-demining/

[9] Land Mine Monitor 2007 p 27 http://the-monitor.org/media/1677574/executive_summary.pdf

[10] Blagden P M 1995 The changing scene of mine clearance *Second Int. Conf. on Abandoned Land Mines (12–14 October 1998)* pp 18–22 'By 1995 it was becoming obvious that in many of the mine clearance operations the more important factor was the release of ground'

[11] Lardner T 2008/09 The GICHD land release project *J. ERW Mine Action* **12** 82 https://commons.lib.jmu.edu/cgi/viewcontent.cgi?article=1578&context=cisr-journal

[12] Statement by Ambassador Stephen Nellen. Director, GICHD. Seventh Meeting of the States Parties to the Convention and Prohibition of the Use, Stockpiling, Production and Transfer of Anti-personnel Mines and on their Destructionm Geneva, 18–22 September 2006 p 3 https://apminebanconvention.org/fileadmin/APMBC/MSP/7MSP/update_day1/Nellen_7MSP_18Sept06.pdf

[13] Stevens J E 2009 A conversation about land cancellation and release with H. Murphey 'Murf' McCloy *J. Convention. Weapons Destruct.* **13** 6 https://commons.lib.jmu.edu/cgi/viewcontent.cgi?article=1490&context=cisr-journal

[14] Geneva International Centre for humanitarian Demining 2005 A Study of Manual Mine Clearance p 13 https://gichd.org/en/resources/publications/detail/publication/a-study-of-manual-mine-clearance/

[15] Article 48 2007 Final Report. Seventh Meeting of the States Parties to the Convention on the Prohibition of the Use, Stockpiling, Production and Transfer of Anti-Personnel Mines and on their Destruction. Geneva, 18–22 September 2006. p 20. https://www.apminebanconvention.org/fileadmin/APMBC/MSP/7MSP/final_report/7MSP_Final_Report_e.pdf

[16] Norway, Coordinator of the Resource Utilization Contact Group. Applying all Available Methods to Achieve the Full Efficient and Expedient Implementation of Article 5. Meeting of the States Parties to the Convention on the Prohibition of the Use, Stockpiling, Production and Transfer of Anti-Personnel Mines and on their Destruction. Ninth Meeting, Geneva, 24–28 November 2008. Item 11 of the provisional agenda. 02 October 2008 https://www.apminebanconvention.org/fileadmin/APMBC/MSP/9MSP/9MSP-NorwPaper-Landrelease-2Oct2008-en.pdf

[17] Land Release. International Mine Action Standard 08.20. First Edition. 10 June 2009 p v
[18] Land Mine Monitor Report 2009 *Special Ten-Year Review of the Mine Ban Treaty* p 42 http://the-monitor.org/en-gb/reports/2009/landmine-monitor-2009/mine-action.aspx
[19] Land Release. International Mine Action Standard 07.11. Edition 01. Amendment 05. 19 March 2019. p 1 https://mineactionstandards.org/en/standards/document-detail/?tx_imas_document%5Bdocument%5D=262&tx_imas_document%5Baction%5D=show&tx_imas_document%5Bcontroller%5D=Document&cHash=a6bc6b5efdc872050f10f20085c953bf
[20] Land Release. International Mine Action Standard 08.20. First Edition. 10 June 2009 p 3
[21] Glossary of mine action terms, definitions and abbreviations. International Mine Action Standard 04.10. Edition 02. Amendment 10. February 2019. https://mineactionstandards.org/en/standards/document-detail/?tx_imas_document%5Bdocument%5D=264&tx_imas_document%5Baction%5D=show&tx_imas_document%5Bcontroller%5D=Document&cHash=611e82a6d46b2fe28da138e248a187ab
[22] Evans R 2018 Lesson from the past: the rapid clearance of Denmark's minefields *J. Conventional Weapons Destruct.* **22** 4 https://commons.lib.jmu.edu/cisr-journal/vol22/iss1/4
[23] Korver A 1996 Injuries of the lower limbs caused by antipersonnel mines: the experience of the International Committee of the Red Cross *Injury* **277** 477–9
[24] Trimble K and Clasper J 2001 Anti-personnel mine injury; mechanism and medical management *J. Royal Army Med. Corps* **147** 73–9
[25] Harjai M, Agarwal D, Dave P, Jog S and Arora P 2005 Mine blast injuries—our experience *Med. J. Armed Forces India* **61** 143–7
[26] Land Release. International Mine Action Standard 07.11. Edition 01. Amendment 05. 19 March 2019, p 5 https://mineactionstandards.org/en/standards/document-detail/?tx_imas_document%5Bdocument%5D=262&tx_imas_document%5Baction%5D=show&tx_imas_document%5Bcontroller%5D=Document&cHash=a6bc6b5efdc872050f10f20085c953bf
[27] Non-Technical Survey. International Mine Action Standard 08.10. Edition 01. Amendment 04. February 2019, p 5 https://mineactionstandards.org/en/standards/document-detail/?tx_imas_document%5Bdocument%5D=260&tx_imas_document%5Baction%5D=show&tx_imas_document%5Bcontroller%5D=Document&cHash=ba457a7fc58ed288abda162f07b1e264
[28] Technical Survey. International Mine Action Standard 08.20. Edition 01. Amendment 05. 19 March 2019, p 4 https://mineactionstandards.org/en/standards/document-detail/?tx_imas_document%5Bdocument%5D=261&tx_imas_document%5Baction%5D=show&tx_imas_document%5Bcontroller%5D=Document&cHash=c892d90740b32344daf12c4ced93d4d5
[29] Glossary of mine action terms, definitions and abbreviations. International Mine Action Standard 04.10. Edition 02. Amendment 10. February 2019 https://mineactionstandards.org/en/standards/document-detail/?tx_imas_document%5Bdocument%5D=264&tx_imas_document%5Baction%5D=show&tx_imas_document%5Bcontroller%5D=Document&cHash=611e82a6d46b2fe28da138e248a187ab
[30] Geneva International Centre for humanitarian Demining. A Study of Manual Mine Clearance. August 2005, p 9 https://gichd.org/en/resources/publications/detail/publication/a-study-of-manual-mine-clearance/
[31] Article 7. Convention on the Prohibition of the Use, Stockpiling, Production and Transfer of Anti-Personnel Mines and on Their Destruction. 18 September 1997. https://apminebanconvention.org/fileadmin/APMBC/text_status/Ottawa_Convention_English.pdf
[32] Article 7. Convention on Cluster Munitions. 2008 https://apminebanconvention.org/fileadmin/APMBC/text_status/Ottawa_Convention_English.pdf

[33] Wallen C and Torbet N 2020 Mobile technology in mine action: the fulcrum application *J. Conventional Weapons Destruct.* **20.2** 17–21 https://issuu.com/cisr-journal/docs/issue20-2
[34] Creighton M, Karlsen A and Qasim M 2013 Cluster munition remnant survey in laos *J. ERW Mine Action* **172** 12–16 https://issuu.com/folkehjelp/docs/cluster_munition_remnant_survey_in_
[35] Evans R 2019 Minefield clearance and casualties—Holland 1945—military operations research unit report no.7 *J Conventional Weapons Destruct* **23** 12 https://commons.lib.jmu.edu/cisr-journal/vol23/iss1/12
[36] TNA PRO: CAB 106/1023 Military operational research unit report No.7 Battle Study. Minefield Clearance and Casualties. pp 09–10 and p 41
[37] TNA PRO: WO 205/1186 Report on Mineclearance in the Netherlands, 01 November 1945. Recording Methods and Organisation p 9
[38] Darrozzes J, Roussel N and Zribi M 2016 The reflected global navigation satellite system (GNSS-R): from theory to practice *Microwave Remote Sensing of Land Surface. Techniques and Methods* **2016** 303–55
[39] Article 5. Convention on the Prohibition of the Use, Stockpiling, Production and Transfer of Anti-Personnel Mines and on their Destruction. 18 September 1997. https://apminebanconvention.org/fileadmin/APMBC/text_status/Ottawa_Convention_English.pdf
[40] Article 4. The Convention on Cluster Munitions. 30 May 2008. Dublin. Ireland. https://clusterconvention.org/files/2011/01/Convention-ENG.pdf#page=5
[41] ISO 31000:2018. https://iso.org/obp/ui/#iso:std:iso:31000:ed-2:v1:en
[42] Paterson T, Pund B and Ziaee A 2013 Landmines and livelihoods in Afghanistan: evaluating the benefits of mine action. *J. Peacebuilding Develop.* **8** 81 http://dx.doi.org/10.1080/15423166.2013.814969 In Afghanistan the National Mine Action Authority, MACCA, recorded 363 casualties before demining in a selection of communities, and none after demining
[43] *Fenix-Insight Ltd Falkland Islands Malvinas Exploitation Report* March 2015 https://commons.lib.jmu.edu/cisr-globalcwd/1332/
[44] Article 5. Convention on the Prohibition of the Use, Stockpiling, Production and Transfer of Anti-Personnel Mines and on their Destruction 18 September 1997 https://apminebanconvention.org/fileadmin/APMBC/text_status/Ottawa_Convention_English.pdf
[45] Schouten P and Bachmann J 2017 *Roads to Peace. The Role of Infrastructure in Fragile and Conflict-Affected States.* United Nations Office for Project Services (Copenhagen: Danish Institute for International Studies) http://roads-to-peace.org/PDF/DIIS%20UNOPS%202017%20Roads%20to%20Peace%20report.pdf
[46] Filippino E and Paterson T 2005 Mine action lessons and challenges: is mine action making a difference or avoiding the question? *J Mine Action* **9** 23–24 https://commons.lib.jmu.edu/cisr-journal/vol9/iss1/11/
[47] International Mine Action Standard 04.10. Glossary of mine action terms, definitions and abbreviations. Edition 02. Amendment 10. 19 March 2019. https://mineactionstandards.org/en/standards/document-detail/?tx_imas_document%5Bdocument%5D=264&tx_imas_document%5Baction%5D=show&tx_imas_document%5Bcontroller%5D=Document&cHash=611e82a6d46b2fe28da138e248a187ab
[48] GICHD 2004 A Study of Mechanical Application in Demining. May p 57. https://gichd.org/en/resources/publications/detail/publication/a-study-of-mechanical-application-in-demining/
[49] Paterson T, Pund B and Ziaee A 2013 Landmines and livelihoods in Afghanistan: evaluating the benefits of mine action *J. Peacebuilding Develop.* **8** 88

[50] Article 2. Protocol on Explosive Remnants of War. Meeting of the States Parties to the Convention on Prohibitions or Restrictions on the Use of Certain Conventional Weapons. Geneva. November 2003 https://documents-dds-ny.un.org/doc/UNDOC/GEN/G03/653/61/PDF/G0365361.pdf?OpenElement
[51] A Guide to Non-Technical Survey. Process Map. Annex A.GICHD. Geneva, 2019 https://commons.lib.jmu.edu/cisr-globalcwd/1292/
[52] ISO 31000:2018 https://iso.org/obp/ui/#iso:std:iso:31000:ed-2:v1:en
[53] International Mine Action Standard 07.11. Land Release. Edition 1. Amendment 5 https://mineactionstandards.org/fileadmin/MAS/documents/standards/IMAS-07.11-Ed1-Am5-Land_Release.pdf
[54] International Mine Action Standard 07.11. Land Release. Edition 1. Amendment 5 p 1. https://mineactionstandards.org/fileadmin/MAS/documents/standards/IMAS-07.11-Ed1-Am5-Land_Release.pdf
[55] Mozambique Mine Action. Contamination and Impact. Land Mine Monitor October 2010 http://the-monitor.org/en-gb/reports/2010/mozambique/mine-action.aspx
[56] Clearing the Mines 2018 Report p 146 http://mineactionreview.org/documents-and-reports/clearing-the-mines-2018
[57] https://mz.undp.org/content/mozambique/en/home/ourwork/environmentandenergy/success-stories/Mozambique_Declared_Mine_free/
[58] Mozambique announces it is now free of all known landmines, 2015 first year in decades without landmine casualties. Anti-Personnel Mine Ban Convention Press Release. Geneva. 01 December 2015. https://apminebanconvention.org/fileadmin/APMBC/MSP/14MSP/2015-12-01-Mozambique_free_of_all_known_landmines.pdf
[59] Technical Note 07.11/02. Key Performance Indicators (KPIs) for Land Release and Stockpile Destruction Operations. Edition 01, 4 February 2021, p 7 https://mineactionstandards.org/filcadmin/uscr_upload/TNMA_07.11.02_Key_Performance_Indicators.pdf
[60] https://dictionary.cambridge.org/dictionary/english/performance
[61] ISO 9001 2015. Definitions
[62] Technical Note 01.10/01. Guide to Technical Notes for Mine Action. Version 4.0. Amendment 01. July 2013. p iv. https://mineactionstandards.org/fileadmin/MAS/documents/standards/TN_01.10_01_2002_Guide_to_TN_for_Mine_Action_V4.0_Amd1.pdf
[63] Federal Aviation Administration. U.S. Department of Transportation. Aviation Instructor's Handbook. Chapter 01: Risk Management and Single Pilot Resource Management. pp 1-1-1-29. https://faa.gov/regulations_policies/handbooks_manuals/aviation/aviation_instructors_handbook/media/03_aih_chapter_1.pdf
[64] Federal Aviation Administration 2009 U.S. Department of Transportation. *Risk Management Handbook.* Change 1 (January 2016). https://faa.gov/regulations_policies/handbooks_manuals/aviation/media/risk_management_hb_change_1.pdf

IOP Publishing

Challenges in Risk Analysis for Science and Engineering
Development of a common language
Tracey J Temple and Melissa K Ladyman

Chapter 9

Risk assessment and management at a heritage site

G Hooper

In the United Kingdom the Ministry of Defence is disposing of much of its estate that is no longer required for defence purposes. Some of that estate has unique historic, architectural, archaeological and artistic value, including the former Royal Gunpowder Mills at Waltham Abbey, which had been in government ownership since 1787. The site features a Scheduled Ancient Monument, 20 listed buildings and a Site of Special Scientific Interest. The Powder Mills have been run as a visitor attraction since 2001. The site poses a number of challenges in terms of the assessment and the management of the risks that occur on the 170 acre site, and various stewardship issues are needed in order to mitigate those risks. These fall into four discreet areas; historic buildings, fauna, flora and people. Either quantitative or qualitative Risk Assessments are conducted in these areas and management plans have been developed in order to control the risks. The built estate comprises some 240 buildings and structures, and a process of prioritising the necessary repair and restoration of these buildings and structures has been developed in the form of a Scheduled Monument Management Plan. For the fauna a management process is in place to control the deer population so that it does not prejudice the well-being of other wildlife. For the flora a management programme has been implemented to reduce the impact of non-native species which may be to the detriment of native ones. The natural environment is therefore controlled through a Woodland Management Plan for the flora and a Wildlife Management Plan for the animals and birds. Strong control measures are in place to ensure that the visiting public, the staff and the volunteers do not put themselves or anyone else at risk. This is achieved through a series of engineering and administrative control measures. The overall aim is to ensure that the rich history of the site and the profusion of unique flora and fauna can be enjoyed by the visiting public whilst at the same time preserving it for future generations.

doi:10.1088/978-0-7503-3643-7ch9 © IOP Publishing Ltd 2022

9.1 Introduction

The Ministry of Defence is steward of the largest historic estate in Government ownership, protected for its respective historic, architectural, archaeological and artistic value. Much of this estate is no longer required for defence purposes and has been, or will be, released for alternative uses [1]. Several have unique historical interest, including the former Royal Gunpowder Mills at Waltham Abbey. It is situated in the county of Essex, about 15 miles north east of the centre of London, just outside the London Orbital motorway, the M25.

Its history goes back about 350 years as a manufacturer of gunpowder, and the use of the site has evolved as the science of explosives has developed (figure 9.1). In the early twentieth century it was expanded considerably to make the propellant cordite and after World War Two it was converted into a research and development establishment. It closed as a government facility in 1991 and in 2001 it was opened to the public as a visitor attraction. The site features many unique historic buildings and is rich in flora and fauna. Most of the site is classified as a Scheduled Ancient Monument (SAM) and a Site of Special Scientific Interest (SSSI). Therefore, in terms of Risk Assessment and management, it poses challenges.

9.2 Risk Management

In terms of Risk Management, the Trustees of this heritage site have two principal tasks, firstly to preserve and hopefully enhance all the various historic buildings, flora and fauna on site and secondly to make sure that the staff, volunteers, contractors and of course visitors are kept safe and healthy (figure 9.2). Risk assessment is carried out on all aspects of the management of the site, and a variety of external regulators monitor the Company's activities. These include Historic and Natural England, the Health and Safety Executive and the Office of Rail and Road.

Figure 9.1. Victorian Gunpowder Incorporating Mills at the Waltham Abbey Royal Gunpowder Mills.

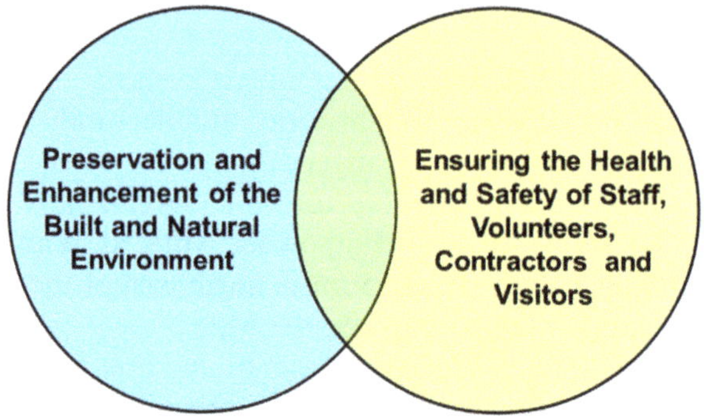

Figure 9.2. The charitable objectives of the Waltham Abbey Royal Gunpowder Mills.

In terms of the Built and Natural environments, there are three management plans in place:

1. A Scheduled Monument Management Plan
2. A Wildlife Management Plan
3. A Woodland Management Plan.

9.2.1 Scheduled monument management plan

There are some three hundred buildings and structures on the site, mostly within the SAM, ranging from ones in current use and in a good state of repair to dilapidated ruins. There is legislation governing how this should be managed [2]. It is not practicable to bring all the buildings back into a serviceable condition, so there is a need to prioritise remedial works in accordance with benefits and risks. This is done in a quantitative way, utilising a scale from regular repair and maintenance to letting the buildings decline in condition. Use is made of a number of indicators including historical, evidential, aesthetic and communal values. Each building or structure is accordingly rated in terms of its significance and its repair category. This is illustrated below for six buildings (figure 9.3).

The Scheduled Monument Management Plan is a living document, and it is updated and revised in line with recommended conservation practices.

9.2.2 Wildlife management plan

This plan is more qualitative in nature. There is a large variety of wildlife on site, principally because a large fraction of the site has been left undisturbed since the end of the Second World War in 1945. Some of these are illustrated below (figure 9.4).

The only species that are actively managed are the deer, where the population is controlled. There are a wide variety of birds; perhaps the most notable being the siskin. Siskins particularly favour the berries of the alder tree; prevalent on the site. This attracts the largest winter flocks of siskins in the UK, some 300 birds.

In March 2015 the Royal Gunpowder Mills commissioned a survey for great crested newts *Triturus cristatus* on the New Hill area of the site. On two occasions

Scheduled Monument Management Plan

	Building	Value				Significance	Repair Category
		Historical	Evidential	Aesthetic	Communal		
1	Quinan Stove	Exceptional	High	Moderate	Moderate	High	1
2	Stamping Mill	Moderate	High	Low	High	High	2
3	Mixing House No 3	High	Moderate	Moderate	Low	Moderate/ High	2
4	Charge Assembly	Moderate	Moderate	Low	Low	Moderate	3
5	Radio Chemistry Laboratory	Low	Moderate	Low	Low	Low	3
6	Small Storage Shed	Low	Low	Low	Low	Low	4

Figure 9.3. Scheduled monument management plan categorising buildings and structures in terms of their significance and repair priority.

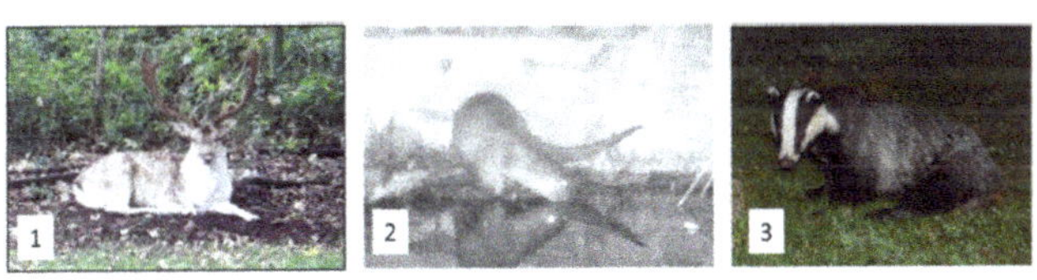

Wildlife Management Plan

1	Fallow Deer
2	Otter
3	Badger
4	Barn Owl
5	Pipistrelle Bat
6	Siskin

Figure 9.4. Wildlife management plan covering the protection of wildlife on the site.

during the survey a female great crested newt was found under a railway sleeper located on the bank of the lake on that site (figure 9.5).

Any disturbance resulting in the killing, injury or disturbance of the low numbers of great crested newts and the disturbance or loss of the terrestrial habitat would be

Figure 9.5. New Hill Lake location of Great Crested Newts.

Figure 9.6. Woodland management plan covering native and invasive species of trees.

an offence under environmental legislation [3, 4]. Mitigation measures have to be put in place to ensure that sufficient replacement habitat is created to ensure that the long-term viability of the small great crested newt population is maintained, and that any newts present within the area that may be lost to development are captured and relocated to areas of protected habitat under license from the regulator; Natural England.

9.2.3 Woodland management plan

There is a risk that the natural habitat of the Royal Gunpowder Mills site will be overtaken by invasive plant species (figure 9.6). In this context the natural habitat is that which was established three hundred years ago to make gunpowder. Alder is the wood most preferred to make the charcoal for gunpowder, however it is in danger of being over-run by non-native sycamore. That needs management and control, and this takes the form of a Woodland Management Plan. This is quantitative in nature in that it specifies the percentage of tree species to be either planted or removed. Back some fifty years ago the thinking was to drain the Marshy area around the Gunpowder Mills and lower the level of the water table. This was subsequently

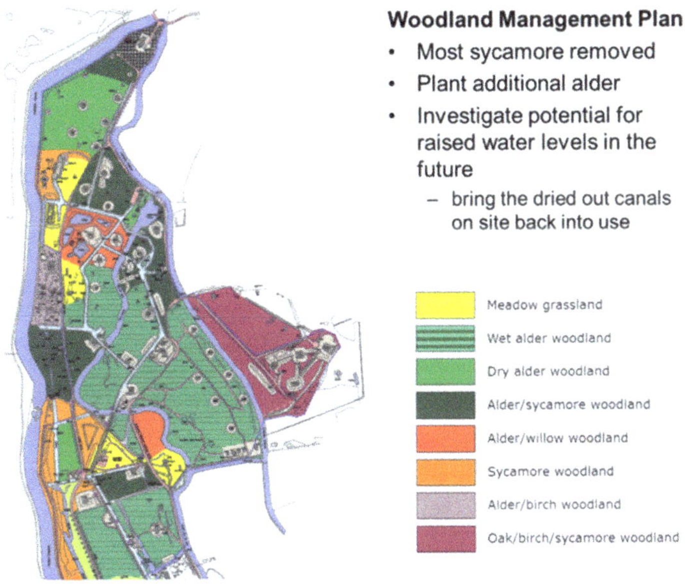

Figure 9.7. Zones of the Woodland Site showing the locations of native alder and invasive sycamore trees.

found to have undesirable side effects, including encouraging invasive sycamore and discouraging native alder, thereby disturbing the ecology of the site (figure 9.7). Plans are underway to deal with this, including the possibility of bringing dried out canals back into use and raising the water table again.

9.3 People safety

Equally important as Risk Management and assessment for the built estate, flora and fauna is identification of the hazards facing our staff, volunteers, contractors and visitors. There are numerous ones on the site. There are deep lakes and canals with sunken hazards. There are two railways with attendant safety hazards, and there is specific legislation dealing with safety management on narrow gauge railways [5]. There is a host of military hardware; there are engineering workshops and also there are a number of dilapidated buildings, several containing asbestos and/or being contaminated with explosives (figure 9.8).

We as occupiers of the property are subject to a duty to take such care as is reasonable in all the circumstances to see that staff, volunteers, contractors and visitors will be reasonably safe for the purposes for which they are invited or permitted to be there [6].

In addition, we can be liable to anyone on our premises, including trespassers, who comes to harm as a result of a danger of which we were aware or suspected and where it was reasonable for us to do something about that danger [7]. This requires a robust safety management system to be in place.

Figure 9.8. The wide variety of health and safety hazards on the site.

All these potential hazards need to be identified and assessed. To aid in this respect the Royal Gunpowder Mills has a comprehensive health and safety management system. This is required for a number of reasons:

- It is a legal requirement; the company has no option but to comply with the law and also the requirements of the Charity Commission for England and Wales.
- The Company wants to send everyone home in as good a condition as when they arrived at the site.
- There is an important issue of business and reputational damage; if there is a serious accident and the Company has not followed the law, it will be prosecuted and/or closed down.

There is a hierarchy of documentation covering all aspects of Health and Safety, and central to this is the procedure to be followed in assessing risks (figure 9.9). Most good Risk Assessment methodologies, and the Royal Gunpowder Mills has a good one, use a matrix to help determine the seriousness of the risk (figure 9.10).

Both consequence and likelihood are scored on a scale from 1 to 5, and the scores are multiplied together, giving a risk score of 1 to 25. If the product of the scores for likelihood and consequence is six or less, the risk is in the green zone, which is satisfactory. If the score is between eight and twelve, then action needs to be taken to reduce either the likelihood or the consequence by tighter control measures. If the risk score is 15 or over then the activity is either stopped immediately or it does not start in the first place.

Every activity on the site is subject to Risk Assessment. That ranges from site maintenance such as cutting down vegetation, through providing entertainment for the visitors in the form of launching small rockets, to safe operation of the site during the COVID pandemic (figure 9.11). Since in the latter instance the consequence of providing inadequate precautions can be a fatality, the Company

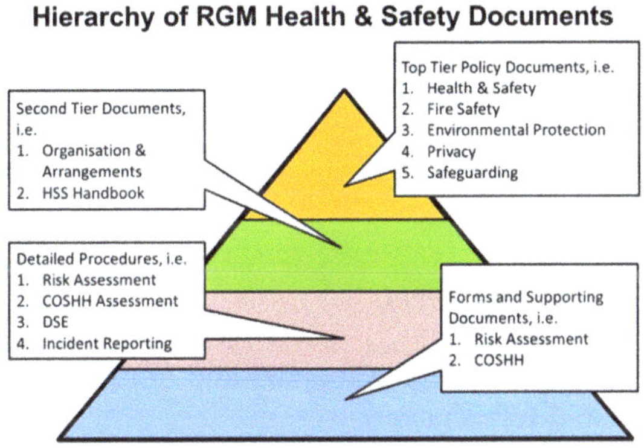

Figure 9.9. Hierarchy of health and safety documentation ranging from top level policies to forms and supporting documentation.

RISK MATRIX					
Consequence / **Likelihood**	Negligible (1)	Minor (2)	Medium (3)	Major (4)	Severe (5)
Almost Certain (5)	5	10	15	20	25
Likely (4)	4	8	12	16	20
Possible (3)	3	6	9	12	15
Unlikely (2)	2	4	6	8	10
Very Unlikely (1)	1	2	3	4	5

Figure 9.10. Matrix used to quantify risks in terms of their likelihood and consequence.

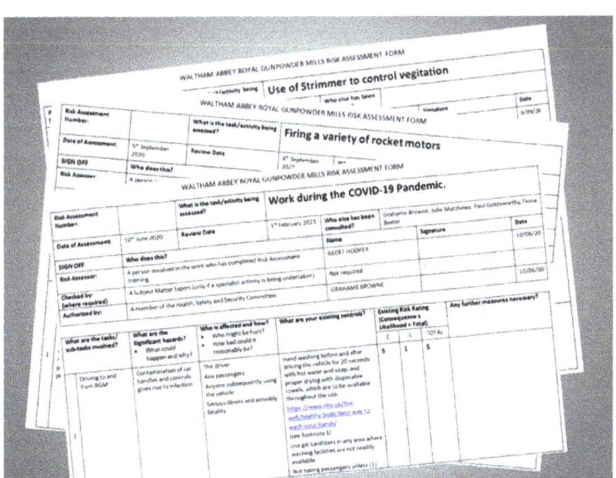

Figure 9.11. Examples of Risk Assessments carried out on the site.

needs really robust controls in place to reduce the likelihood of infection, and there are rafts of control measures to make sure that this is the case. The following activities are currently the subject of Risk Assessments:

- Driving to and from the site
- Entering and exiting the site
- Meeting colleagues
- Handling tools and equipment
- Office working
- Use of work area and handling supplies and office equipment
- Taking breaks and refreshment
- Driving Royal Gunpowder Mills vehicles on site
- Opening and closing the area where operation are taking place.

In common with many other organisations a variety of control measures have been put in place. No single one of these individually is sufficient to protect against the virus. Each layer has holes—representing the measure's flaws—which allow the virus to find a way through but collectively they give a reasonable degree of protection. This is known as the Swiss Cheese Model, and has been applied to a number of risk situations as well as COVID (figure 9.12).

Given the rapidly changing scenario and with it the Government advice and rules, this is necessarily a living document. At the time of writing, England is still subject to restrictions and the site is partially closed to visitors, but staff and volunteers are maintaining the site.

Finally, in common with most visitor attractions the Company must try to avoid vexatious claims. Carisbrooke Castle on the Isle of Wight is surrounded by a moat. It could be argued that the site operator would not need to tell people to be careful not to fall into a moat. The law, however, sees the situation differently; a pensioner won a substantial pay-out from English Heritage because there was not a warning notice telling people of the potential dangers.

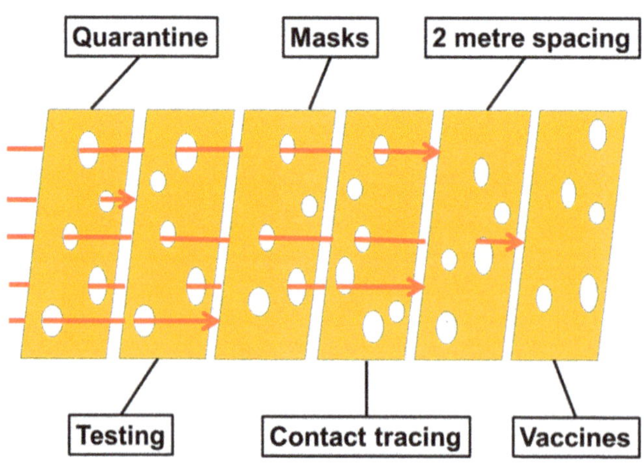

Figure 9.12. 'Swiss Cheese' model showing multiple layers of protection against the COVID-19 virus.

9.4 Conclusion

A complex series of Risk Management and assessment procedures are needed in order to ensure that the risks associated with running a heritage site are adequately controlled. These range from prioritisation of repairs to buildings and structures to proportionate management of the flora and fauna. Additionally, the health and safety of staff, volunteers, contractors and of course visitors need to be, and is, assured.

References

[1] MOD Heritage Report 2013 to 2015—Heritage in the Ministry of Defence
[2] Ancient Monuments and Archaeological Areas Act 1979 (as amended)
[3] Conservation of Habitats and Species Regulations 2010
[4] Wildlife and Countryside Act 1981 (as amended)
[5] Health and Safety (Enforcing Authority for Railways and Other Guided Transport Systems) Regulations 2006 (as amended)
[6] Occupiers' Liability Act 1957
[7] Occupiers' Liability Act 1984

Chapter 10

Amazonian non-timber forest products and ways forward to ensure their sustainability at low risk: Acai case study

K K L Yamaguchi, J Campos Zeballos, S Janzen, E B F Galante and V F Veiga-Junior

The Acai fruit consumption from the Brazilian Amazon region has been rapidly expanding in the national and international context. This Non-Timber Forest Product (NTFP) plays an important role for sustainability of the region, as its fast market expansion is reducing pressures from land-use change. However, research on the Acai industry, as well as on the risks faced by the industry today and in the future, is lacking. This chapter therefore presents factors to consider in an integrated Risk Assessment, as a first step towards such analysis. The two-folded approach of global review and local case study reveals that there is an important gap in assessing risks to emerging NTFP industries. The global literature review found no paper on Risk Assessments and NTFPs, and only one paper specifically on Acai and risk. This lack of research may result in inadequate preparedness of the industry to future challenges. The case study of the Middle Solimões municipality (Coari, Codajás, Tefé and Alvarães) in the state of the Amazonas (Brazil) shows the dependence of local livelihoods on the Acai market as well as the strong price fluctuations linked to seasonality and productivity. By contributing towards integrated Risk Assessments of the NTFP industry and especially the Acai industry, this chapter makes an important contribution to filling the research gap and informing developments of the industry.

10.1 Introduction

At first glance, the Amazon region can be observed as a beautiful homogeneous green carpet of huge trees within 5.5 million square kilometres with marvellous waterways that meander through the woods until they reach the giant rivers. This

doi:10.1088/978-0-7503-3643-7ch10 © IOP Publishing Ltd 2022

perception brings together some essential aspects that, in a way, make the destruction of large areas of forest permissive. After all, it may appear that with this green mantle being so extensive, there would be no harm in removing some of its edges. With a closer look, one can see that there is nothing homogeneous in the region. Among several tremendously different microbiomes, such as bamboo forests, types of forest with unique ecosystems are common in the Amazon region, such as *campinarana* (white sandy forests), *terra firme* (lowland forests) and *Igapó* (flooded forest). Thus, commercially important and well-known species such as guarana (*Paullinia cupana var.* Sorbilis) and rubber tree (*Hevea brasiliensis* (Wild. Ex A.Juss) Mull.Arg) are not uniformly distributed, being found in greater quantities only in very specific micro-regions. Rubber trees for instance are mainly found along the Western region and guarana along the Central one.

The Amazon is a hotspot of biodiversity, and its tropical fruits represent important crops [1]. Acai is one of these outstanding examples of great importance to the region, with highly appreciated berries, valuable palm hearts presenting multiple uses to the riverside families. In some regions, such as the Western Brazilian Amazon, it is estimated that Acai contributes to 17% of households' rent [2]. In the Amazon region, both Acai species (*Euterpe oleracea* Mart. and *Euterpe precatoria* Mart.) are considered hyper-dominant species, being amongst the most commonly observed trees in the region [3].

E. precatoria alone accounts for 1.32% of all trees of the entire Amazon [3]. However, a study on the genetic structure of this species in the Brazilian Amazon identified three different populations with dissimilar characteristics [4]. In this sense, even this widely diffused species presents variations inside the region. Being native of the Brazilian state of Amazonas, Acai from *E. precatoria* species is popularly known as 'Açaí-da-Terra-Firme' (lowland), 'Açaí-do-Amazonas', 'Açaí-solteiro' (single Acai).

Local populations have long consumed Acai berries. The berries can be consumed fresh, with dried shrimp, but are most frequently prepared as a sweet dish, in the form of a frozen pulp [5]. Having high energy values and possessing a high concentration of vitamins and mineral salts, Acai is considered a 'superfood' for being very nutritious and energetic [6, 7]. In recent years, its national and international markets have increased immensely [5]. In ten years, according to IBGE (2020), Brazil moved from a production of 124.4 thousand tons of Acai berries, in 2010, to 1.4 million tons in 2019 [8].

The increasing demand created has put pressure on production that moved several populations to extensively exploit the Acai, creating jobs and development, but also with several adverse effects, such as contamination. Additionally, the plants and industries are exposed to natural hazards. To ensure the sustainability of the industry, it is important to understand the Acai industry in a disaster risk context through an integrated Risk Assessment. This chapter therefore presents an analysis of these factors applied to the *E. precatoria* species, the most common Acai species in the Central-Western Amazon region.

10.2 Background

The Acai industry is one of the strongest in the Brazilian Amazon, with some 92% of the country's production occurring in the northern region. In 2019, Acai accounted for 48.3% of the food-related NTFP's market, which generated a total of around US $ 100 million in Brazil, only as a silvicultural product. The majority of the NTFP is generally consumed by the internal market [9]. However, exports have also increased, with Acai exports from Brazil increasing from 2.3 to 3.9 thousand tons just between 2018 and 2019. The total value also increased, from US$ 7.1 million to US$ 10.6 million [8]. Several certification processes such as Forest Stewardship Council (FSC), Organic Product in Brazil, Fair for Life and Union for Ethical Biotrade (UEBT) have been used in order to improve the productivity and open external markets, with better prices [10].

With the increased demand and production, the berries complement revenues from non-timber forest products (NTFP), which contribute to the livelihoods of more than 6 million households in the Brazilian Amazon [2]. Valued at 1.29 R$ per kg in 2019 [11], Acai is more productive than other NTFP. This important source of NTFP supports extractivism livelihoods and increases the value of standing forests [2, 12]. This is important considering the identified 'tipping points' at 40% of the total deforested area [13], leading to irreversible changes in the Amazon. As other activities, such as logging, soy production expansion, and cattle ranching, are increasingly leading to large-scale land-use changes jeopardising extractivism activities [1, 2, 14]. Acai plays an important role as an exception to the trend and for the sustainability of the region.

However, in order to ensure the sustainability of the Acai industry and the continued withstanding of other pressures, it is important to understand the industry in a disaster risk context and how current and future risk may affect the people, the Acai palm trees and the Acai industry, rather than wait for the industry to experience harm. This move from disaster management to disaster Risk Assessment is a crucial paradigm shift that needs to happen in the NTFP industry to allow for preventive action to reduce disaster risk as well as disaster impacts.

Hazards such as droughts and floods are indeed increasingly being observed in the region. With the intensification of El Niño and La Niña, these extreme events are projected to increase in frequency [15, 16]. During hot and dry years, Acai production experienced declines [17], and the palms may be competing for water with other trees. Flooding, on the other hand, hinders the establishment of new individuals, as young palms are eliminated by high waters [18]. Pests, such as insects (grasshoppers, leafhoppers, mealybugs, aphids and mites), fungi, bacteria and protozoa, are also a hazard to consider. The most common pathology is anthracnose, caused by *Colletotrichum gloeosporioides*, which has been related to excess water in flooding of rivers and poor nutrition of the land [19]. This occurrence of hazards, together with poverty in the region, make municipalities of the Amazonas state susceptible to adverse impact. Studies suggest an average vulnerability index value of 0.406 on a scale of 0–1 [16]. Located at the centre of the Amazonas state, Manaus sits at the interface of two of the world's biggest rivers: Solimões (with white waters) and Negro (with black waters) rivers form the so-called Amazonas river. At the Manaus harbour, the level of the Negro river varies from 13 to 30 metres every year. The last ten years saw six of the ten greatest floods detected in Manaus in over one hundred years, since 1903. Presently, the Alert System to Critical Events and the Hydrologic Alert Systems, in Portuguese SACE and SAHs, respectively, predicted that the 2021 floods of June/July, would be very close to the maximum value already detected (29.97 m, in 2015) [20]. On 30 April 2021, 12 of the 13 cities from the southwestern Amazonas state had already declared an emergency situation.

Considering the hazards in the region as well as the vulnerability of the local population, it becomes evident that the Acai industry needs to be assessed from a disaster risk perspective. A Risk Assessment can serve to identify hotspots of risk and as entry points to address disaster risk as a way to ensure the sustainability of the Acai industry. Risk, as represented in figure 10.1, consists of three components: Hazard, Exposure and Vulnerability, where Hazard is the '*potential occurrence of a natural or human-induced physical event or trend or physical impact that may cause loss of life, injury, or other health impacts, as well as damage and loss to property, infrastructure, livelihoods, service provision, ecosystems, and environmental resources*'; exposure includes all the people, assets and ecosystems that a hazard can impact; and vulnerability is the '*propensity or predisposition to be adversely affected*' [21].

In order to bridge the communities of NTFP industries with the disaster risk community, this chapter explores the relevance of assessing the disaster risk of NTFP industries, using the example of Acai in the Amazon. After reviewing the state of the art through a literature review, this paper presents factors to consider in an integrated Risk Assessment as a first step towards such analysis and a way forward to inform management decisions regarding the Acai industry.

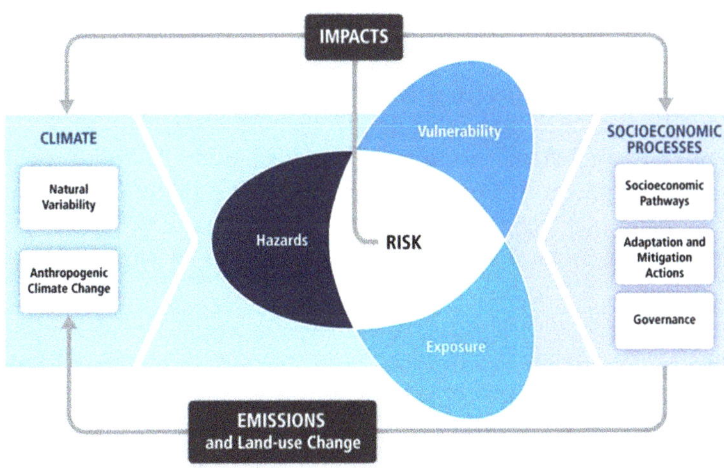

Figure 10.1. Conceptual framework of risk, showing the three main components hazard, exposure and vulnerability. Source: IPCC: 2014 [21].

Table 10.1. Search terms used in the literature review using both SCOPUS and Scielo.

Search	NTFP	Acai-specific
#1	TITLE-ABS-KEY ((risk AND assessment AND ((non AND wood AND forest AND products) OR (non AND timber AND forest AND products))))	TITLE-ABS-KEY (risk AND assessment AND Acai)
#2	TITLE-ABS-KEY ((multi AND hazard AND ((non AND wood AND forest AND products) OR (non AND timber AND forest AND products))))	TITLE-ABS-KEY (multi AND hazard AND Acai)
#3	TITLE-ABS-KEY ((drought OR flood) AND ((non AND wood AND forest AND products) OR (non AND timber AND forest AND products)))	TITLE-ABS-KEY ((drought OR flood) AND Acai)

10.3 Method

Attempts to understand the social vulnerability of the Acai industry in the Amazon were made. For example, a series of interviews were carried out in five cities within the Amazon by the Amazon Federal University. The interviews and visits aimed to collect information to thoroughly comprehend the work conditions and economic dependency in the collection and extraction process of the Acai. The research was descriptive, exploratory and correlated with the literature.

The research activities took place from December 2018 to December 2019 in Coari, Codajás, Tefé, Anori and Alvarães cities with 78 subjects. Visits were made to the commercial, production (industrial processing) and farmers for direct observation of the commercial and extraction process.

This research has ethical and legal Brazilian aspects in accordance with Resolution no. 466/2012 of the National Health Council (CNS) (BRASIL, 2012) and presents permission from the Research Ethics Committee (CEP) of the Federal University of Amazonas under CAAE: 97 439 918.5.1001.5020.

The interviews set a baseline to analyse the social vulnerability further.

In a second step, the current state of the art regarding Risk Assessments addressing NTFP and Acai production was reviewed. A literature review was conducted to this end on SCOPUS and Scielo, using three search terms in both platforms targeting papers related to Risk Assessments of NTFP in general, and more specifically, of Acai. The search was conducted in March 2021, using the following English search terms:

10.4 Results

10.4.1 Metadata analysis

The literature review resulted in a very limited number of papers. The three searches across both SCOPUS and Scielo resulted in a total of 128 papers, as listed in table 10.2. NTFP related papers were significantly more represented with 107 papers, compared to 20 covering Acai. Additionally, the results show that while

Table 10.2. Number of hits resulting from the three literature searches conducted in SCOPUS and Scielo.

	SCOPUS		Scielo		
Search and topic	NTFP	Acai	NTFP	Acai	Of which relevant
#1 Risk Assessment	31	11			1
#2 Multi-hazard assessment	1				
#3 Drought or flood assessment	74	9	1		8
Total	106	20	1	0	9

Table 10.3. Number of hits resulting from the literature searches with the relative country of the study.

Region (total count)	USA	Brazil	Ethiopia	Russia	Mexico	Others
	4	3	3	2	2	11

Acai covers nearly equal share Risk Assessments and specific hazard and drought assessment (11 and 9 papers respectively), NTFP are mainly reported in the context of drought and flood assessments (with 74 papers, of which one was repeated). With only one hit for NTFP, multi-hazard assessments are the most poorly represented assessment considered in this study.

Looking at NTFP specifically, the literature review found relevant papers from different regions across the world. The papers were not relevant for the analysis; however, they show the attempts made to analyse the risk of NTFP in different regions of the world. Table 10.3 shows that NTFP is mostly studied in the USA, followed by Brazil.

The results furthermore show that the Risk Analysis definition varies across papers. From search #1 [22], appearing with both search terms, is the only paper relevant for the Acai system. It gives a baseline for vulnerability assessment and possible scenarios for Acai expansion. Additionally, it suggests considering how Acai is expanding in future analysis and if it can be a risk for the forest ecosystem.

In the same search, other analyses targeted the understanding of one of the risk components. Birhane *et al* (2020) analysed the impact of climate change on Baobab trees (*Adansonia digitata* L.) in Ethiopia [23]. The analysis is the closest attempt to a Risk Assessment for NTFPs. Exposure was measured under climate change scenarios, and socio-economic data were collected to understand the ecosystem services that Baobab trees provide to the communities.

Moulana *et al* [24], Rockwell *et al* [25], and Torres *et al* [26] analysed the growing market of the Jernang (*Daemonorops draco* (Willd.) Blume), the Brazil nut (*Bertholletia excelsa* Bonpl.), and the different types of agave (*Agave* L.) to produce mezcal from the management of natural resources perspective and access to markets. Latawiec *et al*, explored if large-scale restoration plans might push agriculture settlements to areas currently covered by natural vegetation [27].

Regarding papers targeting drought and flood events impacts on the Acai industry, Brum *et al* analysed more in depth the impacts of flood on the growth of the Acai palm [18].

10.4.2 Analysis of Acai industry as an entry point of Risk Assessments

The analysis of the Acai industry allowed schematically putting together the Acai production chain, illustrated in figure 10.2. Taking a risk perspective, i.e., considering the aspects of hazard, exposure and vulnerability, allowed identifying several aspects of the Acai production chain to be analysed for each component of the Risk Assessment.

The Acai palm forest and the Acai industry will be directly impacted by hazards. The Acai yield, the production of the Acai pulp, and the recovery of the waste are direct indicators of the hazard impacts in the system. The international and the national market are components indirectly impacted by hazards.

In terms of hazards, principally, changes in the precipitation patterns and temperature will impact the Acai trees development at every stage. Those changes also shift areas where typical pests for the Acai tree can develop, and could create better environments for them to develop. Issues as flood, drought and pest, and their impact on Acai were found in the literature review. As mentioned before, Braum and Souza explored the impacts of floods on the Acai population trees [18]. De Castro *et al* 2019 [28], de Oliveira *et al* 2019 [29], and Silvestre *et al* 2017 [30], explore the impacts of drought events in the Acai palm tree from different perspectives. No research was found analysing the implications of flood and drought addressing the entire system.

Regarding the vulnerability, there are a number of elements such as natural resources management, social context and developments, and technology innovation that will play a role. Figure 10.2 shows the natural resources components in green, and social components in orange. Although NTFPs present an option to link environmental conservation with sustainable economic development objectives for rural communities [22], and Acai palm trees have been studied for their hydraulic capacity and from and ecological perspective [31, 32], the forest health must be assessed in terms of forest cover, diversity, protected area.

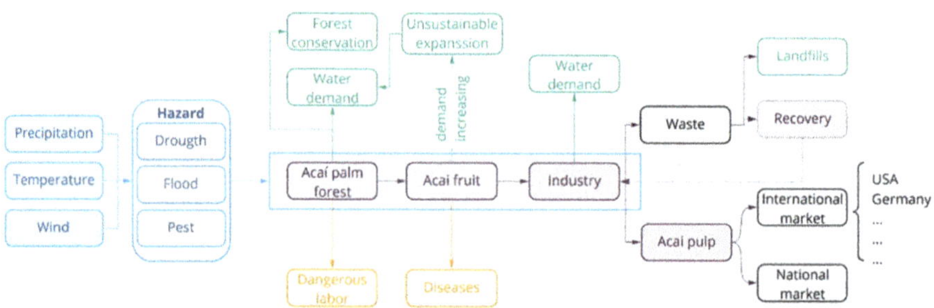

Figure 10.2. Schematic representation of the Acai chain, illustrating the relevant risk components along the chain. Components in purple represent the industry's exposed elements.

Figure 10.3. Picture illustrating the cutting of Acai fruits, Acai using peconha for stability (a) and Acai fruit (b).

10.4.3 Analysis of Acai industry at Solimões river

Social vulnerability depends on the people. Amongst others, their dependence on income generated from Acai, their education level and their degree of poverty will affect social vulnerability. Additionally, previous experience with drought or flood events as well as an understanding and perception of future risks must be considered. Because of the precarious conditions involved in Acai picking, the number of hospital beds and distance to hospitals represent an important factor. Indeed, the fruits are picked by climbing up the palm tree, using a 'peconha' (a sort of fabric to increase stability of the bare feet), a machete and footwear (figure 10.3). With the limited utensils and protection, injuries are thus frequent. Furthermore, Acai because it is common to 'jump' from tree to tree, children are generally used to avoid Acai breaking the thin palm tree trunk. Injuries furthermore arise due to the distant journeys involved in the Acai chain. Acai is currently collected in mostly distant places, which are difficult to access, with transport being done only by small boats. The journey increases the susceptibility to accidents and represents a problem for the supply chain.

Finally, expertise and traditional knowledge (and the lack thereof) play an important role in shaping vulnerabilities. After the preparation of the Acai concentrate, the pulp of the two species that occur in the Amazon is often mixed. The preparation of the pulp concentrate still relies on traditional knowledge in several locations; however, with the sensorial variations so accentuated, mixing the two species can result in the evaluation of a lower quality product. The Acai industries in the Middle Solimões river region (Coari, Codajás, Tefé, Anori and Alvarães) are small in size and are generally artisanal or semi-industrial (making use of a pulping machine). The amount of water added in the fruit pulp extraction process is an important parameter that directly influences the value that will be used as a basis for commercialisation. To meet the demands, the detection of adulterations to make the product more concentrated has been recurrent. Furthermore, the storage of the pulp is critical. The purplish colour can undergo variations and turn brown, due to oxidation Acai resulting from inappropriate storage. The lack of traditional knowledge and experience may thus reduce the quality of the pulp, increasing the vulnerability of the producers.

10.5 Discussion

10.5.1 Considerations for integrated Risk Assessment

The chapter's findings confirm that very little attention is being dedicated to disaster risks faced by the NTFP and Acai industries. Having identified only nine relevant papers to the discussion, the results highlight an important gap. As disaster Risk Assessments are critical for informed decision-making and allow for preventive measures rather than post-disaster action, the lack of research in this field may carry important consequences for the NTFP and Acai industry.

As risk is a function of hazard, exposure and vulnerability, one paper identified a couple of factors to consider for an integrated Risk Assessment. Floods, droughts and pests have already been witnessed in the area and represent hazards that may threaten the exposed people, the Acai palm trees as well as the Acai industry. A literature review also evidenced an important gap regarding the NTFP/Acai industry in the context of multiple hazards. The Acai palm trees and the production chain have already been and will likely again be exposed to droughts, floods and pest invasions. The risk profile of the industry should therefore be assessed considering these multiple hazards and their interplay. This emphasis on individual hazards and Risk Assessments is generally dominant. However, there is a recognised need to shifting to multi-hazard Risk Assessments, with a transdisciplinary system analysis that involves multiple stakeholders [33].

In terms of vulnerability, one paper encourages decision makers to consider ecological and social vulnerability as well as industrial processes. Forest health, including forest cover, diversity and share of protected areas, serve as indication for sustainable forest management, which ensures forests renewal and long-term productivity and reduces the ecological vulnerability [34]. In terms of social vulnerability, aspects such as dependence on income generated by Acai, previous experience with hazards, risk perception and hospital beds are to be considered. Finally, the industrial aspects linked to water additions and pulp storage can affect the quality of the Acai pulp.

10.5.2 Further enhancing sustainability of the industry

The Acai industry currently produces mainly Acai pulp, which is frozen and used in preparations from juices, sorbets, porridge, sauces, and jams [35]. However, this process discards an important share of the fruit, including seeds and fibres. Indeed some 80%–95% of the fruit processed is agro-industrial waste [36]. In order to turn the linear industrial process into a circular one and further contribute to the sustainability of the industry, additional uses of the waste should be explored. Developing by-products from the waste would be not only of environmental interest, managing the fruit's waste problem, but could also bring additional economic value to the Acai and the standing forest as a whole [1].

Being widely available throughout the Amazon and with their production rate exceeding the usage rate (it is estimated that residues can account for up to 80% of the fruit weight [37]), the Acai residues present many opportunities. Research is currently looking into using this waste as a source of bioenergy [38], while it's chemical composition could be profitable for the food and cosmetic industry.

10.6 Conclusion

National and international markets of Acai have increased immensely in recent years. As Acai represents an important NTFP that generates extractivism income, the Acai industry expansion is highly welcomed. Yet, the lack of Risk Assessments confirmed in the literature review may affect the sustainability of the industry today and in the future.

Based on the literature review and the qualitative research, the paper identifies flood, droughts and pests as hazards that the people, the Acai palm trees and the industry can be exposed to. The interplay of these hazards in a multi-hazard Risk Assessment may be particularly interesting. As for the vulnerability, the paper identifies ecological and social aspects that will need to be included in the Risk Assessment.

Conducting an integrated Risk Assessment of the Acai industry, as well as NTFP in general, will be critical to inform the industry and decision-making. The understanding of risks faced by the industry allows to take preventive action, so as to reduce disaster risks and thereby enhance the sustainability of the industry in the future.

Acknowledgements

The authors acknowledge the support of the Coordination for the Improvement of Higher Education Personnel (CAPES) and the National Council for Scientific and Technological Development (CNPq) and FAPERJ.

References

[1] Villacís-Chiriboga J *et al* 2020 Valorization of byproducts from tropical fruits: extraction methodologies, applications, environmental, and economic assessment: a review (part 1: general overview of the byproducts, traditional biorefinery practices, and possible applications) *Compr. Rev. Food Sci. Food Saf.* **19** 405–47

[2] Lopes E *et al* 2019 Mapping the socio-ecology of Non Timber Forest Products (NTFP) extraction in the Brazilian Amazon: the case of açaí (*Euterpe precatoria* Mart.) in Acre *Landsc. Urban Plan.* **188** 110–7

[3] ter Steege H *et al* 2013 Hyperdominance in the Amazonian Tree Flora *Science* **342** 1243092

[4] Ramos S L F *et al* 2021 Genetic structure in populations of *Euterpe precatoria* Mart. in the Brazilian Amazon *Front Ecol Evol.* **8** 603448

[5] Pinheiro de Lima A C *et al* 2019 Physicochemical characterization of residual biomass (seed and fiber) from açaí (*Euterpe oleracea*) processing and assessment of the potential for energy production and bioproducts *Biomass Convers. Biorefin.* **11** 925–35

[6] Melo P S *et al* 2020 Simulated gastrointestinal digestion of Brazilian açaí seeds affects the content of flavan-3-ol derivatives, and their antioxidant and anti-inflammatory activities *Heliyon* **6** e05214

[7] Chamberlain J L *et al* 2020 Rediscovering the contributions of forests and trees to transition global food systems *Forests* **11** 1098

[8] IBGE 2020 https://sidra.ibge.gov.br/pesquisa/pam/tabelas

[9] IBGE 2019 Produção da Extração Vegetal e da Silvicultura, Rio de Janeiro, **34** 1–8 https://biblioteca.ibge.gov.br/visualizacao/periodicos/74/pevs_2019_v34_informativo.pdf

[10] Brazilian Federal Government, Padrões de Sustentabilidade na Cadeia de valor do Açaí https://gov.br/agricultura/pt-br/assuntos/agricultura-familiar/publicacoes/projeto-mercados-verdes-e-consumo-sustentavel/outras-publicacoes/padroes_sustentabilidade_acai-1.pdf/view

[11] Conab. Conab—Açaí—Análise Mensal—Dezembro 2020 https://conab.gov.br/info-agro/analises-do-mercado-agropecuario-e-extrativista/analises-do-mercado/historico-mensal-de-acai/item/14918-acai-analise-mensal-dezembro-2020

[12] Homma A K O *et al* 2006 Açaí: Novos desafios e tendências *Amazônia: Ci. Desenv.* **1** 7–23

[13] Nobre C A *et al* 2016 Land-use and climate change risks in the Amazon and the need of a novel sustainable development paradigm *Proc. Natl Acad. Sci. USA* **113** 10759–68

[14] Chandrasekhar A n.d. Why is the Amazon still at risk despite 'zero deforestation' commitments?—SWI swissinfo.ch https://swissinfo.ch/eng/why-is-the-amazon-still-at-risk-despite--zero-deforestation--commitments-/46314800

[15] Garcia B N *et al* 2018 Extreme drought events over the Amazon Basin: the perspective from the reconstruction of South American Hydroclimate *Water* **10** 1594

[16] Menezes J A *et al* 2018 Mapping human vulnerability to climate change in the Brazilian Amazon: the construction of a municipal vulnerability index *PLoS One* **13** e0190808

[17] Tregidgo D *et al* 2020 Vulnerability of the Açaí palm to climate change *Hum. Ecol.* **48** 505–14

[18] Brum H D and Souza A F 2020 Flood disturbance and shade stress shape the population structure of açaí palm *Euterpe precatoria*, the most abundant Amazon species *Botany.* **98** 147–60

[19] Batista T F C *et al* 2007 Ocorrência de antracnose em frutos de açaí, *Euterpe oleracea*, em Muaná, Pará *Fitopatol. Bras.* **32** 360

[20] Boletim de Monitoramento Hidrometereológico da Amazônia Ocidental, Boletim 17 2021 http://cprm.gov.br/sace/

[21] Intergovernmental Panel on Climate Change (IPCC) 2014 *Climate Change 2014: Impacts, Adaptation, and Vulnerability. Part A: Global and Sectoral Aspects. Contribution of Working Group II to the Fifth Assessment Report of the Intergovernmental Panel on Climate Change* ed C B Field *et al* (Cambridge: Cambridge University Press)

[22] Lewis J A 2008 The power of knowledge: information transfer and açaí intensification in the peri-urban interface of Belém, Brazil *Agroforest Syst.* **74** 293

[23] Birhane E *et al* 2020 Vulnerability of baobab (*Adansonia digitata* L.) to human disturbances and climate change in western Tigray, Ethiopia: conservation concerns and priorities *Global Ecol. Conserv.* **22** e00943

[24] Moulana R *et al* 2021 Risk analysis of Jernang plant (*Daemonorops draco*) in Aceh Province *IOP Conf. Ser.: Earth Environ. Sci.* **644** 012039

[25] Rockwell C A *et al* 2015 Nut production in *Bertholletia excelsa* across a logged forest mosaic: implications for multiple forest use *PLoS One* **10** e0135464

[26] Torres I *et al* 2015 TEK, local perceptions of risk, and diversity of management practices of *Agave inaequidens* in Michoacán, Mexico *J. Ethnobiol. Ethnomed.* **11** 61

[27] Latawiec A E *et al* 2015 Creating space for large-scale restoration in tropical agricultural landscapes *Front. Ecol. Environ.* **13** 211–18

[28] de Castro G L S *et al* 2019 Photosynthetic apparatus protection and drought effect mitigation in açaí palm seedlings by rhizobacteria *Acta Physiol. Plant* **41** 163

[29] de Oliveira H O *et al* 2019 Coupling physiological analysis with proteomic profile to understand the photosynthetic responses of young *Euterpe oleracea* palms to drought *Photosynth. Res.* **140** 189–205

[30] Silvestre W V D *et al* 2017 Differential tolerance to water deficit in two açaí (*Euterpe oleracea* Mart.) plant materials *Acta Physiol. Plant* **39** 4

[31] Aparecido L M T *et al* 2015 Ecological applications of differences in the hydraulic efficiency of palms and broad-leaved trees *Trees* **29** 1431–45

[32] Freitas M A B *et al* 2015 Floristic impoverishment of Amazonian floodplain forests managed for açaí fruit production *Forest Ecol. Manag.* **351** 20–7

[33] Ismail-Zadeh A T *et al* 2017 Forging a paradigm shift in disaster science *Nat. Hazards* **86** 969–88

[34] MacDicken K G 2015 Global Forest Resources Assessment 2015: what, why and how? *Forest. Ecol. Manag.* **352** 3–8

[35] Yamaguchi K K L *et al* 2015 Amazon Acai: chemistry and biological activities: a review *Food Chem.* **15** 137–51

[36] Yuyama L K O *et al* 2011 Caracterização físico-química do suco de açaí de *Euterpe precatoria* Mart. oriundo de diferentes ecossistemas amazônicos *Acta Amaz.* **41** 545–52

[37] Melo O S *et al* 2021 Açaí seeds: an unexplored agro-industrial residue as a potential source of lipids, fibers, and antioxidant phenolic compounds *Ind. Crops Prod.* **161** 113204

[38] Bufalino L *et al* 2018 Local variability of yield and physical properties of açaí waste and improvement of its energetic attributes by separation of lignocellulosic fibers and seeds *J. Renew. Sustain. Energy* **10** 053102

Chapter 11

Development of a risk matrix for low-cost engineering solutions: a systematic approach

L Roberts, R Critchley, D Salberg, S Bloodworth-Race and R Hazael

Using a matrix can be a useful tool to identify and select risk mitigation strategies. The development of this selection matrix is often extremely difficult, the matrix itself will normally be a subjective assessment of the available facts and individual experiences, and it will seldom provide an objective comparison. It will however enable a reference for a decision-making process and provide an understood reference to enable risk mitigation decisions to be made. Ideally the development of the matrix will be underpinned by a scientific approach and each risk mitigation or control measure will be supported by credible and reliable evidence.

This chapter utilises a case study approach to demonstrate the process in a challenging real-world scenario. The case study looks at methods to assure the suitability of demining personal protective equipment (PPE). This is an example of an extreme hazard with a very real risk of death, and which necessitates mitigation measures which are low cost, field suitable and can be completed quickly and easily by a wide group of users, geographically dispersed and who must be able to implement any proposal with little training. As such it is an extreme scenario for the selection of risk mitigation techniques.

The example also shows how the development of a selection matrix can be a useful tool which is independent of any selection of mitigtion or control measures. In this specific example demining organisations on different continents or with different drivers and pressures may select different approaches (or combinations of approaches) however the development of the matrix remains consistent and provides a single comparative reference.

Whilst the approach can be altered the systematic approach should include an understanding of:

The hazard, the risk, and the need for control/mitigation

The available techniques, technologies and strategies

doi:10.1088/978-0-7503-3643-7ch11 © IOP Publishing Ltd 2022

The real-world constraints of the risk environment; including the financial limitations and human factors which will impact implementation.

With all the above studied and understood a comparative matrix can be developed to assess the suitability and effectiveness of any mitigation/control measures.

11.1 Introduction

11.1.1 The risk to be assessed

Personal Protective Equipment (PPE) is used in a variety of situations to protect the user either personally, or when in a vehicle or building from a specified threat (Arifurrahman *et al* 2020, Bloodworth-Race *et al* 2021). There are numerous papers that discuss the many blast and fragmentation effects of improvised and conventional explosive devices (DaSilva *et al* 2020, Rao *et al* 2020) that have occurred in civilian, defence and security sectors. As well as studies that investigate the various materials that show promise in PPE uses (Bruton *et al* 2020, Critchley *et al* 2021). In the development of a scientific approach to the Risk Assessment development we are basing our methodology on the PPE used in the Humanitarian demining charity sector. By 1996, the International Committee of the Red Cross (ICRC) estimated there were 110 million landmines scattered across 70 countries (ICRC 1996) which shows their steady rise due to their efficacy and therefore becoming a common feature in historical and current conflicts. This number is an estimate due to several factors including that the states that deploy landmines do not keep records, and, dependent on the operating environment, mines can be scattered by aircraft, further adding doubt to their location (ICBL 2019). When state records are available, they are in many cases inaccurate or incomplete (Land Mine Free World 2020). This can be attributed to the well documented reduction in record keeping observed during— and post-conflict (Minor 2012). Even when information on the location of landmines is dependable at the time of recording, natural phenomena such as erosion, flooding, earthquakes, and sandstorms displace the mines themselves or their physical location markers (ICBL 2019).

11.1.2 How current PPE responds to blast and fragmenting incidents

When deminers are working in a contaminated environment they must wear PPE in line with the correct standard (UNMAS 2013). The most common pieces of PPE used are the facial visor, which provides protection to the face as well as the soft fibre 'apron' that provides protection to vital areas including the torso and pelvis region. When worn correctly, the visor absorbs part of the blast wave pressure in a ductile manner and reflects and diffracts some of the pressure front around the head (Tan *et al* 2017). The authors produced data by modelling a polycarbonate (PC) visor attached to a helmet on a stationary human head and neck. The inclusion of a modelled neck allows for displacement of the head in a more natural manner on contact with a detonation wave. The blast was modelled to be 'similar to the pressure wave experienced from a 28.4 g TNT charge explosive at a standoff distance of 0.81 m' (Tan *et al* 2017) which is comparable to the blast from a typical anti-personnel landmine. It was observed that not all the pressure is deflected and absorbed, and some residual pressure propagates through the PC to the face, however, this is approximately a tenth of the pressure experienced when they conducted the modelling without a visor. This propagation through the visor is to be expected; Courtney *et al* (2012) used 6.35 mm samples of various transparent armour to demonstrate that a face shield will attenuate but not stop a blast wave of 1.173 MPa. This value is significantly lower than 1.5 MPa[1] that a standard landmine will output at a working distance of 0.6 m (UNMAS 2013) and so it is expected that the head will experience some overpressure. Zhang and Pintar (2010) identified a trend where overpressure impulse and peak overpressures experienced by the skull when wearing a helmet is significantly higher than without a helmet. This was investigated by designing a model of a standard sized (ICBL 2019) landmine represented by 250 g of TNT placed 0.5 m from the head. This higher pressure observed is likely to be due to the close proximity of the inner surface of the helmet to the skull. This extra surface increases the overpressure by allowing blast wave reflection to occur in this space.

Using a modelled 250 g TNT charge and a concrete wall, Leppänen (2006) demonstrated that at standoffs below approximately 4.5 m, fragmentation reaches the target before the blast wave. This is in agreement with analysis of demining injuries that generally occur at standoffs of ~0.6 m where it is observed that fragmentation makes contact with the visor first, prior to the blast wave (Smith 2003). This is observed following demining accidents where fragmentation makes contact with the headband-secured visor, however the visor is subsequently removed by the blast wave. Smith argues that despite the visors being removed in the vast majority of cases, crucially, it remains in place for the fragmentation phase of the detonation. In the research it is concluded that the visor removal is a beneficial feature which avoids transferring the pressure to the head and avoids the whiplash type effect described above. This could be considered analogous to a crumple zone of a car or the designed failure of a cycling helmet.

After subjecting mannequins wearing different types of facial PPE to the blast of 50–200 g of C-4, Nerenberg *et al* (2000) has an opposing opinion to Smith. They state that 'it is required that a protective system remain over the head and face throughout the blast event'. They conclude that a visor attached to a helmet coupled with an integrated chest plate has the highest performance in preventing injuries. One argument they propose for this view is that it protects against the 'the intense short-lived fireball', which would be the heat of the expanding gases. In their study they rigidly fix the mannequins in the kneeling position commonly adopted by deminers. This rigid affixed position prevents the natural displacement of the body following blast loading. This suggests the

[1] Calculated as $Z = \frac{d}{\sqrt[3]{m}} = \frac{0.6}{\sqrt[3]{0.24}} = 0.965$, where: Z = scaled distance; d = standoff distance; m = mass of explosive. A reading of $Z = 0.965$ gives a peak incident (side-on) overpressure of 1.5 MPa (Kingery and Bulmash 1984).

effects of hot gases experienced by the rigid mannequin would be greater than a real deminer would experience because the mannequin cannot displace away from the explosion compared to the more fluid reactions of real deminers. If Nerenberg's hot gas argument is correct, it could still be argued that reducing the pressures subjected to the head as stated by Smith (2003) could be more important than mitigating burns. It is the role of the PPE to provide an adequate level of protections as designated by the appropriate standard when used in conjunction with an appropriate Safe Operating Procedure (SOP) and an appropriate Risk Assessment (RA).

11.1.3 PPE review

Smith (2003) in a review of demining PPE requirements, states that it is not necessary to attach the visor to the head using a chin-strap despite some PPE manufacturers offering this option (Nerenberg *et al* 2000). The deflection mechanism is likely to be different as it is fixed at two points as opposed to one. The strap is likely to be offered due to convention but given that blast visors are somewhat niche and not as well researched as other safety visors or helmets there is no evidence to support the practice. There is however evidence that a chin-strap increases the incidence of injury by transferring the blast load directly to the neck causing a whiplash type effect. There are recorded instances of a visor strap being pulled off the face causing significant damage to the deminer's nose. There is also an incident of a metal clip cutting into a deminer's throat on impact (Global Conventional Weapons Destruction Repository 2020a).

Regardless of which mechanism is optimal in the protecting the deminer against a detonation, it is clear that the correct use of PPE, in whatever configuration, does typically reduce the incidence of injuries (Global Conventional Weapons Destruction Repository 2020a). This is likely to be simply because of the PPE acting as designed but also could show that there is a link between a culture of PPE use and good Risk Management including adherence to other safety related SOPs.

Despite the clear benefits of having access to and safely using PPE; when the PPE does not act as designed, it can prove to be more hazardous than not wearing PPE (Global Conventional Weapons Destruction Repository 2020b). Visor failure, in particular, can be a source of additional fragmentation and therefore cause more injuries than not wearing a visor; references to this effect have been made by an advocate for Humanitarian Mine Action (Smith 2019). Outside some accounts by the Global Conventional Weapons Destruction Repository's Humanitarian Demining Accident and Incident Database (2020a) there is only limited information concerning what circumstances and conditions the visors have undergone prior to failure is available. This again is attributed to the reduction in record keeping observed during and post-conflict (Minor 2012).

International Mine Action Standards states that 'eye protection shall be capable of retaining integrity against the blast effects of 240 gm of TNT at 60 cm' (UNMAS 2013). Where there has been a PPE incident where the visor did not perform as expected, either a defect was present pre/post manufacturing there was potential degradation in response to the conditions experienced in use and/or storage. Deminers will visually inspect the visors, however, this is generally to determine the amount of surface scratches which obstruct vision rather than for structural degradation. Currently, no non-destructive tests are used by demining organisations to ensure that the visors provided to personnel will function when needed. This lack of testing is an area which should be considered for additional risk mitigation in the form of a regular pre-use test for demining visors.

Due to the unique nature and varied location of the deminer's role, the tests will need to fulfil some unique requirements. The tests will need to be:

1. **Low cost** to ensure that demining charities can justify the expenditure to stakeholders.
2. **As simple as possible** to require limited training, again for cost, but also due to time limitations. This simplicity should also incorporate logistical requirements—they should be simple to move between countries in terms of size but also by not incorporating any controlled goods e.g., chemicals.
3. **As unambiguous as possible** with the ideal standard being a simple pass or fail criteria, as opposed to requiring subjective judgement.
4. **Universally understood** regardless of linguistic and cultural factors or literacy levels.
5. Ideally **non-destructive**; conditional ageing will be as unique to the individual visor and therefore testing individual items to assure the batch will be difficult (not to mention the loss of a visor due to testing will act to de-incentivise the test from occurring).

The proposed tests will be laid out in a reproducible manner for a laboratory to be able to test and evaluate the work.

11.2 Literature review of non-destructive tests to assess residual stress/damage to the visor

The following tests are drawn from published literature and identify peer-reviewed options which could provide the desired additional mitigation. These tests will be considered within the selection matrix.

11.2.1 Test 1—direct illumination test: author designed method adapted from Horn *et al* (2017)

This method creates shadows of the visible surface scratches for closer inspection, and highlights invisible crazing or microcracking if present. Figure 11.1 shows this microcracking or crazing.

11.2.1.1 Equipment required

For this test a light source such as a small battery-operated torch is required. A plain white or light-coloured work surface, or alternatively, a piece of white paper is also needed.

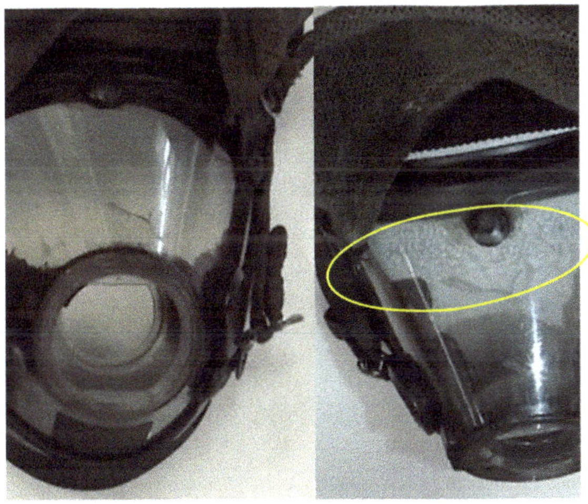

Figure 11.1. The image on the left shows a heat damaged facepiece where only the hairline crack is visible. The same facepiece is shown on the right under the direct illumination method where microcracking/crazing can be identified across the width of the visor (15–20 cm) and is highlighted by the yellow circle. Adapted from Horn *et al* (2017).

11.2.1.2 Instructions

1. Ensure that the PC visor is clean of any dirt or dust by washing in water and mild soap.
2. Ensure you are in a shaded area, or preferably undercover to reduce the amount of ambient light.
3. Hold the visor over a plain white or light-coloured work surface, or alternatively, onto a piece of white paper.
4. Shine a light source through the visor in the same direction as the wearer would be looking. Shadows of the surface scratches, and potentially micro-cracks or crazing will appear onto the surface allowing for inspection.
5. Vary the distance between the visor from the surface and vary the distance between the light and the visor.
6. Ensure that all angles of the visor are tested by rotating it left and right.
7. Ensure you are inspecting the work surface directly as opposed to looking through the visor which may make identifying anomalies more difficult.

The equipment setup and the shadows of the surface scratches is shown in figure 11.2.

11.2.1.3 Test result analysis

If any evidence of hairline cracks or crazing are present then the visor has failed the test. Crazing typically precedes failure in polymer materials, which are exposed to excessive stresses and can be identified by a whitening of this area. A crack or crazing shows decreased structural integrity, which indicates a reduction in impact strength. This reduction in impact strength cannot be quantified using this test alone therefore the visor is in an unknown state and must be decommissioned.

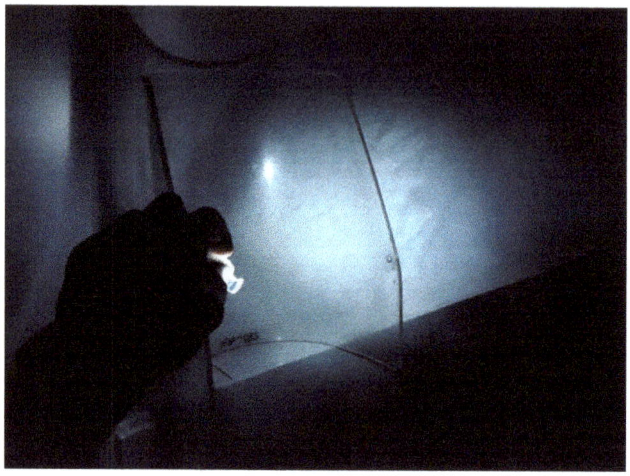

Figure 11.2. Showing the direct illumination method setup. A low ambient light has been used for the purposes of contrast for the photograph but this is not required. The shadow of the surface scratches can be seen on the white surface, which is distinct from the crazing seen in figure 11.1. This would be an example of a visor that would pass the direct illumination test due to no crazing being observed.

11.2.2 Test 2—low-level illumination test: background and discussion

Following investigation using the direct illumination test on visor specimens it was observed that casting light at very shallow angles on the visors showed a notable visual phenomenon in some samples. Figure 11.3 shows the 'glistening' or a very recognisable 'shine' that was observable in these visors. In the figure, the distinctive curve of this mark is clearly observable, but when inspecting the visor by holding it up to the light, or by using the direct illumination method this mark is nearly invisible. These marks were not previously identified by members of this research department by simple visual inspection suggesting that these can easily be overlooked. The visor donors had marked some of the affected visors as 'polished—repaired—usable' suggesting that these areas had not been identified previously (Heafitz 2006).

When the shallow angle test was repeated on the other samples it was noted that all the repaired polished visors displayed this phenomenon and none of the others exhibited this effect. Whilst the exact mechanism of what is occurring within the PC layer is not yet known, and taking into account the small sample size; it can be concluded with some certainty that polishing PC visors induces a change in the physical structure. This induced change in the structure is highly likely to have a lower performance in terms of blast and fragmentation protection. A literature research into other materials did not produce results showing similar phenomena to what has been observed in this instance. This method has been shown to identify instances where light behaves in a different manner and displays a shine or glistening which indicated the visor is in an unknown state (figure 11.3).

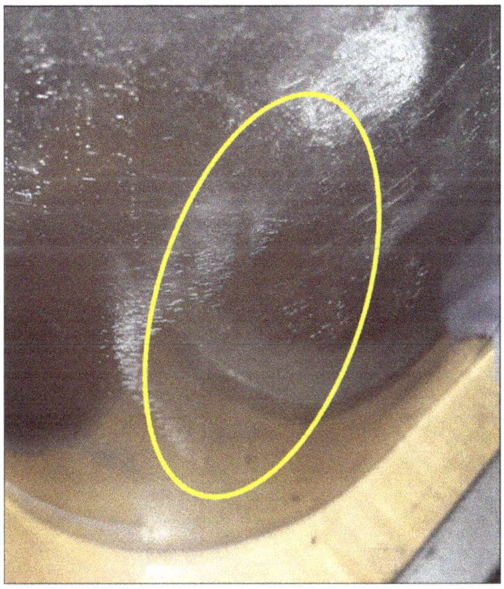

Figure 11.3. Image shows the glistening phenomenon which is only observable under low angle light. The area is approximately 5 cm in length and is indicated by the yellow oval. The other markings visible in the picture are surface scratches and are clearly visible at all times.

11.2.2.1 Equipment required

For this test a light source such as a small battery-operated torch is required.

11.2.2.2 Instructions

1. Ensure that the PC visor is clean of any dirt or dust by washing in water and mild soap.
2. Place the visor upright and shine the light source at a shallow angle down through the PC.
3. Inspect the visor for the characteristic shine or glistening as in figure 11.3.

11.2.2.3 Test result analysis

Due to the lack of data concerning this novel glistening/shining effect any evidence of the presence of this effect means the PPE is in an unknown state and its performance characteristics are unproven. Therefore, if any glistening/shining is identified, then the visor should be decommissioned. This conservative approach should be the case until scientific data can be gathered to provide a more informed, evidence-based decision.

11.2.3 Test 3—birefringence test: background and discussion

This type of testing is based on the induced birefringence property that PC and any translucent material displays. This effect occurs due to the changing speed of light through a clear solid dependent on the amount and the direction of externally applied or internal stresses (Yeager 2018). Internal stresses usually occur in PC components as a result of residual stress from manufacturing, but we hypothesise that a change in internal stress would also occur due to cracking and/or degradation.

Birefringence is observed when polarised light passes through the stressed PC and separates into two wave fronts which move at different speeds which is a function of the amount of stress in each direction (Yeager 2018, Murphy *et al* 2020).

The Stress–Optic Law states that birefringence is proportional to the difference between two stresses in that particular area (Yeager 2018). This is because the indices of refraction are also proportional to the stresses (Murphy *et al* 2020). Industry uses this relationship to perform a qualitative test on manufactured components (Juergens 2002, Yeager 2018). By being able to identify the difference between the indices of refraction at any point of the component, the difference in stresses can be calculated. This is done by the equation:

$$\sigma = \frac{\delta}{(t - C_B)} \tag{11.1}$$

where:

σ = stress (MPa)
δ = retardation value (nm)
t = thickness of material (mm)
C_B = Brewster's constant (84 for PC).

When PC is observed through polarising filters, the two phases of the light waves passing through the sample interfere and produce a visible colour, or a colour 'fringe'. By measuring the colour, either qualitatively by eye, or quantitatively by a polarimeter, the retardation value can be determined which allows completion of the stress calculation (Juergens 2002). A polarimeter is a calibrated instrument which removes subjectivity in judgement and allows accurate stress calculations. These are relatively high in cost, are not physically robust and require specialist training (Yeager 2018) and therefore not suitable for the visor field tests.

The areas of lowest stress in the material appear in order of increasing stress; black, grey and then white. As retardation levels and stress increase, the colours transform through a largely repeating pattern with the intensity of the colours decreasing as the stress increases (Yeager 2018). It is noted however that this is difficult by eye and states the 'results can be quite subjective and allow for only a rough estimate of retardation and stress' however for the application required it can be argued that this level for accuracy is sufficient (Yeager 2018).

Yeager (2018) in a review of medical manufacturing techniques advocates a quick test in which pictures of fringe patterns which are acceptable and those which are unacceptable are displayed at outputs of manufacturing key points to allow a quick evaluation of internal stresses. It therefore follows that this qualitative technique could be used as a field test. As such Test 3 (Birefringence test) makes the two separate indices of refraction associated with a stressed material visible by eye.

11.2.3.1 Equipment required

For this test a light source such as a small battery-operated torch is required as well as polarising filters.

11.2.3.2 Instructions

1. Ensure that the PC visor is clean of any dirt or dust by washing in water and mild soap. Place the visor upright.
2. Place the polarised light source behind the visor shining towards the observer.
3. The observer should wear a pair of polarised glasses. This will allow the birefringence to be observed.
4. Compare the birefringence with the reference photo of a new, unstressed visor. Periodic photographs of each visor should be taken through the polarised lens to document any change in stresses over time.
5. Identify and record any changes in birefringence.

11.2.3.3 Test result analysis

Should the material be deemed to be unduly stressed it should be decommissioned. Whilst this is currently a subjective measure, it is expected that further academic work, along with field trials will more accurately identify how normal birefringence and therefore a 'pass' appears. This should give more objective indications for when a visor is deemed to fail the test.

11.2.4 Test 4—yellowing test: background and discussion

When PC is exposed to outdoor conditions, the material degrades both its physical and visual properties which is caused by UV light and exposure to humidity and oxygen (Diepens 2009). In order to determine whether a simple, cost effective test for degradation can be devised it is necessary to understand what reactions occur and how they are initiated. Exactly which reactions take place when PC is outdoors remains under debate, this stems mainly from the fact that many of the studies (Factor and Chu 1980, Pryde 1985, Factor *et al* 1987, Tjandraatmadja *et al* 2002, Diepens 2009, Diepens and Gijsman 2011, Wypych 2020) into ageing were conducted under different conditions. Some of the environmental factors investigated included varying the humidity and a different range of UV wavelengths (both accelerated UV ageing and natural UV) leading to observing different results.

There is evidence for the existence of two distinct chemical mechanisms that are responsible for the photodegradation of PC; namely the photo-Fries rearrangement, and photo-oxidation (Tjandraatmadja *et al* 2002, Diepens and Gijsman 2011). The sunlight received at ground level is mostly comprised of 300 nm and above wavelengths but with limited energy at approximately 295 nm. Diepens and Gijsman (2011) showed that when PC is subjected to short wavelength (<300 nm) that the photo-Fries rearrangement is more likely to occur. Above this region it is photo-oxidation that is the dominant effect. This means at sea level the dominant mechanism is likely to be photo-oxidation but with limited, yet significant levels of degradation caused by the photo-Fries. As such Test 4 (Yellowing test) identifies when the visors have become an unacceptable shade of yellow and therefore show UV damage.

11.2.4.1 Equipment required

This test requires the bespoke manufactured acetate strips described in the following sections.

11.2.4.2 Instructions

1. Hold the visor and the acetate test strip to natural light.
2. Identify using figure 11.4 what yellow number can be viewed thought the visor.

Figure 11.4 shows a schematic of what the test strip may look like with numbered test patches. Any yellowing in the visor will block the appearance of some of the test patches. Identify which is the first yellow patch that can be seen through the visor. This is the measured yellowness level that the visor displays.

11.2.4.3 Test result analysis

Currently this test is not viable without more work to determine what level of yellowness is acceptable. An analysis of actual aged visors and artificially aged visors that display yellowness should be done. Once the appropriate level of yellowing that indicates an unacceptable degradation is determined, the test strip can be calibrated to this level. Any visor that reaches that particular reading is deemed to fail the test.

11.3 Subjectivity of scientific testing

The visor tests are designed to remove the element of subjective approach to Risk Assessment. Numerous Risk Assessments rely on subjective judgements although there is literature to suggest a more scientific approach to Risk Management and decision making is being considered (Ladyman *et al* 2019). This is to allow rapid assessment but also to reduce user error given that the test operators are likely to be competent in demining but have very little competence in material testing. This area of competence and judgement making was studied by Crouch *et al* (2002) where they compared the results of five different drug detecting field tests used by the police across different states in the US. Half of the 400 randomly selected urine samples from arrestees were tested by lab technicians and the other half by the police officers. Both parties were trained and were experienced in the urine collection method. Whilst the test was designed to be used with very limited training i.e. reading the label being sufficient, the lab technicians were inherently experienced in clinical techniques. The results proved that the field tests were accurate, and that the main source of error was the subjectivity of the tests. The trained lab technicians had an

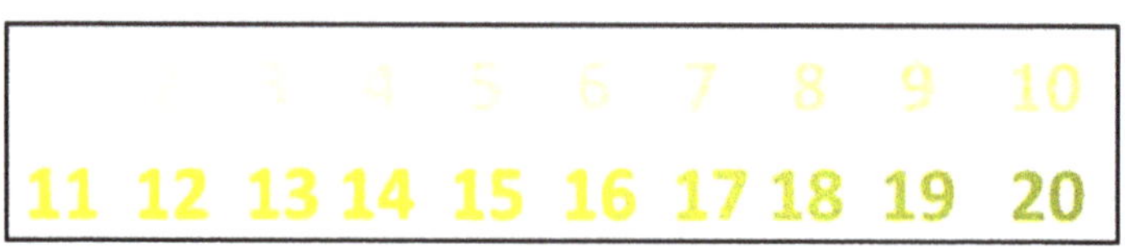

Figure 11.4. A schematic of the acetate strip. Each number increases in yellowness. The first number that can be observed through the visor would be the yellowness figure reading.

error rate of 0.8% whereas the police officers had an error rate of 2.5%, which is more than three times that of the experienced personnel.

Assuming the visor tests can be fully verified, the unambiguous 'yes' or 'no' assessment then removes the decision making from the operator or supervisor on the ground. Removing the team's empowerment to make a judgement, and replacing it with a pre-defined rigid SOP is generally viewed as a negative management strategy. Kahreh *et al* (2011) argues that an effective and popular strategy is to give employees breadth in their job descriptions and the associated authority to carry out their role autonomously. This breadth however only reaches as far as the employee's competence and experience (Kahreh *et al* 2011). The test conductors are likely not going to have the training or experience to make decisions relating to material science and therefore the ambiguity and decision making is required to be reduced to a minimum level.

If this testing regime is adopted, it would be considered a business process change. Using the manufacturing industry as a case study, de Treville *et al* (2005) argues that implementing a new change as a SOP is beneficial as it facilitates the gathering of data. If several different independent demining teams in different locations and working conditions each undertake a test without a pre-defined SOP, then they would likely reach different conclusions. It would not be possible to meaningfully measure the difference in visor shattering incidence and therefore identify how effective the tests are in practice. For an SOP to be implemented it is useful to have a pass/fail result as opposed to a scoring system which then facilities clear instructions on whether to dispose or retain the visor.

11.4 Discussion of tests that can inform the Risk Assessment

11.4.1 Test 1—direct illumination test: evaluation and discussion

The direct illumination test fulfils the criteria defined in the Aims section. It is low cost, given that it only requires the provision of a small torch. It is a simple test with negligible training time required. It is unambiguous because any evidence of crazing/microcracking is considered a fail and therefore there can be subjective assessment of an 'acceptable' level of crazing. It is also universally understood requiring no text. Whilst this is based on a peer reviewed method, and has legitimacy within the firefighting community, to be deemed as a recognised method for use in the demining community it will need to be further tested in a larger sample. When 36 donated visors were inspected using this method no crazing/microcracking was observed, meaning the test efficacy is currently inconclusive however this may be due to the relatively small sample size. Further testing would ideally happen by the demining organisations in the field who have access to more visors to determine whether any heat damaged crazing/microcracking specimens exist.

11.4.2 Test 2—low-level illumination test: evaluation and discussion

The low-level illumination test also fulfils the criteria defined in the Aims section for the same reasons as the direct illumination method. The main obstacle for this method becoming a legitimate candidate to be an accepted test in the demining community is determining what physical and chemical state the glistening phenomenon represents.

We propose the hypothesis that two or three separate initiators cause the glistening phenomenon caused by polishing. The first mechanism we propose is that the application of localised heat created by friction of the sanding may induce a loss of ductility as identified by Horn *et al* (2017) and demonstrated by Fawcett (2009). There is some evidence that photo-oxidation is a surface phenomenon and its products eventually reduce the rate of degradation by forming a protective surface layer (Diepens 2009). The polished visors are then reused in the field and are subjected to further UV without the benefit of the protective layer therefore increasing the rate of degradation. These two mechanisms may act synergistically or independently. There is a possibility that a third mechanism may be involved based on the chemicals used for polishing. Polishes are often abrasives and may have deleterious effects on the chemical structure, although it is difficult to provide an evidence-based argument on what effect this may have based on limited records of what polishes are used.

To hypothesise what the physical change in the PC may be, it may be useful to examine why the effect is only observed at low angles of light. The use of light at low angles is exploited in the dark field illumination method (Li *et al* 2013). This method is used in quality control in manufacturing, especially in inspection of small parts or electronics This method illuminates edges, cracks, engraving or imperfections in a surface by reflecting light from that indentation. Because the light used is at a low angle, a smooth unindented surface appears dark as light does not reflect into the camera, while indentations cause reflections and are highlighted (Batchelor 2012).

We hypothesise that the glistening illumination of the visor at low light angles indicates that more surfaces exist within the PC to allow the reflections to occur. An increased amount of surfaces would be consistent with the existence of small cavities or internal cracking leading to decreased performance.

11.4.3 Test 3—birefringence test evaluation and discussion

The birefringence test fulfils most of the criteria defined in the Aims section. It is low cost because again, the most expensive item required is a small torch. A cost effective and convenient source of polarising filters to create polarised light is the lenses sourced from polarising 3D glasses. These can be attached to torches to create polarised light and can be worn to provide the second filter required in the setup as seen in figure 11.5. The lenses from the 3D glasses can also be attached to cameras to allow photographs of the birefringence. This means the equipment setup and process is simple and requires little training. Some skill and training will be required in the interpretation of results, which is why it does not fulfil the unambiguity requirement. Like the previous two tests it does not require any text or reading.

A limitation to this technique that polarised light must only pass through one 'wall'. For example if a clear piece of tubing was analysed, the light would interfere twice and produce an aberrant result (Yeager 2018). The visor is a good candidate for this type of analysis due to its semi-circle shape when viewed from above and allows full analysis of its whole area. Figure 11.6 illustrates this concept.

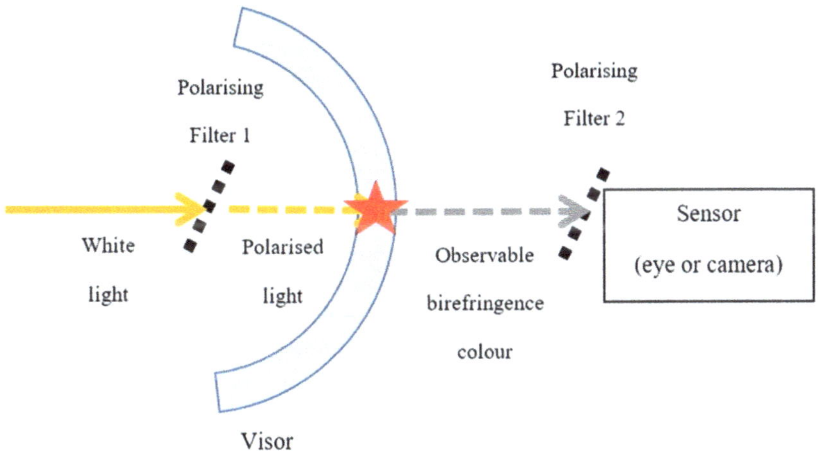

Figure 11.5. The equipment setup for the birefringence test method. The recommended configuration for use in a field setting would be to affix a polarising filter taken from a pair of 3D glasses onto a torch which would act as 'Polarising Filter 1' in the diagram. Wearing another pair of 3D glasses would act as 'Polarising Filter 2'. Original diagram.

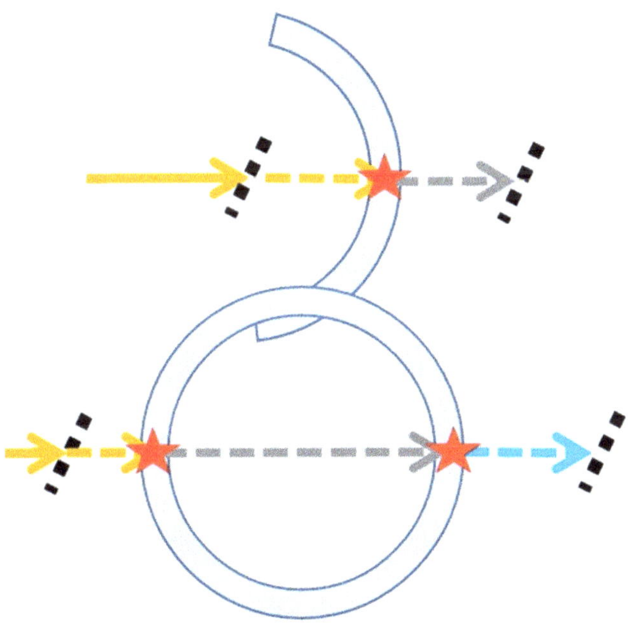

Figure 11.6. A schematic diagram illustrating why birefringence is only suitable for analysis through one surface. The top diagram shows a hollow semi-cylinder (viewed from above), which is similar in shape to the visors. White light (orange arrow) is directed through a polarising filter and the polarised light enters the single wall of PC. The coloured light due to birefringence (green arrow), which can be analysed by viewing it through another polarising filter. The bottom diagram has the same setup but the hollow cylinder shape inherently causes the light to travel through two PC surfaces. Instead of analysis the correct light (greey), the light is further split and causes an aberrant result represented by the blue arrow. Original diagram.

Areas that display a change in thickness, e.g. corners of a structure also cannot be reliably analysed by this technique (Janeschitz-Kriegl 1977). This is because the colours displayed are a result of interference between the two polarised phases, which are a function of the time taken to travel through the medium (Yeager 2018). If the thickness is not uniform then the different phases may take different times to travel through the medium, leading to interference at different stages (Janeschitz-Kriegl 1977). This would produce a different and, in the context of this analysis, an incorrect colour fringe. Again, the visor is suitable for this analysis because its thickness is uniform. Whilst shining light at different angles will create different path lengths through the PC there is no requirement to ensure that the angle of light to the PC is precise or consistent. The stress fringes, or the changes in stress, will still be visible, despite small changes in colour (Janeschitz-Kriegl 1977), which is an acceptable for this level of precision.

The validity and efficacy of birefringence analysis can be seen by the investment in this area by industry. It is used in a quantitative and qualitative manner in manufacturing components that undergo stress. This can be achieved in parts that are not transparent by creating a birefringent coating that moulds around the component to be tested (Vishay Precision Group 2011, Gowda 2014). A historical application of birefringence was in optometry in order to quality control glass spectacles by verifying that the glass did not have areas of high stress that would eventually lead to a cracked spectacle (Cook 2008). Whilst this technique is no longer used due to more advanced lens mounting technology and the use of more ductile glass material, it further demonstrates how birefringence can be used to qualitatively stress test in this manner.

11.4.4 Test 4—yellowing test: evaluation and discussion

This test is both low cost and is simple to use with a negligible training requirement. The test does not fully fulfil the ambiguity criteria due to its somewhat subjective assessment. The test in its current form uses Arabic numerals (i.e. Western style numbers) which are used in the majority of the world (Pixner *et al* 2011). Notable mine contaminated exceptions to this are in Cambodia amongst other states who have their own numbering systems. The test could be reproduced to use different symbols, although this causes the complexity and the cost to rise. A ‘dot’ based system similar to what is used on dice could be used universally but this increases the probability of miscounting in larger numbers.

It is clear using a subjective visual observation that none of the visors donated to the authors exhibit yellowing. This is likely due to robust management of a one year decommissioning routine, and therefore the visors do not reach those levels of degradation. As an indication of timescales prior to observing yellowing, Factor *et al* (1987) use yellowed PC in their study which had been aged for four years. A study by a commercial PC glazing company state that unprotected PC shows a discernible yellowing at approximately two years (SABIC 2008). These are merely indications, and their samples of PC have different thicknesses and conditions to the visors but support the fact that yellowing may take longer to manifest itself.

This therefore shows that whilst the yellowing test is viable it may have limited application in a situation where the visor decommissioning regime is robust. The visors would likely have failed some of the other tests outlined in the methods section prior to reaching yellowness levels which would be discernible to the eye. However due to its low cost, speed and low training requirement it is recommended the test is adopted as it may be useful in objectively identifying poor practice.

One potential issue of using this test is its subjectivity and the potential for variance between judgements of yellowness. The test is designed to remove as much subjectivity by offering several graduated yellow sections. Despite this, there is potential for the yellowness of the PC to lie between two graduations meaning less agreement in judgement and more error. Lighting also needs to be considered which is difficult to control in the field therefore it is suggested the measurements should be taken at the same time of day.

The assessment agreement issue has been discussed in a study by Mentes *et al* (2006), where they attempt to objectively judge urine colour to quantify dehydration in patients as a function of yellowing. They use a standardised eight colour chart designed by Armstrong *et al* (1994) under the same light conditions. Mentes' team established the 'interrater reliability' i.e. a measure of agreement between assessors, by using four coloured water specimens that had been previously rated by the lead researcher. This judgement was based on defined criteria which was disseminated and taught to the assessors. It is not mentioned how many assessors were used, but they do state they used research assistants for this task. This suggests a small sample was used and that they would have high degree of education, which may skew the results. The Kappa statistic, a measurement of agreement and therefore interrater reliability (Cohen 1960) ranged from 0.42 to 1.00 which is measured from a scale of −1 to +1. The range observed in the test is deemed between 'moderate' to 'perfect agreement' and is just within the boundaries of being acceptable for use in medical decision making (McHugh 2012).

It would be beneficial to have more confidence in agreement rates and it is likely that increasing the training of judgement criteria and a more graduated scale could be used e.g. a scale of 100 colour increasing in yellowness.

In addition to the lack of agreement in determining what the dominant mechanisms of yellowing are, there is a further complicating factor, which is the environmental humidity level. High humidity has been observed to reduce the rate of yellowing in PC films. Pryde (1985) recorded that by exposing 287 nm light to PC film, the samples held in high relative humidity conditions slowed the onset of yellowing compared to dry samples. Pryde (1985) attributed this reduced yellowing to a reaction between the hydrogen in the environmental water molecules and the excited products of photo-Fries. This would inhibit propagation of further reactions reducing the rate of chain scission.

In high humidity conditions, Factor and Chu (1980) identified a thin layer of photo oxidised PC formed which they hypothesised protects the inner layers of the PC from being affected by UV as the rate of ageing decreased once this layer was established. Therefore the environmental conditions must be considered when designing an experiment into ageing of PC (Wypych 2020). Humidity varies by geographical location, time of day and time of year, therefore this is not a factor that can be pragmatically controlled. An uncontrollable humidity disparity would also occur either side of the visor due to the deminer's breath. This may be a critical factor in determining whether the yellowing test is viable.

11.5 Conclusions

The issue of how to test for degradation in PPE visors was investigated and four different tests suitable for use in the field were identified. Five distinct criteria were identified that the test would need to fulfil to be of use to a demining organisation. The extent of how much the tests fulfilled the criteria was conducted and a summary is below in table 11.1.

It could be considered that the final requirement criterion is 'effectiveness' i.e. how effective the test is at identifying degradation. This cannot be answered with confidence without further testing and the further work which has been outlined in this section. Once further testing is achieved, an evidence-based decision can be made as to which tests should be invested in and adopted by demining organisations. Therefore the appropriate SOPs and RAs can rely on low costs, rapid tests to aid the user in determining whether the PPE fulfils the requirements of the documentation before use by the deminer.

Table 11.1. Showing a summary of how the various tests developed in this work fulfilled the criteria in order to be of use to a demining organisation.

	Direct illumination	Low angle illumination	Yellowing	Birefringence
Low cost	X	X	X	X
Simplicity	X	X	X	X
Unambiguous	X	X		
Universally understood	X	X		X
Non-destructive	X	X	X	X

Abbreviations

3D	Three-dimensional
ICRC	International Committee of the Red Cross
PC	Polycarbonate
PPE	Personal Protective Equipment
RA	Risk Assessment
SOP	Safe Operating Procedure
TNT	2,4,6-Trinitrotoluene
UN	United Nations
US	United States
UV	Ultra Violet Light

References

Arifurrahman F, Critchley R and Horsfall I 2020 Experimental and numerical study of auxetic sandwich panels on 160 grams of PE4 blast loading *J. Sandwich Struct. Mater.* **23** 3902–31

Armstrong L, Maresh C, Castellani J, Bergeron M, Kenefick R, LaGasse K and Riebe D 1994 urinary indices of hydration status *Int. J. Sport Nutr.* **4** 265–79

Batchelor B 2012 *Machine Vision Handbook.* 1st edn (London: Springer)

Bloodworth-Race S, Critchley R, Peare A, Hazael R and Temple T 2021 Testing the blast response of foam inserts for helmets *Heliyon* **7** e06990

Bruton K, Critchley R, Bloodworth-Race and Hazael R 2020 Lignocellulosic natural fibers in polymer composite materials: benefits, challenges and applications *Encyclopedia of Materials: Plastics and Polymers* vol 3 (Amsterdam: Elsevier) pp 353–69

Cohen J 1960 A coefficient of agreement for nominal scales *Educ. Psychol. Meas.* **20** 37

Cook O 2008 *Birefringence: The Dispenser's Forgotten Friend* https://2020mag.com/article/birefringence-the-dispensers-forgotten-friend

Courtney E, Courtney A and Courtney M 2012 Blast wave transmission through transparent armor materials *J. Battlef. Technol.* **15** 19–22

Critchley R, Hazael R, Bhatti K, Wood D, Peare A, Johnson S and Temple T 2022 Blast mitigation using polymeric 3D printed auxetic lattice structures *J. Protect. Struct.* **13** 496–586

Crouch D, Hersch R, Cook R, Frank J and Walsh J 2002 A field evaluation of five on-site drug-testing devices *J. Anal. Toxicol.* **26** 493–9

da Silva L A, Johnson S, Critchley R, Clements J, Norris K and Stennett C 2020 Experimental fragmentation of pipe bombs with varying case thickness *Forensic Sci. Int.* **306** 110034

DeTreville S, Antonakis J and Edelson N M 2005 Can standard operating procedures be motivating? Reconciling process variability issues and behavioural outcomes *Total Qual. Manag.* **16** 231–41

Diepens M 2009 Photodegradation and stability of bisphenol a polycarbonate in weathering conditions *PhD Thesis* Technische Universiteit Eindhoven, Eindhoven

Diepens M and Gijsman P 2011 Outdoor and accelerated weathering studies of bisphenol A polycarbonate *Polym. Degrad. Stab.* **96** 649–52

Factor A and Chu M 1980 The role of oxygen in the photo-ageing of bisphenol-A polycarbonate *Polym. Degrad. Stab.* **2** 203–23

Factor A, Ligon W and May R 1987 The role of oxygen in the photo-ageing of bisphenol-A polycarbonate: GC/GC/high-resolution MS analysis of Florida-weathered polycarbonate *Macromolecules* **20** 2461–8

Fawcett C 2009 Blast testing of visors used for humanitarian demining *J. ERW Mine Action.* **13** 29

Global Conventional Weapons Destruction Repository 2020a Humanitarian demining accident and incident database James Madison University

Global Conventional Weapons Destruction Repository 2020b Humanitarian demining accident and incident database: DDAS accident 313 James Madison University

Gowda H 2014 *Photostress on an Aircraft Linkage System* [Photograph in Lecture] Bangalore University, 4 April https://slideshare.net/hareeshang/unit-6-birefringent-coatings

Heafitz A 2006 Visor scratch repair and prevention *J. Mine Action.* **10** 96–8

Horn G, Kesler R, Regan J and Madrzykowski D 2017 *A Study on the Effect of Repeat Moderate Intensity Radiant Exposures on SCBA Facepiece Properties* Grant/Contract Reports (NISTGCR) National Institute of Standards and Technology, Gaithersburg, MD

ICBL (Int. Campaign to Ban Landmines) 2019 Landmine monitor 2019 *Int. Camp. Ban Landmines Cluster Munition Coalition* **21**

ICBL 2020 *Anti Personnel Landmines* http://Antipersonnellandmines

ICRC (Int. Committee of the Red Cross 1996 Anti-personnel mines: an overview

Janeschitz-Kriegl H 1977 Injection moulding of plastics: some ideas about the relationship between mould filling and birefringence *Rheol. Acta* **16** 327–39

Juergens R 2002 *Optical Design and Analysis Software II Conf. 2002 (7–11 July 2002)* (Seattle, WA: Society of Photo-Optical Instrumentation Engineers Conf.)

Kahreh M, Ahmadi H and Hashemi A 2011 Achieving competitive advantage through empowering employees: an empirical study *Far East Res. Centre.* **6**

Kingery C and Bulmash G 1984 Air blast parameters from TNT spherical air burst and hemispherical burst *U.S. Army Engineer Research and Development Center*

Ladyman M K, Temple T J, Piperakis M M, Fawcett-Hirst W, Gutierrez-Carazo E and Coulon F 2019 Decision framework for the environmental management of explosive contaminated land *Sci. Total Environ.* **690** 730–8

Land Mine Free World 2020 *Landmines* https://landminefree.org/2017/index.php/support/facts-about-landmines

Leppänen J 2006 Concrete subjected to projectile and fragment impacts *Int. J. Impact Eng.* **32** 1828–41

Li L, Wang Z, Pei F and Wang X 2013 Improved illumination for vision-based defect inspection of highly reflective metal surface *Chin. Opt. Lett.* **11** 021102

McHugh M 2012 Interrater reliability: the kappa statistic *Biochem. Med.* **7** 276–82

Mentes J, Wakefield B and Culp K 2006 Use of a urine color chart to monitor hydration status in nursing home residents *Biol. Res. Nurs.* **7** 197–203

Minor E 2012 Towards the recording of every casualty *Oxford Res. Group.* **9** 4

Murphy D, Spring K, Fellers F and Davidson M 2020 *Principles of Birefringence* https://microscopyu.com/techniques/polarized-light/principles-of-birefringence

Nerenberg J, Islam S, Makris A and Dionne J 2000 Comparative study of different lightweight head protection systems with full-face visors for humanitarian deminers *J. Conventional Weapons Destruct.* **4**

Pixner S, Zuber J, Heřmanová V, Kaufmann L, Nuerk H and Moeller K 2011 One language, two number-word systems and many problems: Numerical cognition in the Czech language *Res. Develop. Disabil.* **32** 2683–9

Pryde C 1985 Photoaging of polycarbonate: effects of selected variables on degradation pathways *ACS Symp. Ser.*

Rao P, Painter J, Appleby-Thomas G, Critchley R, Wood D, Roberts and Hazael R 2020 Fragmentation studies by non-explosive cylinder expansion technique *Int. J. Impact Eng.,* **146** 103714

SABIC 2008 *Superior Resistance to Outdoor Weathering* https://sfs.sabic.eu/wp-content/uploads/resource_pdf/1345451400-45917995-SABIC-SFS-2744_Lexan_Thermoclear_Lemon.pdf

Smith A 2003 DDAS: IMAS and PPE requirements *JMU J. Mine Action.* **7**

Smith A 2019 *Humanitarian Mine Action* http://nolandmines.com/

Tan L, Tse K, Tan Y, Sapingi M, Tan V and Lee H 2017 Face shield design against blast-induced head injuries *Int. J. Numer. Meth. Biomed. Eng.* **33** e2884

Tjandraatmadja G, Burn L and Jollands M 2002 Evaluation of commercial polycarbonate optical properties after QUV-A radiation—the role of humidity in photodegradation *Polym. Degrad. Stab.* **78** 435–48

UNMAS 2003 *IMAS 04.30: Glossary of Mine Action Terms, Definitions and Abbreviations*

UNMAS 2013 *IMAS 10.30: Safety and Occupational Health—Personal Protective Equipment*

Vishay Precision Group 2011 Introduction to stress analysis by the PhotoStress® method *Vishay Precision Group Technical Note*

Wypych G 2020 *Handbook of UV Degradation and Stabilization.* 3rd edn (Toronto: ChemTech Publishers)

Yeager M 2018 Photoelastic stress analysis of polycarbonate medical parts *Bayer Mater. Sci.*

Zhang J and Pintar F A 2010 A finite element study of blast overpressure on the skull with and without helmet *ASME Summer Bioengineering Conf. Proc.* SBC2010–19083 pp 799–800

Chapter 12

Assessing the academic and experiential risk to students from the transition to hybrid and online delivery

M Ladyman, T Temple and A Temple

The COVID-19 pandemic forced higher education establishments to convert their teaching and deliver by the use of online platforms, such as Microsoft Teams or Zoom. The initial transition was rapid, and largely successful thanks to available technology, but with teaching returning to normal in the UK many providers are faced with the decision of whether to continue to provide remote online teaching. While current resources, policies and course content were sufficient for an emergency transition to an online format, they may not be able to meet the required standards and quality in the long term. In addition, students had minimal expectations of online tuition and were appreciative of the efforts made to ensure course completion. Alternatively, if courses are advertised as available for hybrid or fully online cohorts, student expectations will understandably rise, and therefore providers will have to adapt to new modes of delivery. To assess the risks to students from the transition to hybrid and online delivery, a Risk Assessment was developed using the mandatory higher education standard and quality descriptors and Postgraduate Taught Experience Survey. The Risk Assessment focused on the risk to student failure and poor student experience and demonstrated that while it is feasible to deliver level 7 qualifications in a hybrid or online format, significant mitigation would be required for both hybrid and fully online cohorts. From the Risk Assessment, it was found that while mitigations for hybrid and online delivery are similar, it may be more challenging to achieve equality of opportunity and experience for hybrid cohorts where some students are face-to-face and some are remote online. It is likely that academic providers will have to decide how to approach the desire for hybrid and online teaching on a course-by-course basis by weighing student experience against required resource, practical components, and the risk to the academic provider.

doi:10.1088/978-0-7503-3643-7ch12

© IOP Publishing Ltd 2022

12.1 Introduction

Traditionally, university providers deliver higher education in a face-to-face setting within a physical building where student cohorts are together in one teaching space [1]. However, the COVID-19 pandemic forced higher education establishments to convert their teaching and deliver by the use of online teleconferencing platforms. This means that all taught lectures were delivered virtually, often using presentation slides with a member of the academic staff delivering technical aspects, answering questions and engaging the students in the online workshops and group work. This approach was seen as an opportunity to get the students through the course, avoid delaying graduation and meeting the university providers' responsibilities. For the COVID-19 situation this change was implemented in a rapid timeframe, e.g. in days, and fully implemented in weeks, so that the regulatory mandate of 'lockdown' could be adhered to. Therefore, any direct or indirect impacts of delivering this type of teaching with regards to the quality of the education and student experiences were not necessarily considered [2–6, 7].

Two years since COVID-19, teaching online has progressed in that university staff have learned how to deliver online and have continually evolved with the virtual platform technical up-dates [1]. However, as face-to-face teaching is returning to normal, hybrid delivery and online teaching for higher education is becoming a necessity as students expect and want this type of flexible learning. Remote learning at this education level is well established and works well online for students who specifically want to study part-time, for example the Open University have offered this method for many years [8]. Hybrid learning is when courses are synchronously delivered both in the classroom and online live at the same time for remote students. This style usually offers visuals of the lecturer via camera and an online platform where live chat can take place and provides a virtual sense of belonging [2, 9, 10]. Hybrid learning has become an essential option for some students who have been self-isolating or are unable to travel to the university, therefore becoming a 'want' situation rather than a compulsory requirement. According to a study from an education services project in 2021 students were asked if the COVID-19 pandemic lockdowns had made them think differently about how to learn and study. Almost half of the students surveyed said it changed their feelings for the better and it would change their priorities when choosing a study programme and many would now consider a fully online course. The same study canvased thoughts on what students would do if a course did not offer the option of online study and over 60% said they would look for another online course [11]. Surveys like this suggest that the forced change to online learning is proving to be a preferred option and is making way for flexible and inclusive learning.

To enable hybrid or online teaching at higher education to be successful for the future it is essential to undertake a Risk Assessment prior to academic registration on an annual basis as technology and student preferences continue to evolve. The assessment would evaluate the academic and experiential risk to students so that key provisions are included in the delivery and to implement mitigation when and where required. The Risk Assessment process can be achieved by undertaking a qualitative approach that is aligned to key academic higher educational standard and quality criteria [12]. Therefore, this chapter provides a generic example of a Risk Assessment of a level 7 science and engineering course. A level 7 qualification is normally studied face-to-face at a university (referred to as the 'provider') and the programme or course includes various entry and exit routes, e.g., Post Graduate Certificate, Post Graduate Diploma and Master of Science. Typically, a level 7 education is classed as postgraduate study, meaning that usually a student has an undergraduate qualification to demonstrate the ability to proceed to the next academic level [13]. The course usually consists of taught and practical modules with designated learning hours that contribute to the associated academic award. The provider must align their practices to higher education standards for level 7, for this example the hazards incorporated in the Risk Assessment used criteria from the Quality Assurance Agency for Higher Education (QAA) and survey questions from the Postgraduate Taught Experience Survey (PTES) [13, 14]. The QAA is an independent body that checks standards and quality in UK higher education. It conducts quality assessment reviews, develops reference points and guidance for providers, and conducts or commissions research on relevant issues. PTES, is a UK sector wide survey that collates student experiential feedback to gain insight on their teaching and learning experiences of taught postgraduate students.

It is clear that higher education is responding to student expectations by adapting traditional delivery styles to include more technology that enable hybrid and online options for learning. However, when providers decide to move in this direction it is important that it maintains the standards and quality required at level 7 and offers a good student experience. Therefore, to support providers successfully transitioning to hybrid and/or online delivery a Risk Assessment was developed to evaluate the risk to students of failing or receiving a lower quality experience at level 7.

12.2 Method development

12.2.1 Risk Assessment development

A Risk Assessment for transitioning from traditional style face-to-face classroom-based teaching to hybrid synchronous teaching or fully live online teaching during a level 7 course was developed. First steps in the development was to identify suitable assessment criteria to ensure higher education providers can maintain quality and standards required for a level 7 qualification in this new format. This was achieved by extracting the core practices criteria (core standard and core quality) from the UK's QAA for higher education and summarised questions from the student experience PTES Survey in seven categories. The combination of the two sources created a list of hazards that ensured all aspects of level 7 education was assessed from adequate top-level policy, course design, resources, research community, IT provision, admissions, and staff competency etc (table 12.1).

To score against the Risk Assessment criteria a matrix was developed defining likelihood and severity in terms of the impact to a student undertaking a level 7 qualification by hybrid or online delivery. Severity was defined on a scale of one to five where 1. was 'negligible' impact, which was comparable to traditional teaching

Table 12.1. Risk Assessment for transition to hybrid and online teaching of level 7 qualifications.

Hazards		Hybrid synchronous delivery			Live online remote delivery		
Core standard criterion		Risk	Risk response and mitigation	Mitigated Risk	Risk	Risk response and mitigation	Mitigated Risk
1	The threshold standards for qualifications are not consistent with the relevant national qualification's frameworks for delivery mode.	4,4	Modify internal provider policy and individual course intended learning outcomes and realign to framework. Create different intended learning outcomes specific to online students.	1,1	4,5	Change internal provider policy and individual course intended learning outcomes and realign to framework.	1,1
2	Students who are awarded qualifications do not have the opportunity to achieve standards beyond the threshold level that are reasonably comparable with those achieved in other UK providers.	4,3	Ensure all students have equitable access to progressive feedback. Ensure extra-curricular activities are available online as well as face to face. Modify assessment rubrics to ensure online students are not disadvantaged.	2,2	4,3	Ensure all students have access to progressive feedback. Ensure extra-curricular activities are provided in an online community. Modify assessment rubrics to be appropriate for online.	1,2
3	Where a provider works in partnership with other organisations, arrangements to ensure that the standards of its awards are credible and secure are not effective.	4,2	Ensure external partners have appropriate training and suitable equipment for hybrid delivery. Undertake audit of delivery and manage any issues regularly.	1,2	4,2	Ensure external partners have appropriate training and suitable equipment for online delivery. Undertake audit of delivery and manage any issues regularly.	1,2
4	Reliable, fair and transparent external expertise, assessment and classification processes are impaired by the delivery mode.	4,1	Engage with and train external examiners to be aware of requirements for hybrid learning and any other changes.	2,1	4,1	Engage with and train external examiners to be aware of requirements for online learning and any other changes.	2,1
Core Quality							
5	The admissions system is not reliable, fair and inclusive for all students.	4,5	Include technical requirements for equipment e.g. internet speed, camera, adequate working space in admissions criteria.	1,1	4,5	Include technical requirements for equipment e.g. internet speed, camera, adequate working space.	1,1

6	The provider does not have sufficient appropriately qualified and skilled staff to deliver a high-quality academic experience.	**4,4**	Train academic and technical staff to use equipment and relevant software for hybrid teaching. Create opportunities for staff to share and discuss lessons learned for hybrid delivery.	**1,4**	**4,4**	Train academic and technical staff to use equipment and relevant software for online. Create opportunities for staff to share and discuss lessons learned for online delivery.	**1,4**
7	The provider has not designed and/or delivered high-quality courses appropriate to the mode of delivery.	**4,2**	Update internal policy on course design for hybrid delivery. Staff training on how to design hybrid delivery. Adapt teaching style for hybrid delivery i.e. ensuring equal engagement opportunity with online and in person students, might include additional staff member online.	**1,2**	**4,2**	Update internal policy on course design to include online delivery. Staff training on how to design online. Adapt teaching style for online delivery i.e. using appropriate interactive tools and activities to ensure engagement.	**1,2**
8	The provider does not have sufficient and appropriate facilities, learning resources and student support services to deliver a high-quality academic experience.	**4,2**	Provide automatic integrated classrooms with recording facility for hybrid delivery with internet, cameras etc. Communicate requirements to support services for a joined-up experience. This should include stream-lined provision of on –line resources i.e. one place for everything. Make provision for students who require support.	**1,2**	**4,2**	Develop bespoke teaching facilities for online delivery with recording facilities. Communicate requirements to support services for a joined-up experience. This should include stream-lined provision of on –line resources Make provision for students who require support.	**1,2**
9	The provider does not actively engage students, individually and collectively, in the quality of their educational experience.	**4,1**	Ensure both the online and face-to-face students have a student representative as point of contact. Have regular student liaison meetings in a hybrid format.	**1,1**	**4,1**	Ensure that an online student representative is assigned. Have regular student liaison meetings online.	**1,1**

Table 12.1. (*Continued*)

10	The providers fair and transparent procedures for handling complaints and appeals are not accessible to all students.	**4,1**	Have duplicate online bookable spaces for drop-in sessions where students can talk to a person in a private and secure environment. Create dedicated email addresses and clear procedures for complaints that are easily accessible.	**1,1**	**4,1**	Create online bookable spaces where students can speak to a person about complaints in a private and secure environment. Have clear and transparent online complaint processes with emails and electronic forms for complaints that are easily accessible.	**1,1**
11	Where the provider offers research degrees, it cannot deliver these in appropriate and supportive research environments.	**4,3**	Create literature-based projects with dedicated supervisors or balance timetable for students to undertake practical elements. Adapt intended learning outcomes accordingly. Create hybrid research community spaces for staff and students to work collaboratively.	**2,2**	**4,3**	Create literature-based projects with dedicated supervisors or non-practical projects. Adapt intended learning outcomes accordingly. Create online spaces where regular project meetings and wider group meetings can take place for online students.	**2,2**
12	Where a provider works in partnership with other organisations, the arrangements to ensure that the academic experience is high-quality irrespective of where or how courses are delivered and who delivers them are not effective for the mode of delivery.	**4,2**	Ensure provider classroom has all equipment and necessary materials to maintain control of delivery with other organizations and training.	**1,2**	**4,2**	Agree same online platform is available to everyone and appropriate connections. Provider staff to control online training and give support.	**1,2**

13	The provider cannot support students to achieve successful academic and professional outcomes.	4,5	Create suitable online spaces so that students can have one-to-one support discussions with academic supervisors. Create online tutorial spaces and train staff to engage with students online. Map updated course ILO's to external accredited bodies which can provide memberships, leading to Charterships etc. Create connections with external partners online so that students gain access to professional networks.	2,2	4,5	Create suitable online spaces so that students can have one-to-one support discussions with academic supervisors. Create online tutorial spaces and train staff to engage with students online. Map updated course ILO's to external accredited bodies which can provide memberships, leading to Charterships etc. Create connections with external partners online so that students gain access to professional networks.	2,2
Student Experience							
14	Teaching and Learning: Lack of engagement and intellectual stimulation for students and learning materials and support inappropriate for mode of delivery.	4,3	Train staff to engage with online students same as those in the classroom. Ensure access to resources is equitable for all students. Provider ensures course design includes diverse teaching styles such as interactive activities, group work, forums etc… and that appropriate equipment is available.	2,2	4,3	Train staff to engage with online students. Ensure suitable resources available online and in accessible spaces. Provider ensures course design includes diverse teaching styles such as interactive activities, group work, forums etc… and that appropriate equipment is available.	2,1
15	Engagement: Students feel nervous to ask questions and do not have a balanced workload. Insufficient opportunities to provide feedback.	4,2	Train staff to engage equally with online students. Develop other ways that the students can ask questions, i.e.written questions or anonymous polls. University has suitable online resources, such as cameras, polling and other interactive capability. Ensure duplicate feedback mechanisms are in place.	2,2	4,2	Train staff to engage with online students as they would with those in the classroom. Treat all students the same i.e., all students have cameras to overcome this fear. Ensure electronic feedback capability is available and consistent response mechanisms are in place.	2,2

Table 12.1. (*Continued*)

16	Assessment and Feedback: Assessment is impaired by mode of delivery making it unfair or having insufficient feedback.	4,4	Ensure online assessments are the same as face-to-face i.e. open book with the same durations and the same style of feedback. Alternatively, all students take online exam.	2,1	4,4	Assessments must be suitable for online, and students have clear communication on how to undertake them. Provider must have suitable resources to support online assessments.	2,1
17	Dissertation or major project: Insufficient support from supervisors when planning and undertaking projects.	4,2	Ensure that supervisors are trained and duplicating the same support for all project students. Create online spaces where regular project meetings can take place for online students.	2,2	4,2	A change may be needed in supervision style for online students to ensure efficient meetings etc. Create online spaces where regular project meetings can take place for online students.	2,1
18	Organisation and management: Organisation and management is impaired by mode of delivery.	3,2	Use multiple communication methods to ensure that online students do not miss information. Create accessible communication strategies and disseminate regularly. Train staff to undertake effective communication.	1,1	3,2	Create suitable and consistent online communication methods. Train staff to communicate regularly.	1,1

19	Resources and services: Access to all required subject specific IT resources is impaired for mode of delivery.	**4,4**	Undertake a comprehensive feasibility assessment of resource and software requirements for remote students. Ensure software licenses are available for all students remote and on-site irrespective of their location. Create a virtual space for all students to access required software and on-demand training.	**1,1**	**4,4**	Undertake a comprehensive feasibility assessment of resource and software requirements for remote students. Ensure software licenses are available for all students irrespective of their location. Create a virtual space for all students to access required software and on-demand training.	**1,1**
20	Skills development: Impaired ability to improve research skills, such as independent and creative learning.	**4,2**	Duplicate suitable online spaces so that students can have one-to-one support discussions with academic supervisors. Create online tutorial spaces and train staff to engage with students online. Create suitable online spaces for students and staff to interact and work co-operatively.	**3,2**	**4,2**	Online students may not experience a research community in the same way as in person because there is not the same vibrant, active online community as there is in a workplace. Create suitable online spaces for students and staff to interact and work co-operatively.	**3,2**

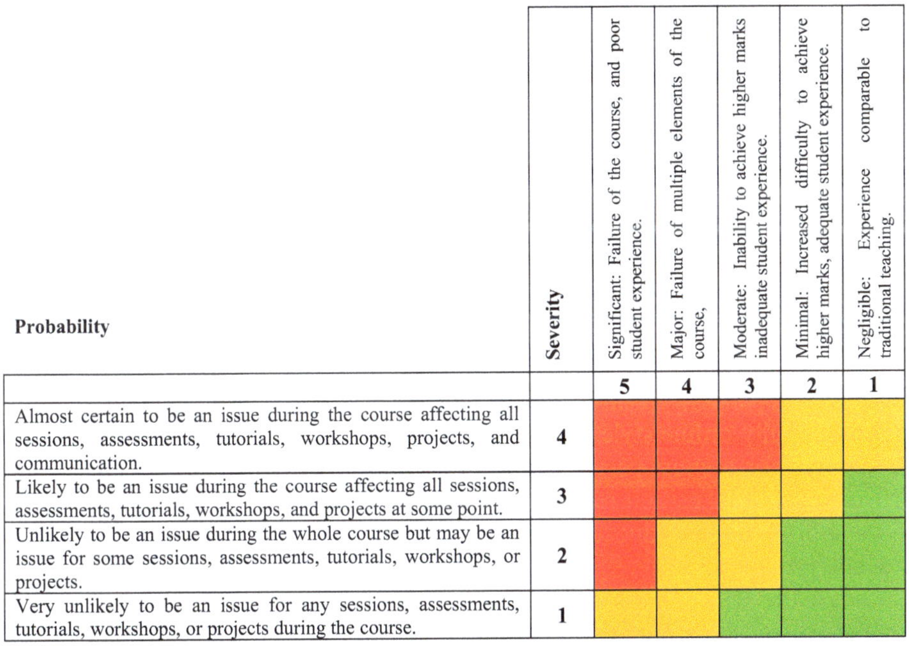

Figure 12.1. Risk matrix defining probability and severity for the risk of failure and poor experience for students studying a hybrid or fully online level 7 qualification. Red= high risk, yellow = moderate risk, and green = low risk.

and 5. was 'significant' i.e. failure of level 7 award (figure 12.1). Probability was defined on a scale of one to four, related to how much delivery of the course would be disrupted where 1. was 'very unlikely' to be affected and 4. 'almost certain' to be affected (figure 12.1). The likelihood and severity definitions were included in a matrix which enabled individual scoring of each category, followed by cross-referencing to determine the risk category as high (red), moderate (yellow) or low (green).

Following the scoring results were entered into a table that enabled comparison of hybrid and online delivery, with a column to propose mitigation strategies.

12.2.2 Risk Assessment methodology

The assessment (table 12.1) was completed collaboratively by a group of academic staff based on their experience of delivering higher education using all three styles. The 20 identified hazards were assessed using the developed matrix to assign a risk category (high, medium, low). Following this a mitigation strategy was proposed and the hazard was rescored.

12.3 Results

Results of the qualitative Risk Assessment for transition to hybrid or online delivery based on hazards identified from the QAA core standard and core quality criteria and the PTES metrics for student experience at level 7 are shown in table 12.1.

12.4 Discussion

The Risk Assessment for the transition of higher education to hybrid or remote online teaching was based on academic standards, quality [12] and student experience metrics [14]. The assessment focused on the risk to students of failing, or receiving lower marks due to the teaching style, and the risk of poor student experience compared to traditional on-campus education for science and engineering level 7 qualifications. The assessment focused on level 7 qualifications in particular which are usually the highest level of taught education and bridge the gap between undergraduate study and independent research or the transition into a workplace. Therefore, at level 7 qualifications tend to include more practical elements such as laboratory and field demonstrations, and usually require completion of an independent research project which is potentially challenging to deliver in a hybrid or fully online style. While it has been possible to transition to the online environment for a short period during the COVID-19 pandemic, it is clear that a more permanent transition requires careful consideration to ensure students receive the same high-quality education and student experience that they expect from an academic provider.

Overall, the Risk Assessment demonstrated that it is feasible for an academic provider to transition to hybrid or online teaching, but not without considerable mitigation of current teaching and processes to meet the required academic standards and maintain quality. Overall, the risk to student success was similar for both hybrid and online modes of delivery, with all twenty hazards being assessed as at least a moderate risk (yellow) if no mitigation was applied i.e. the provider immediately delivers hybrid and online teaching with no changes to policy or course content. It is appreciated that the current level of technological and experiential readiness of providers may significantly differ depending on the mitigations already introduced during the recent pandemic. Therefore, individual providers may find that their organisational risk is higher or lower in some areas.

When considering mitigation strategies, it was found that they would be more challenging for hybrid cohorts due to the necessity of providing equal experience and opportunities to both students undertaking a face-to-face and remote online course. In general, mitigation to meet the required level 7 'core standards' is highly achievable by adjusting the syllabus to remove practical elements, which are not mandatory. However, it was found that it may be more challenging to fully replicate the quality of teaching or the student experience for remote online students. The following sections outline the high (red) and moderate (yellow) risks to hybrid or remote online delivery and suggest top-level mitigations

12.4.1 Highest risks to standards, quality and student experience

12.4.1.1 Core standards

The highest risks arise when there is a high probability of a significant number of students failing their level 7 qualification. This may be due to serious issues that affect all elements of a course i.e. inadequate provider policy, poor student communication, inappropriate assessment and inaccessibility of essential resources that a student could not independently mitigate e.g. if remote online students do not have access to software required for an assessment they will be unable to take the assessment and more than likely fail. These issues arise most acutely in the 'core standards' section of the Risk Assessment (Hazards 1–4) which cover hazards relating to an inability to meet the required standard for level 7.

The highest risk in the 'core standards' section arises from the Intended Learning Outcomes (ILO) not being adapted to meet the new style of teaching, making it impossible for students to achieve their award when studying online. For example, if ILO's include practical experimental elements which cannot be assessed or achieved online due to safety or lack of facilities then students would have no chance of passing. Therefore, ILO's must be adapted to be suitable for the entire cohort. This is challenging for the hybrid cohort as it is likely to disadvantage either the online or the on campus students depending on how it is implemented. For example, re-writing ILO's so they are suitable for online teaching may mean removing desirable practical elements that attracted campus-based students in the first place would no longer be available. However, offering campus-based students' additional non-assessed practical elements disadvantages the online students, they would have to agree to this prior to registration and be aware that no hands-on experience will be offered. Students would need to consider how this may impact their future careers, as often employers require evidence of practical experience in industry critical practical skills and lacking this could impair their employability. This is equally relevant for assessments and marking criterion, which if not adapted may impair online students' ability to achieve higher standards than required by the threshold for level 7 as highlighted by the high risk for the second 'core standards' hazard. This could also be mitigated by ensuring that feedback mechanisms for online students are comparable to campus-based students, and that any extra-curricular activities are equally available to online students.

12.4.1.2 Core quality

In the 'core quality' section three hazards (5, 6 and 13) were identified as high risk for students. This means insufficient course management, delivery and student support may be significant enough to result in students failing their level 7 qualification. For example, additional requirements may need to be included in the admissions process for remote online students to ensure they meet the academic standard and have appropriate resources to successfully undertake the course (Hazard 5). Students may need to prove that they have an internet connection that meets a minimum speed requirement and that they have a device with a camera and sufficient processing power. The remote location of the students may also require consideration to ensure local internet requirements do not block essential academic resources such as the providers teleconferencing software, virtual learning environments or library resources etc. Lack of ability to access these resources would put a remote online student at a significant disadvantage compared to classroom-based students, or online students in other locations. The provider may decide that while students can study remotely, there are limitations on where they can study from i.e. within the same national boundary as the provider or an agreed supranational area.

Another high risk to student failure is when the provider does not have sufficiently skilled staff to deliver course content in a hybrid or online style (Hazard 6). This may be due to the inability to effectively use equipment and software required for online or hybrid delivery i.e. not being able to use multiple software and hardware at the same time such as switching between cameras, shared screens and interactive tools. Or this could be due to inexperienced staff delivering in a hybrid or online style resulting in remote online students unable to hear key information, for example staff moving away from stationary microphones and cameras. These issues can be mitigated by providing mandatory staff training and additional support, such as training in how to use hybrid classroom facilities and essential teaching software.

An important and high risk arises when the provider does not have sufficient support to enable students to pursue a rewarding, professional career and achieve positive academic outcomes [12]. Part of this is, the student expectation that the provider should regularly connect with them to discuss their academic progression and any pastoral needs to ensure they reach their full potential. For this to happen successfully regular discussions should take place either in one-to-one setting and in groups. However, students may not feel comfortable talking about this or about sensitive personal issues in an online environment. Also, there should be opportunities for impromptu discussions between staff and students, although these might be minimised online due to student's perception that they may be recorded, not private or not enough time. This can be mitigated by creating scheduled times online for students to drop-in and discuss any issues privately and securely with trained staff. Another part of this hazard is the ability to achieve professionally relevant outcomes to increase their employability post-graduation. Many providers already invite external organisations such as those that offer accreditation to visit student face-to-face, and this should be transferred to hybrid or online provision to ensure equal opportunity to all students.

12.4.1.3 Student experience

In the final section of the Risk Assessment concerning 'student experience', course design (Hazard 14), assessment and feedback (Hazard 16) and resources (Hazard 19) were identified as posing the highest risk to student success and experience.

Teaching hybrid or online requires a very different approach compared to traditional teaching methods to ensure the best possible learning environment. Therefore, almost all elements of a level 7 qualification currently taught face-to-face would require adapting to increase the remote online student experience. Lower engagement would not necessarily translate to poor academic performance for all students as some may be prepared to study independently or may prefer a less interactive teaching style. However, this can make it more difficult to formatively assess students' progress and may mean academics miss opportunities to help students achieve higher marks. Previously discussed mitigation such as ensuring use and training of appropriate equipment and resources would enhance the online student experience, and although it is unlikely that the face-to-face experience could be replicated online, capitalising on the unforeseen benefits of hybrid and remote online could create an equally positive experience.

Assessments are essential to taught qualifications and may take many forms, though in the last decade there has been a shift from predominantly examined qualifications to a diversity of assessments such as dissertations, group projects, oral assessments and peer-review [15]. This shift is beneficial to hybrid or online delivery as oral or written assignments can be equally as successfully in a classroom or online environment providing the issues identified in sections 12.1 and 12.2 of the Risk Assessments are mitigated. However, exams are particularly difficult to transfer to an online environment for remote students and are usually required to be proctored or designed to be open access to minimise academic dishonesty (cheating) and maintain academic standards [16]. Proctoring i.e. remote monitoring of students by an external company while they sit an exam, is a solution if examination is deemed essential, but can be far more stressful for students and incurs a cost for the provider. In addition, to ensure equality all students in a hybrid cohort would have to sit a proctored exam, which may seem unfair to face-to-face students. This can be fully mitigated by redesigning assessments to avoid setting exams as there is no requirement for students to be assessed by examination. In addition, for hybrid cohorts equal opportunities for success must be provided to both face-to-face and remote online students ensuring appropriate online environments are provided. This links to the wider issue of resource provision (Hazard 19) that has been discussed extensively in other sections with regards to ensuring staff have sufficient resources, but that also extends to student resources. This may include licenses to enable remote use of essential software, access to journals and books through the library as well as access to virtual learning environments as discussed elsewhere. Prior to offering remote online provision through hybrid or fully online delivery providers should undertake a full feasibility assessment of whether currently used resources can be made available for remote students. Resources deemed as essential may need to be made available remotely, which may incur additional cost, or may impose limitations on delivery. A decision can be made as to whether access to non-essential courses which may include consideration of additional costs or restrictions. These mitigations would enable a remote student to successfully undertake a level 7 qualification, but may impair the quality of experience received as students are less likely to be offered impromptu access to local resources such as laboratories, bespoke equipment and staff experience.

12.4.2 Moderate risks to standards, quality and student experience

12.4.2.1 Core standards

Hazards 3 and 4 within the 'core standards' section both concern reliance on partnerships and other organisations to meet the standards required for level 7. For example, universities frequently employ external partners to deliver technical or industry focused content specific to credit-bearing assessments. These partnerships are highly valuable for level 7 students to bridge the gap between academic theory and real word application and to promote development of professional transferrable skills. However, the awarding organisation (provider) remains accountable for the academic standards of the external partners, and must therefore ensure appropriate arrangements are made to manage and monitor the external partners (Hazard 3). When transitioning to hybrid or online delivery, these arrangements are likely to require modification to ensure the quality and academic standard of external partners is maintained. However, due to the minimal number of learning hours delivered by external partners, the severity for students is low and the risk easily mitigated by updating policy and ensuring monitoring and audit processes are in place. The provider should also make appropriate training and equipment available for the external partner if necessary. Similarly, when external examiners are employed to impartially review assessments, procedures and delivery they may not be aware of the requirements for hybrid or online delivery. Although, this is unlikely to directly affect student's opportunity to pass as all assessments and course materials would be approved prior to provision to students. This can be effectively mitigated by ensuring that selected external examiners meet the required academic and technical requirements including awareness of hybrid and online teaching standards.

12.4.2.2 Core quality

The majority of hazards in the second section ('Core Quality') of the Risk Assessment taken from the QAA higher education framework were assessed as a medium risk. This means while students are unlikely to fail the course due to these issues, the quality of delivery and resources may be impaired. The medium risks fell into three broad categories: (1) engagement; (2) quality of delivery and (3) provision of resources.

Engagement relates to the provider creating an academic research culture that enhances the quality of individual student experiences, such as encouraging students to feedback to the provider on the quality of their experience (Hazard 9) and creating supportive research environments (Hazard 11). While many existing processes can simply be replicated in an online environment for remote online cohorts, it may be much more difficult to ensure equity in hybrid cohorts. Therefore, it would be beneficial to have two student representatives in hybrid cohorts, one for the classroom-based students and one for the online students to ensure any particular concerns of both groups can be communicated and resolved. For remote online student cohorts there would always be a student representative, same as if face-to-face and if the entire cohort is online only one student representative is likely to be required. In addition, level 7 qualifications in science and engineering will usually involve undertaking an independent research project requiring students to critically analyse primary literature and generate technical data in a specific topic area. The academic requirement could be met for remote online delivery if ILO's were adapted to allow for desk-based projects involving collating literature and survey responses rather than a practical project. However, doing this may disadvantage the remote students by excluding them from the benefits of being exposed to the thriving research culture in many university laboratories, workshops and research groups. In these environments, students have the opportunity to interact with other postgraduate students and researchers that is far less likely to arise when studying remotely. This enables them to develop practical skills essential to future employment. In addition, practical environments cultivate and nurture sough after skills such as creativity, curiosity, and critical independent thought.

The second theme identified in the 'core quality' section was the quality of teaching delivered by internal academic staff (Hazard 7) and external teaching partners (Hazard 12). The hazards associated with not meeting the required academic standard for an external partner were discussed in section 12.4.1 (Hazard 4), and the identified mitigations such as auditing and training would also promote high-quality delivery. However, additional mitigations such as ensuring the external partner has the appropriate equipment to deliver online or in a hybrid style, or fully supporting the external partner by hosting the online provision on their behalf would ensure a high-quality experience. Implementing these mitigations would significantly reduce the likelihood of low-quality delivery, but there is always a risk when responsibility is shared with an external provider outside the control of the provider. It should be easier to mitigate the quality of internal staff delivery (Hazard 7) by ensuring that internal policy for course and assessment design is updated to include the delivery to remote online students, and that staff are encouraged to take advantage of additional training on how to design for online or hybrid classrooms. For example, when studying remotely students may feel less engaged with the rest of their cohort and additional mechanisms for encouraging participation such as regular polling, anonymous question forums and interactive group software such as whiteboard facilities. These all require careful management by staff if they are to be fully leveraged and encouraging staff to try new approaches in safe and supportive environments could enable a successful transition to hybrid or remote online delivery.

Ensuring the high-quality of course design is only half the battle, as excellent course design can be hampered by poor provision of resources, which is the third broad category of risk identified under the 'core quality' section. Resources include both the student-facing equipment and software (Hazard 8) and the underpinning processes such as for complaints and appeals (Hazard 10). Online and hybrid delivery is far more likely to be successful if the provider ensures that staff have suitable equipment such as fully integrated classrooms for hybrid teaching that automatically initiate cameras, microphones, screensharing and recording. Equally, for fully online delivery suitably equipped and soundproofed teaching spaces should be provided, and in both cases an infallible internet connection with at least one backup is essential such as a wired connection with a Wi-Fi backup by a different network. Equally, mitigation is also possible to ensure underlying services meet the needs of remote online students such as ensuring they have equal access to information online, and to staff. For hybrid cohorts, this may require staff to be mindful of sharing information equally with both remote and face-to-face students.

In most cases, the medium risks can be mitigated to low risks with the possible exception of providing appropriate and supportive research environments for remote online students which remained a residual risk after mitigation. This was due to the limitations of online environments particularly for facilitating serendipitous meetings, and its inability to fully immerse students in existing university research culture which is one of the greatest benefits of undertaking a level 7 qualification.

12.4.2.3 Student experience

Four of the seven hazards in the 'student experience' section presented a medium risk, and while all risks can be mitigated to a lower level, it is unlikely that the current student experience could be fully replicated online. This disparity is likely to be more pronounced for a hybrid cohort where face-to-face students may benefit from spontaneous ideas and impromptu meetings with other students and staff which cannot be replicated online. Many of the mitigations suggested for hazards in the 'core standards' and 'core quality' section would improve the online student experience, such as creating online 'drop-in' times and active online student communities. Three of the medium risks arise primarily due to the potentially isolating nature of online remote study. Lack of engagement (Hazard 15) with remote online students in a hybrid cohort may make them feel left out, and for both hybrid and online cohorts the nature of delivery may make them feel nervous to ask questions or seek help. Lack of support during the project phase of a science and engineering level 7 qualification (Hazard 17) may mean students do not benefit from the existing research culture at the university. Students may also feel like they are interrupting 'normal' activities unless measures are taken to fully incorporate remote online students into research groups. Finally, remote online learners (in hybrid and fully online cohorts) may have an impaired opportunity to accrue research and transferable skills beneficial to their future careers (Hazard 20). These three hazards are interlinked and can be summarised as a potential risk to the students' 'sense of belonging'.

As stated in the QAA Advice and Guidance on Enabling Student Achievement creating a 'sense of belonging' can aid retention and provide a safe space for students to seek help from academic and support staff, as well as their peers [17]. An established university creates this 'sense of belonging' from the pre-existing academic community of research staff and students and provides physical spaces such as libraries, cafes and outdoor meeting areas that promote networking and collaboration as well as providing non-academic respite such as through sports clubs. While it is possible to create hybrid or fully online events such as research seminars, group meetings and social meet-ups, these may be less attractive to campus-based students who might prefer to engage in more practical activities. It must be noted that some students may not perceive this as an issue and may find remote online learning is more suitable for their circumstances making level 7 qualifications more accessible and inclusive. It may be that the current research and academic culture can be enhanced by hybrid and online approaches, such as developing globally collaborative research networks with virtual faculties enabling greater cross-collaboration and multi-disciplinary research than ever before. In the future it may be possible for a fully online student in the UK to work collaboratively with laboratory-based students on the other side of the world to gather practical data.

Finally, student experience can be severely hampered by simple issues such as poor course organisation and communication [Hazard 18]. This can often be a challenge for traditional learning environments where students have different platforms to navigate to find their assessment results, their feedback, university policy, course materials and course delivery. This becomes even more challenging for hybrid cohorts where different information may be provided to the face-to-face and online students such as physical classrooms versus digital meetings links. For fully online cohorts the issue may be compounded by the isolation of remote learning where they cannot just 'drop-in' and speak to a member of staff. Therefore, it is essential to prepare materials and information well ahead of time and meet student expectations. For example, if students are told a recording will be available 'after' the lecture, it would help for the student to know whether 'after' means minutes, hours or days.

12.5 Risks to the provider

The Risk Assessment developed for this chapter focused on the risks to students of failing or receiving lower quality education through a hybrid or remote online approach but did not cover the risks imposed to the provider. While this could be the subject of an additional Risk Assessment, and would require scrutiny to develop rigorous criterion, it is evident that for some courses hybrid or online delivery would pose a risk. For example, when delivering online in any capacity, the provider loses control of the material at the point of receipt i.e. students could record lectures on their mobile devices, or have other people in the room with them. While not an issue for open access information, when delivering secure or commercially sensitive information the provider must be able to control the information meaning that some courses may never be able to be delivered online. This also includes export control, where information not available in open source can be considered an export and may require licenses if it contains information that could be used in the proliferation of Weapons of Mass Destruction.

In addition, one of the tenets of higher education is the exposure to a thriving research environment, which is supported and enhanced by students attending taught courses. If a significant proportion of students receive their higher education remotely it may impair the research culture at the institution, resulting in a loss of reputation and over time potentially a reduction in research students. The body of higher education providers may need to consider when and why hybrid or online courses are offered to maintain university culture or accept that higher education may need to evolve to meet the expectations of a new generation of students.

Finally, academic providers have a duty of care to their students to support their mental and physical health. While there are many inherent benefits in the flexibility of a hybrid approach, or the inclusivity of an online cohort, the impact of full-time online education on physical and mental health is not fully understood. In addition, the impact of hybrid or online delivery may be different for everyone and some students may thrive in an online environment, while other students may find it much more challenging. Academic providers may need to redefine 'duty of care' to be inclusive of the needs of online and hybrid cohorts.

12.6 Conclusion

A Risk Assessment was developed to assess the risk to students failing, or receiving poor experience, during a level 7 qualification delivered in a hybrid or fully remote online style. The Risk Assessment was partly based on the QAA core standards and quality to ensure holistic coverage of the practices required by all UK HE regulatory jurisdictions. This was combined with the PTES survey prepared by Advance HE to enable assessment of risk to student experience, often a deciding factor for prospective students. This chapter demonstrates assessment of potential risk and prosed top-level mitigation for identified risks, however it did not include the risk control and monitoring steps. When taking this approach forward, these steps should be included to ensure continual improvement for the organisation.

Overall, the Risk Assessment in this chapter demonstrated that delivery of level 7 qualifications in a hybrid or online style could be successfully achieved, although significant changes to policy, student and staff behaviours and course design would be required to achieve equal opportunity for success (in hybrid environments) and to provide a good student experience.

Some high-risk hazards (such as not meeting level 7 standard (Hazards 1–4)) can be rapidly and effectively mitigated by adapting course ILO's to ensure they provide equal opportunity to remote online and classroom in hybrid delivery, or to ensure they are suitable for fully remote online cohorts. Providers may find additional challenges in changing ILO's where courses are accredited by external professional bodies which require demonstration of practical skills. Though challenging, this is not insurmountable as solutions could be to offer a different version of a course to remote online students where students are made aware of the differences prior to registration. Internal resource issues (Hazards 8 and 19) can also be rapidly solved with appropriate application of funding to install high-quality equipment, and making teaching software and tools accessible to all staff and students.

Another subset of risks are those such as quality of course content and staff competence (Hazards 6, 7, 12, 14, 15 and 16) which can be mitigated by providing training and resources (such as time and IT support), but rely on staff and partner uptake. Internally some training can be mandated, though forcing change is often ineffective and therefore it is preferential to secure staff buy-in. It is far more challenging to control the quality of teaching by external partners as they may be less willing to undertake training or may be unwilling or unable to transition to a hybrid or fully online style. The same is true to risks pertaining to provider processes relying on clear and effective communication (Hazards 5, 9, 10 and 18), while policy can be adapted and technology can be leveraged to enhance student education and experience, it is only as effective as how well it is used. If staff and students are unaware of the processes, or do not use them, there will remain a risk and it is likely to be more acute for remote online students who have no physical location where they can find answers.

The third category of risks presents the lowest risk to student failure, but the highest risk to student experience and relates to the 'sense of belonging' felt by being part of a cohort, a research group and an academic institution (Hazards 11, 13, 17 and 20). Technology has made it possible to take part in events and meetings remotely, and continues to improve, however it cannot yet mimic spontaneous meetings in campus courtyards or ad-hoc technical discussions post-lecture. However, this is predicated on replicating the classroom experience online, which at best may not be appropriate and at worst may not be possible. Therefore, over time, developments in technology and improved parity in access to technology (such as more affordable fibre broadband) may actually provide even more benefits than currently found in universities. Therefore, the challenges identified should not be perceived as a barrier, but as an opportunity to ensure the success of hybrid and online delivery.

References

[1] The Quality Assurance Agency for Higher Education 2022 Made digital: exploring student engagement and performance during the shift to digital teaching, learning and assessment.

[2] Al-Nasa'h M, Al-Tarawneh L, Abu Awwad F M and Ahmad I 2021 Estimating students' online learning satisfaction during COVID-19: a discriminant analysis *Heliyon* **7** e08544

[3] Jogezai N A, Baloch F A, Jaffar M, Shah T, Khilji G K and Bashir S 2021 Teachers' attitudes towards social media (SM) use in online learning amid the COVID-19 pandemic: the effects of SM use by teachers and religious scholars during physical distancing *Heliyon* **7** e06781

[4] Melgaard J, Monir R, Lasrado L A and Fagerstrøm A 2021 Academic procrastination and online learning during the COVID-19 pandemic *Proc. Comp. Sci.* **196** 117–24

[5] Yu-Fong Chang J, Wang L H, Lin T C, Cheng F C and Chiang C P 2021 Comparison of learning effectiveness between physical classroom and online learning for dental education during the COVID-19 pandemic *J. Dental Sci.* **16** 1281–9

[6] Gumede L and Badriparsad N 2022 Online teaching and learning through the students eyes —uncertainty through the COVID-19 lockdown: a qualitative case study in Gauteng province, South Africa *Radiography* **28** 193–8

[7] Irawan A W, Dwisona D and Lestari M 2020 Psychological impacts of students on online learning during the pandemic COVID-19 *KONSELI: J. Bimbingan Konseling (E-Journal)* **7** 53–60

[8] Sharples M *et al* 2016 Innovating Pedagogy makers *Inst. Educ. Technol.* **26** 42 https://ec.europa.eu/jrc/sites/default/files/2014-nmc-horizon-report-eu-en_online.pdf

[9] Vernadakis N, Antoniou P, Giannousi M, Zetou E and Kioumourtzoglou E 2011 Comparing hybrid learning with traditional approaches on learning the Microsoft Office Power Point 2003 program in tertiary education *Comput. Educ.* **56** 188–99

[10] Lemoine P A, McElveen J P, Waller R E and Richardson M D 2021 *Is Online Learning the Future of Global Higher Education?: The Implications from A Global PandemicAdvancing Online Course Design and Pedagogy for the 21st Century Learning Environment* (IGI Global) pp 28–44

[11] Capranos D, Dyers L and Magda A J 2021 *Voice of the Online Learner 2021: Amplifying Student Voices in Extraordinary Times* (New York: Wiley Education Services)

[12] Quality Assurance Agency 2014 *UK Quality Code for Higher EducationPart A: Setting and Maintaining Academic Standards* https://qaa.ac.uk/docs/qaa/quality-code/qualifications-frameworks.pdf

[13] QAA 2018 *UKSCQA UK Standing Committee for Quality Assessment The Revised UK Quality Code for Higher Education* pp 1–5 www.ukscqa.org.uk

[14] AdvanceHE 2020 *Postgraduate Taught Experience Survey (PTES) 2020 Welcome* https://warwick.ac.uk/students/dialogue/ptes/staffresources/introducingthesurvey/ptes_questionnaire_2020.pdf

[15] Quality Assurance Agency and UK Standing Committee for Quality Assessment *UK Quality Code for Higher Education Advice and Guidance Assessment*

[16] Remote online exams in higher education during the COVID-19 crisis, OECD Education Policy Perspectives, OECD iLibrary https://oecd-ilibrary.org/education/remote-online-exams-in-higher-education-during-the-covid-19-crisis_f53e2177-en; jsessionid=sSy7Xoz0lSbsXM942-SgJXD5.ip-10-240-5-134

[17] Quality Assurance Agency and UK Standing Committee for Quality Assessment *UK Quality Code for Higher Education Advice and Guidance Enabling Student Achievement*

Chapter 13

Practical application of a decision framework to mitigate environmental risks posed by the treatment of explosive contaminated wastewater

W Gilroy-Hirst, J Padfield, M Ladyman, F Coulon and T Temple

The increased demand for Insensitive High Explosives (IHE), i.e. those less sensitive to accidental impact, in militaries around the world has required munitions manufacturers to adapt their processes to accommodate changes to the explosive formulations. IHE formulations may combine up to three compounds, often a mixture of traditional legacy explosives and compounds previously unused in explosive formulation. Therefore, when explosive manufacturers start to process with IHE, they must often review the entire processing lifecycle, including the environmentally responsible disposal of side products and excess explosive material to ensure current procedures are suitable for the new formulation. Failure to assess the risk of changing standard procedures on the environment may have significant impacts to business if the existing procedures cause environmental impacts such as pollution, which may lead to fines, or 'cease' notices from environmental regulators.

Therefore, in this chapter a four-stage decision framework was applied to an explosive manufacturing facility to determine whether the facility would be able to process IHE using the existing disposal procedures without negatively impacting the environment. The decision framework was based on the manufacturers existing environmental management system and extended the assessment to evaluate the safety, compliance, reputational and financial risks that a change to disposing of IHE using existing procures would impose on the facility. The aim of the decision framework was to identify significant impacts on the environment that imposed a significant risk on the business and use an evidence-based approach to propose short-term and long-term disposal solutions in line with environmental best practice. The decision framework facilitates an evidence-based approach for making decisions relating to environmental risk as well as providing a transparent and defensible record of how that decision was reached that can be applied to any activity that has the potential for an environmental impact regardless of industry.

doi:10.1088/978-0-7503-3643-7ch13

© IOP Publishing Ltd 2022

13.1 Introduction

There are multiple historical cases of environmental contamination caused by the manufacture of munitions. For example, the Cornhusker Army Ammunition Plant (AAP) in Nebraska was found to have leached significant amounts of 1,3,5-trinitroperhydro-1,3,5-triazine (RDX) to a nearby aquifer (Fleming *et al* 1996). The RDX contamination was detected in 1981 and known to have migrated three miles from the site by 1983, causing at least three reported cases of cancer among the local population (2d Supp. and Benson 2001, Reddy *et al* 2005). If the operational risks of the Cornhusker plant had been identified and properly communicated to decision makers earlier, it is likely that preventative action would have been taken and the subsequent environmental, reputational and financial impacts reduced (Ramachandran *et al* 2021).

Insensitive munitions make use of less-sensitive IHE fills and/or munition design features to reduce the likelihood that the munitions will detonate outside of their designed mode, making it safer for operators in storage, transport and use. One such IHE formulation consists of RDX, 2,4-dinitroanisole (DNAN), and 3-nitro-1,2,4-triazol-5-one (NTO) and is being considered as a drop-in alternative for the current melt-cast ordnance processing in EU manufacturing facilities. The formulation will be referred to as IHE-X throughout this chapter.

Of the three individual constituents of IHE-X, none have yet been assigned limits of concentrations permitted in discharge water in EU legislations. Similarly, there are currently no discharge limits for explosives in water in the EU. However, some assessments have taken place in the United States that can be used as a guide, namely drinking water values (US Department for Health and Human Services 2012). For example, RDX has been assessed and assigned a Drinking Water Exposure Level (DWEL) and a lifetime health advisory limit for humans and could be used to inform limits in the EU (United States Environmental Protection Agency 2018). Although, the enshrinement of the precautionary principle and the 'polluter pays' principle in EU laws mean that an activity that has the potential to cause environmental contamination (such as discharge of wastewater) must be proven to be safe before it can be carried out, and that liability for the damage caused lies with the person or organisation that produced the source of pollution (Balzarolo *et al* 2011).

While RDX can be considered to have the greatest toxicological impact due to its low LD_{50} in rats, it makes up a relatively small percentage of the IHE-X formulation (US Department for Health and Human Services 2012, Felt *et al* 2013). DNAN has become an often-used replacement for commonly used 2,4,6-trinitroanisole (TNT) in melt-cast compositions; while it has only 90% of the explosive power of TNT, it shows a much greater degree of insensitivity, and is still melt-castable with little or no modifications required to existing filling plants (Davies and Provatas 2006, Taylor *et al* 2017). By contrast, NTO has the least potential for toxicological impact of the three explosives based on its LD_{50} toxicity to rats but is known to cause visual discolouration in water at low concentrations (London and Smith 1985, Tennant *et al* 2019). Discolouration of water can be considered as an environmental impact in its own right due to its visual impact as well as the reputational impact caused by the perception of pollution. The expected concentrations of each individual explosive material in solution in wastewater,

Table 13.1. Comparison of expected concentrations of RDX, DNAN and NTO in wastewater processing, their toxicity values as measured by LD50 in rats and potential for visual pollution.

	RDX	DNAN	NTO
Chemical structure	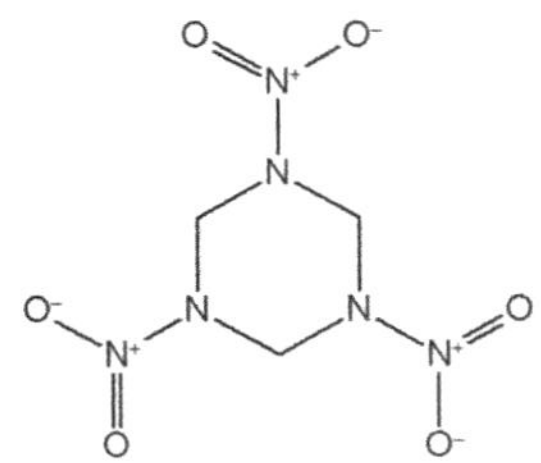	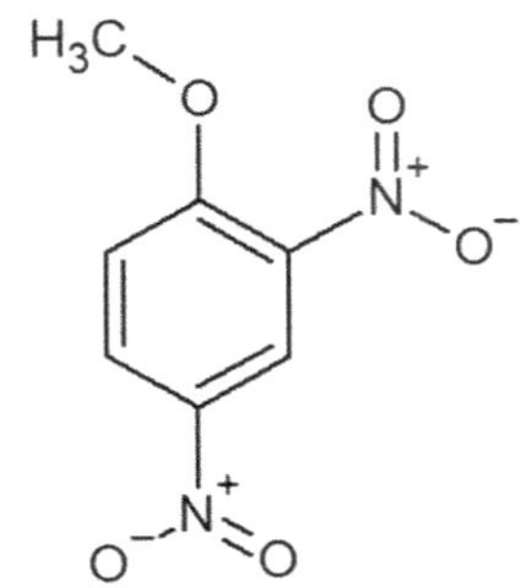	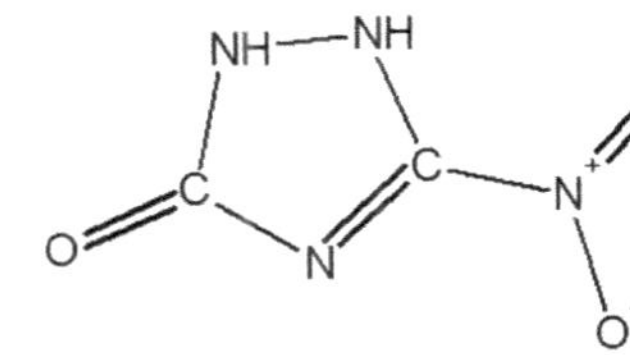
Expected concentration in wastewater (mg L^{-1})	37[1]	137[1]	1413[1]
LD_{50} in rats (mg kg^{-1})	17–118[2]	199[3]	>5000[4]
Visual water discolouration	No	No	Yellow above 100 mg L^{-1} [5]

[1] Felt *et al* (2013).
[2] US Department for Health and Human Services (2012).
[3] Reddy *et al* (2020).
[4] London and Smith (1985).
[5] Tennant *et al* (2019).

taken from a study by the US Army Engineering Research and Development Centre (ERDC) along with potential toxicological effects, is shown in table 13.1. Whilst there has been some investigation of the environmental issues posed by the individual constituents of IHE-X, including toxicity to fauna and fate and transport in soils, the combined effect of the mixed composition is still under investigation (Walsh *et al* 2014, Taylor *et al* 2015, Walsh 2016).

Due to the Polluter Pays principle, common throughout EU law, there is the potential for other pollution events to attract fines and/or remediation costs even if they were not considered as such when they occurred. This could lead to the facility being shut down and as such is an existential risk to the business. In addition, safety and reputational concerns arising from activities can have a negative impact that may not be captured using the preceding two risk areas alone.

13.2 Site description

A decision framework was used to determine the environmental impact and subsequent business risk imposed on a munitions processing facility by replacing a traditional legacy explosive with IHE-X in a melt-cast process. In the process, a large melter-mixer is used to melt the composition and ensure that it is thoroughly mixed and homogenous before filling of munitions into prefabricated casings depending on munition type. Once filling is complete the mixer can be reused if the next batch uses the same type of composition otherwise it must be washed with high pressure steam to prevent contamination between compositions. This process is carried out every 3 months on average and generates approximately 220 L of contaminated water with each wash.

The processing facility is located in an area with a warm temperature, steppe, hot summer climate (CSa) when classified using the Koppen–Geiger climate classification method (Kottek *et al* 2006). The site has a thin topsoil layer (2–5 cm) covering impermeable metamorphic rock. The site is situated amongst scrubland used for arable farming and livestock grazing as shown in the labelled site map in figure 13.3. There is an ephemeral stream running through the centre of the facility dividing the administrative buildings from the processing buildings. The stream flows to the north-east for 6 months of the year dependent upon rainfall but no groundwater is present. There is a discharge point into the stream, but this is currently not used. The trees, farmland and stream can all be considered as environmental receptors in this case.

The process diagram in figure 13.1 outlines the melt-cast munitions filling process that is used at the processing facility from receipt of IHE-X to disposal. The proposed changes, shown by the dashed lines, are a result of the need to clean equipment to prevent cross-contamination between the two types (i.e. TNT-based or DNAN-based) of formulations processed in the mixer. The processing facility is particularly concerned with the potential environmental impacts from the wastewater generated by processing IHE-X at their facility and the resulting risks to the business.

As a result of the environmental concerns raised by the processing facility, the aim of this work was to mitigate risks of IHE-X processing wastewater, including possible treatment and or disposal at the munitions manufacturing facility. This was

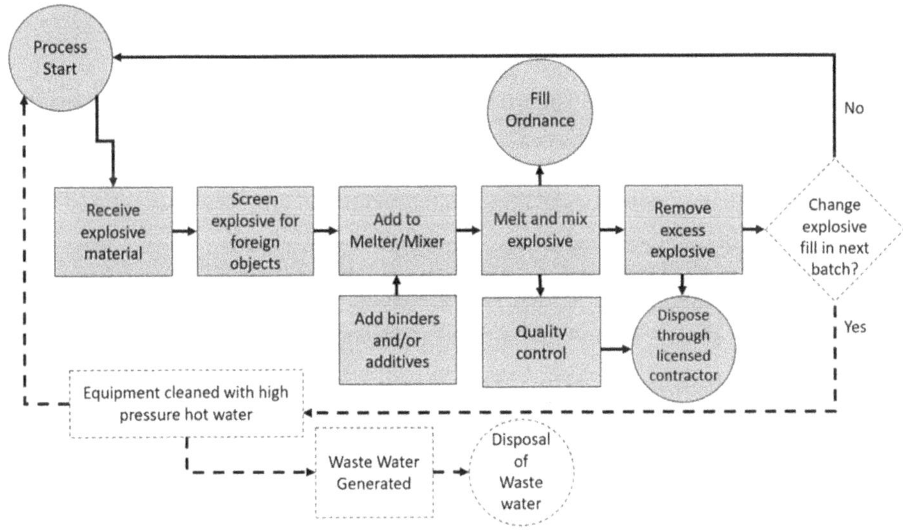

Figure 13.1. Flow diagram of the current melt-cast explosive process shown in grey and required changes to the process for the IHE-X drop-in alternative shown in dashed lines.

to reduce the likelihood of environmental impacts and to ensure that they do not disrupt business capabilities. This was achieved by the application of a decision framework to provide an evidence-based, transparent justification for decision making across three iterations to produce an actionable result. Consideration was given to local laws and discharge limits, the potential for shipping waste offsite for disposal and the possibility of treatment of wastewater onsite using an adsorption-based system. In each iteration the suitability of current methods was assessed, and any knowledge gaps identified and filled before potential mitigations were proposed.

13.3 Decision framework

The decision framework used in this scenario had previously been applied to three scenarios within the defence sector and provides a template with guidance on the content required for each step of the process (Ladyman *et al* 2019). The application of this framework to a manufacturing facility allows its effectiveness to be tested and changes proposed to enhance it. For example, a flow diagram has been constructed that details the steps required for each stage in the process (figure 13.2) that is intended as a supplement to the original template.

In this case the decision framework tool was applied iteratively to the same issue three times to address issues found by the previous version until a workable solution was identified, this is a progression from the original work which only applied the framework once.

In Stage 1 of the framework, the environmental risks resulting from the proposed activity are detailed from the existing organisational environmental Risk Assessment. These risks are then described in terms of the specific business risks that have the highest priority to the organisation carrying out the assessment.

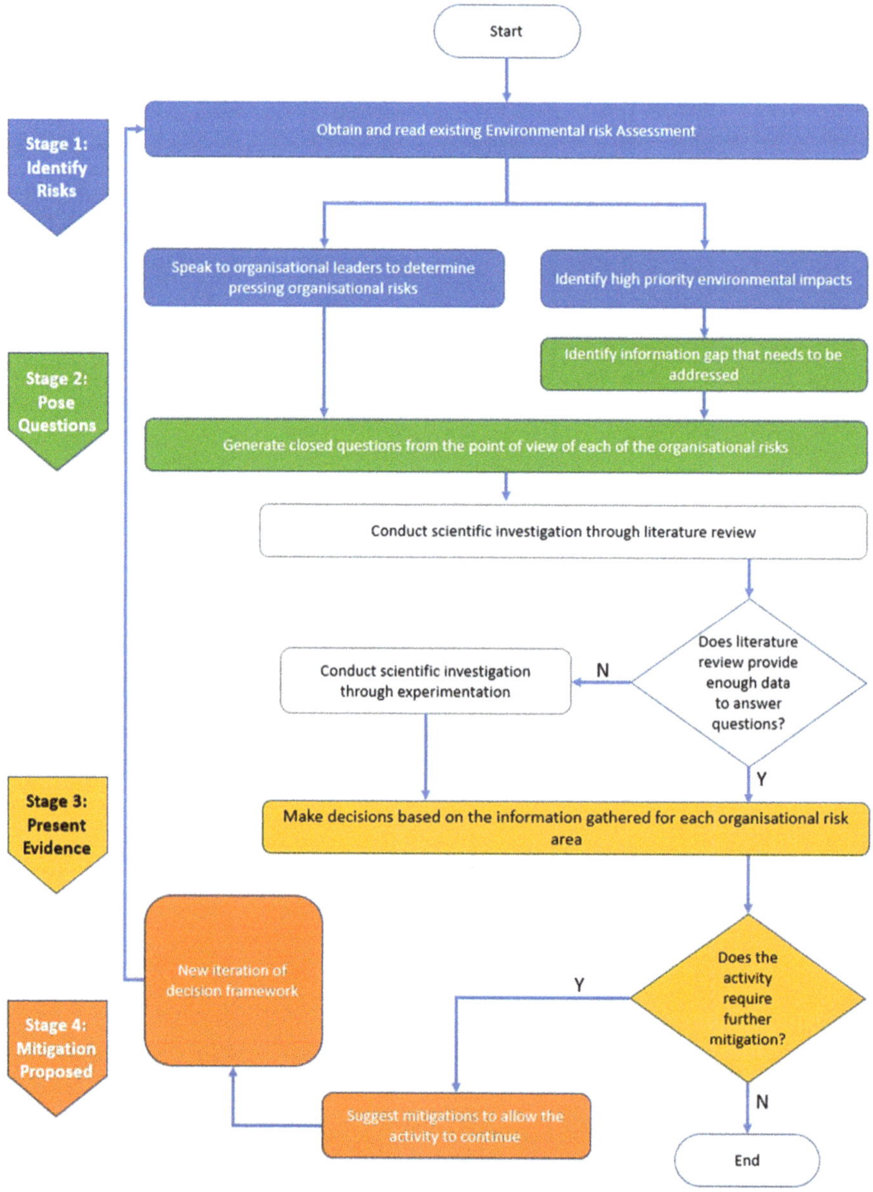

Figure 13.2. A flow diagram of the four stages that are used when applying the decision framework.

In Stage 2, questions are posed that allow the frequency, and severity of impacts from the proposed activity to be assessed. Questions are also posed to allow the potential risk to operations from the proposed activity to be assessed in line with the business risks identified in Stage 1. The questions must be specific, and unique to the assessment being undertaken. Care was taken to ensure that questions generated were able to be answered using empirical data and would give a binary

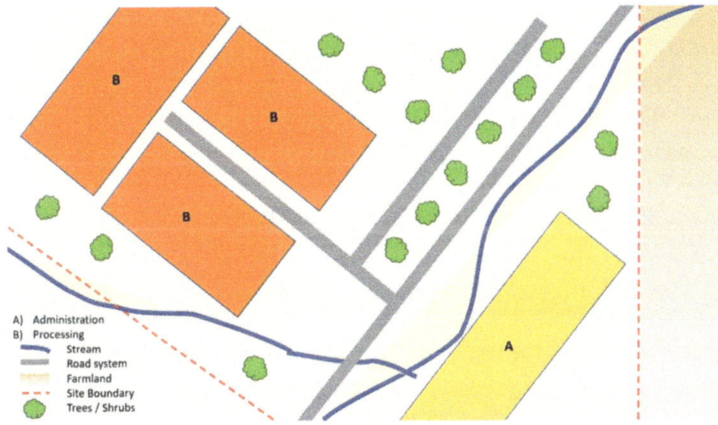

Figure 13.3. A simplified map of the processing facility including potential environmental receptors, i.e., stream, farmland, trees and shrubs including the stream running the through site, local flora and farmland.

result as to whether additional mitigation activity was required for the process to continue.

An investigation was then carried out to answer the questions posed. In some cases, this would take the form of a literature investigation of relevant environmental laws or scientific literature; however, the information was not always available in which case primary scientific investigation was required to obtain the evidence required.

The types of documentation analysed as part of the literature review were not restricted and included, books, peer reviewed journal papers, technical reports and primary legislation. In all cases, the information was accessed in English. In the case of legislation only extant documents were considered.

However, legislation from international sources was used in place as a guide for limits that could be imposed in future. Legislation was accessed using the online EU law repository EUR Lex.

Other topics of interest included the physicochemical and toxicological properties of IHE-X components and their expected concentrations in wastewater from other manufacturing facilities. Information was also collated on how each material had been shown to be removed from solution in water using different methods. During the literature review, each method was investigated, and a Strength, Weakness, Opportunity and Threat (SWOT) analysis conducted as to their suitability for the munitions manufacturing facility.

In Stage 3 of the decision framework, results of the investigation were used to answer the question posed in Stage 2, where a separate decision was made for each of the business risk areas included in the framework. A binary decision was then recorded on the continuation of the activity in question as to whether the activity could continue in its current form. The evidence used to reach these decisions, from literature review or experimental work, was then collated to create a defensible argument should the risk lead to impacts.

The outcomes of Stage 3 informed the activity in Stage 4. In each case, where an activity was found to be able to continue without mitigation then no further action was required. However, if an activity was determined not able to continue in its current form, then mitigations were suggested. The pathway of decisions and mitigations that led to the risk being deemed acceptable can be collated as a transparent record of decision making.

13.4 Results and discussion

13.4.1 Decision framework 1—discharge of IHE-X contaminated wastewater

During the investigation, the potential for an environmental risk arising from the proposed change of use of their manufacturing facility. A method was required to ensure potential environmental risks were mitigated in a logical and defensible way; therefore, the decision framework was used. The decision framework enables a holistic consideration of how environmental risks impact business by identifying the risk categories that are most relevant to the processing facility including (a) compliance with legislation, (b) reputational impact, (c) financial penalty and (d) staff safety. The original decision framework recommended three categories however safety was considered significantly important to the processing facility to warrant its own distinct category.

In order to start processing IHE-X, a legally compliant and effective disposal method for the wastewater produced must be in place. The initial proposed change in process was to discharge wastewater contaminated with IHE-X directly to the stream on-site as this had been agreed as an acceptable method of disposal for legacy explosives (TNT/RDX) by the relevant authorities. This method of disposal became redundant after previous changes to the manufacturing process, however, all the infrastructure remains on-site making this a practical and cost-effective solution. Therefore, as per stage 1 of the framework, the prioritised environmental risks were taken from the existing site environmental management strategy and assessed against the four risk categories.

This was done in the background of the contextual site model (figure 13.3) and specific local operating and climactic conditions. Aspects to consider included the ephemeral nature of the stream running through the site that would quickly dilute discharged wastewater in winter months but would quickly evaporate in summer causing a build-up of material in the soil surface for later release downstream or transport through underlying soil and geology.

The processing facility recognised the potential of contamination to the stream on-site and set about determining their duty to comply with all relevant legislation to ensure that there was no damage that could result in a financial or reputational loss to the company as a result. This resulted in several unknowns that has to answered in terms of the potential risk to staff, the extant legislation, the potential for visual contamination through discolouration of the stream and fines that could result if an incident occurred. It was identified that investment was required to gather the data needed through a contracted literature review as well as analysis of wastewater from scoping activities.

It was identified in the literature search that discharge of IHE-X wastewater would breach the national implementation of the Water Framework Directive due to a lack of evidence that the water was safe to do so. This is embodied by permitting regulations which are required for wastewater to be discharged, regardless of content. This presents a challenge because there being no defined safe limit for RDX, DNAN or NTO in the environment by the legislation. Therefore, following the precautionary principle of needing to prove an activity is safe before it begins is recommended. As a result, drinking water exposure limits from the United States could be used as a guide for future limits should they be imposed (United States Environmental Protection Agency 2018).

There is a reputational risk from visual pollution and degradation of the natural environment surrounding the plant that has the potential to result in a loss in confidence from the local population (Fleming *et al* 1996, Tennant *et al* 2019). This loss of confidence has the potential to be a long-term impact by increasing local opposition to the running and future expansion of the facility. Financial risks resulting from contravention of legislation include fines of up to €52,000 and 2–5 years imprisonment for those found liable. There were also safety risks in the form of direct injury to the workforce if they come into contact with the contaminated wastewater based on the toxicological information gleaned from the open literature as shown in table 13.1.

To assess the risk of the activity compared to the limits identified, an estimate was made on the concentration of each IHE-X constituent based on the observation of crystals forming in collected wastewater. This approach was necessitated by the difficulty in transporting explosive samples for analysis. Since RDX has the lowest solubility of the IHE-X constituents at 40 mg L^{-1} in water at 25 °C then this will be used as a minimum, with the ratio of other constituents used to calculate minimum values of the other materials. The results of this calculation are shown in table 13.2.

The ephemeral nature of the stream on the site would mean that there would be a concentration of IHE-X materials on the stream bed during the summer months with a bulk release of material when it dissolves in the summer. This has the potential to increase concentrations of NTO above those required for visual discolouration to occur, albeit for short periods until the flowing stream dilutes it.

There is potential to engage with the regulating authority to set discharge limits for each IHE-X material and obtain a permit to discharge the wastewater generated from the proposed activity. However, RMWI seek the best practicable environmental option (BEPO) for their operations. As such, the mitigation option proposed for this activity was the investigation of alternative treatment or disposal methods for explosive contaminated wastewater. The full decision framework can be seen in table 13.3.

Table 13.2. Estimated minimum concentration of each of the individual IHE-X materials found in solution in wastewater from the munitions manufacture facility, based on the presence of crystal solids present.

IHE-X material and percentage of composition	RDX (15%)	DNAN (32%)	NTO (53%)
Minimum concentration in collected wastewater (mg L^{-1})	40	85	141

Table 13.3. First decision framework on the discharge of explosive wastewater to local drains.

	Stage	Potential environmental impact	Business risks: Compliance	Reputational	Finance	Safety
1	**Operational risks arising from potential adverse environmental impact**	Contamination of local stream with wastewater containing IHE-X constituents.	Potential to breach EU and national legislation for water quality, permitting, soil and water contamination.	Visible discolouration of water due to presence of NTO (Tennant *et al* 2019).	Potential for fines and contributions to the cost of remediation activity from pollution.	Exposure to IHE-X wastewater.
2	**Questions arising from Risk Assessment**	Would discharge of wastewater at processing concentrations have a negative environmental impact on local receptors?	What are the regional regulations in place regarding the discharge of contaminated water? What are the discharge limits for IHE-X constituents?	Will there be noticeable discolouration of water after discharge without treatment? What are the concentrations of RDX, DNAN and NTO that cause discolouration of water?	What level of fine can be imposed if there was a release of explosive wastewater?	What is the risk to staff in carrying this out?
	Scientific Investigation Undertaken: Testing to ascertain the concentration and volume of explosive wastewater generated from a cleaning cycle. Desktop study to determine discharge limits for IHE-X components and levels of penalty that can be applied by national legislation.					
3	**Decisions based on results**	Discharge of explosive contaminated wastewater would pose a health hazard due to RDX	**Continue activity? No** There is a lack of local or national discharge limits for IHE-X materials DNAN and NTO.	**Continue activity? No** There is noticeable discolouration of water due to the presence of NTO at concentrations	**Continue activity? No** Breaches of Legislative decree no. 152.2006, the Environmental consolidated Act (ECA) include	**Continue Activity? Yes** Staff do not come into contact with IHE-X wastewater as discharge is immediate.

		exposure, a potential health hazard due to DNAN exposure and a visible nuisance to local residents in the form of discolouration from NTO.	The US EPA has a drinking water equivalent level (DWEL) of 0.1 mg L^{-1} and a lifetime health advisory limit of 0.002mg/L for RDX so wastewater must be treated to below that level. (United States Environmental Protection Agency 2018)	above 100 mg L^{-1} (Tennant *et al* 2019). Potential severe reputational impact. Expected concentrations of NTO in IHE-X wastewater are 1400 mg L^{-1} (Felt *et al* 2013).	sanctions of up to 52k EUR and 2–5 years imprisonment for the discharge of water containing dangerous substances as well as the cost of remediating environmental damage caused. A reliable disposal method is required that will minimise the likelihood of penalties being applied to the organisation.
4	**Mitigation option**	Explore options for disposal of explosive contaminated wastewater including onsite treatment and shipping offsite.	Seek consensus with local authority on discharge limit for NTO and DNAN Investigate treatment and disposal options for explosive wastewater.	Investigate treatment for explosive wastewater to reduce NTO concentrations below 100 mg/l for discharge or alternative disposal options.	Investigate treatment and disposal options for explosive wastewater.
				Immediate solution to collect and ship offsite for disposal by licensed contractor whilst other options are explored.	

13.4.2 Decision framework 2—contracted disposal of explosive wastewater off-site

Due to continuing production activity on the site, an interim solution was needed to dispose of wastewater being generated. It was identified that wastewater could be temporarily disposed of off-site using a contracted organisation for safe treatment and disposal until a more permanent solution was found. The potential environmental risks identified from the organisations initial feasibility study was the potential for a spillage or leak in storage or handling of the IHE-X contaminated wastewater whilst it was still on-site. Wastewater was therefore absorbed onto sawdust to reduce possibility of leaks, or large scale contamination from spillages (1.5 kg of sawdust adsorbing 8 l of contaminated wastewater for a total weight of approximately 10 kg) and this waste was double packaged in plastic bags to prevent leakage. These bags were then packaged in cardboard boxes, two bags per box, for ease of transport. For each washout procedure conducted, 14 boxes of waste (280 kg) are produced from the 220 L of wastewater. This allowed the transport of contaminated water offsite as it was now considered as solid waste and did not contravene the Agreement concerning the International Carriage of Dangerous Goods by Road (ADR) or the resultant EU Directive 95/50/EC (Council of the European Union 1995).

Disposal of the contaminated and packaged waste takes place off-site at a dedicated waste management facility that is equipped to deal with hazardous waste products. Transport to the disposal site takes place with other explosive contaminated waste produced onsite, however an estimate has been made of the shipping costs per box to allow a cost comparison to be conducted as seen in table 13.4.

Whilst this procedure would mitigate the environmental, safety, and reputational risks posed, the financial risk from the cost of disposal remained high. Although currently affordable for the company it runs contrary to the proximity principle in which waste is disposed of as close to source as possible (Okuda and Thomson 2007). As previously mentioned, there was a desire for a BPEO as well as recognising the exposure to cost rises in the contract and a limiting of expansion opportunities. Therefore, an on-site disposal of waste was desired, and it was proposed to investigate methods by which the explosive wastewater could be treated onsite before discharge, thereby mitigating these risks. The process of choosing and implementing an onsite treatment method would take time, and so IHE-X wastewater generated in the interim would either need to be disposed of offsite or stored onsite awaiting disposal, creating further risks. The full second iteration of the decision framework can be seen in table 13.5.

Table 13.4. Breakdown of disposal costs for wastewater off-site by registered contractor, estimated.

Transport step	Approximate cost (EUR)	Additional notes
Disposal at registered facility	26.25	Per box of waste
Transport cost	8.13	Best case assuming a full 16 tonne vehicle at a cost of 13000 EUR
Total per washout procedure	481.32	Assuming 14 boxes of waste produced per washout

Table 13.5. Second decision framework on the disposal of explosive wastewater by an external contractor offsite.

Potentially high impact process: disposal of IHE-X contaminated water by contracted removal offsite					
			Business risks		
Stage	Potential environmental impact	Compliance	Reputation	Finance	Safety
1 Operational risks arising from potential adverse environmental impact	Environmental impacts resulting from an accident during storage or handling of wastewater.	Transport of hazardous waste requires classification under UN Recommendations on the Transport of Dangerous Goods. As well as regional or national legislation such as the EU Directive 95/50/EC (1995) and its transposition into national laws.	Bad press. Could be local, national or international depending on severity of incident.	The cost of the contract to remove waste from the site can be expected to increase. In the absence of an alternative disposal method, the company will have to pay.	Exposure to contaminated wastewater from leaks.
2 Questions arising from Risk Assessment	Can IHE-X contaminated waste be safely disposed of by shipping offsite?	Can IHE-X contaminated waste be classified for transport and disposal offsite?	Can IHE-X contaminated waste be made safer to reduce the likelihood of unintended initiation?	What is the increase in cost if disposal is carried out by an external contractor?	Can IHE-X contaminated waste be made safer to transport by reducing likelihood of leaks.

(*Continued*)

Table 13.5. (*Continued*)

Potentially high impact process: disposal of IHE-X contaminated water by contracted removal offsite					
			Business risks		
Stage	Potential environmental impact	Compliance	Reputation	Finance	Safety
Scientific Investigation: Conduct a literature review for ways IHE-X contaminated wastewater can be made safe and classified for transport and disposal (Fawcett-Hirst *et al* 2021). Seek out hazardous waste disposal agencies to ascertain the cost of IHE-X waste disposal.					
3 Decisions based on results	Disposal offsite is feasible using suitable contracted authorities however the cost of doing so remains high.	**Continue activity?: Y** Contracted organisations are available who specialise in the removal of hazardous waste and would be able to categorise waste for transport to their own facility for disposal.	**Continue activity?: Y** Explosive wastewater can be made safer using sawdust and packaged in bags that would not confine burning to detonation. Reduce exposure to negative press.	**Continue activity? N** The cost for disposal offsite is high *at 481.32 EUR* and there is a desire to reduce this for a more long-term solution. Reduce risk.	**Continue activity? Y** Risk of leaks can be mitigated by absorbing wastewater on sawdust.
4 Mitigation option	Investigate onsite alternatives for explosive wastewater disposal.		Investigate onsite alternatives for explosive wastewater disposal.		

Table 13.6. Third decision framework iteration assessing the risks posed by treatment of explosive wastewater using activated carbon adsorption.

Stage	Potential environmental impact	Compliance	Business risks: Reputation	Business risks: Finance	Business risks: Safety
1 Operational risks arising from potential adverse environmental impact	Water and land contamination if the carbon adsorption system does not reduce concentration of IHE-X to environmentally safe levels	Permit required.	Visible discolouration of water due to presence of NTO (Tennant *et al* 2019)	Potential for fines and contributions to the cost of remediation activity from pollution	Exposure of personnel to IHE-X wastewater.
2 Questions arising from Risk Assessment	Does a carbon adsorption system reduce concentrations of IHE-X to a permissible level?	What are the discharge limits for IHE-X materials? How far can the concentration of IHE-X materials be reduced?	Will there be noticeable discolouration of discharged water after treatment?	How much activated carbon is required to reduce NTO concentration to agreed levels?	What are the risks to the operator(s) from this process?
Scientific Investigation: Adsorption study of NTO on to activated carbon to determine the quantity required to reduce concentration in wastewater to below detectable levels.					
3 Decisions based on results	Removal of IHE-X from solution in wastewater can be done using	**Continue process? No** Reduction of concentration to non-detect levels is possible using	**Continue process? Yes** Reduction of concentration to non-detect levels is possible using	**Continue process? Yes** Process data from the facility indicates that even if the activated carbon cannot be reused, the price per	**Continue process? Yes** Risks posed from manual handling and exposure to carbon can be

(*Continued*)

Table 13.6. (*Continued*)

Stage	Potential environmental impact	Business risks			
		Compliance	Reputation	Finance	Safety
	commercial activated carbons and at an affordable price to the business.	carbon adsorption on this scale (Fawcett-Hirst *et al* 2020).	carbon adsorption. Non-detect levels of NTO will not discolour water (Tennant *et al* 2019). Non-detect levels of RDX are not considered to pose a toxicological impact (United States Environmental Protection Agency 2018).	disposal would be 275–470 EUR depending on the activated carbon product chosen (Fawcett-Hirst *et al* 2020; Felt *et al* 2013). This price is considered affordable for the business compared to the cost of continued disposal offsite of 481.32 EUR.	covered by existing safety protocols.
4 Mitigation option	Permit required to discharge water. Continue monitoring wastewater discharged provide a record and ensure that agreed limits are not breached.				

13.4.3 Decision framework 3—on-site treatment of explosive wastewater using an adsorptive system

The main environmental risk identified in the third iteration of the decision framework closely mirrored that of the initial assessment. There was still an unknown about the ability of commercial carbon adsorption systems to reduce the concentration on IHE-X constituents, particularly NTO to below detectable levels in wastewater before discharge. The original issue of compliance with the Water Framework Directive (WFD) remained however the processing facility was now able to gather information to support seeking a permit. Likewise, the issues of potential financial and reputational loss from visual contamination and resulting fines were able to be answered. With these gaps in information identified, a series of experiments were commissioned.

In this work it was found that there was a gap in the literature for adsorption of NTO and DNAN on to activated carbons. Equilibrium and timed adsorption experiments were conducted using three commercial activated carbons in individual solutions of RDX, DNAN and NTO at a range of concentrations based on the expected concentration from the literature (Fawcett-Hirst *et al* 2020). The resulting data obtained on the adsorption capacity of each activated carbon for each IHE-X constituent revealed the amount of activated carbon that would be required to treat the wastewater generated in the scenario. Equilibrium experiments to determine the adsorption capacity of activated carbons for RDX, DNAN and NTO confirmed that while RDX and DNAN would be removed, the high concentration of NTO would saturate the carbon before a sufficient reduction in concentration could be achieved. Even initial concentration of 70mg L^{-1} saturated the carbons tested by three hours (Fawcett-Hirst *et al* 2020). Therefore, significantly higher mass of carbon would be required in order to treat the volume (~220 L) and concentration (1400 mg L^{-1}) of wastewater expected making the process potentially uneconomical. Additional testing in a column-based treatment method, as would be expected to be used at the facility, is required to determine the value. This risk was further expanded on in terms of the potential for negative perception of resource use and the cost of the activated carbon required by the process. As a result of the experimental work undertaken, and indicative costs of activated carbon materials provided by industry leading suppliers, it was assessed that disposal of explosive contaminated wastewater through carbon adsorption would allow discharge of treated water through normal municipal drainage channels and that the process would be economically viable for the manufacturing organisation compared to continued shipment of explosive wastewater offsite for disposal via contractor. The full results of this iteration of the decision framework can be seen in table 13.5.

As a result of the application of three iterations of the decision framework the manufacturer decided to transfer the wastewater offsite in the short term while further investigating the capital costs of introducing a batch adsorptive treatment capability onsite. The first iteration of the decision framework tool proved useful in providing a structure by which the risks posed to a munitions manufacturing organisation, highlighting gaps in current understanding as to the compliance,

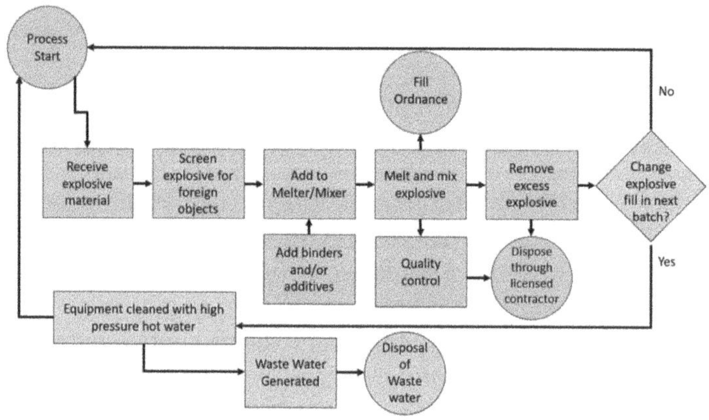

Figure 13.4. Revised flow diagram for the munitions manufacture facility with carbon adsorptive wastewater treatment prior to disposal.

reputational, financial and safety risks to the organisation. A literature review filled these knowledge gaps and led to the second and third iterations of the decision framework that allowed alternative options to be explored.

The second decision framework explored an interim solution for IHE-X wastewater generated to allow production to continue whilst a Best Practicable Environmental Option (BPEO) was explored further. In the third iteration of the decision framework, the application of experimental investigation was key to providing first-hand empirical data to inform the decision-making process and providing a record of how decisions were made. By integrating the disposal of waste into the facilities onsite there is greater control of the process and a reduced reliance on external contractors. In addition, the costs associated with disposal would benefit from economies of scale onsite instead of a linear increase if offsite disposal continued to be used. The resultant changes to the manufacturing process of the site used in this scenario are outlined in the revised flow diagram in figure 13.4 where the dashed lines have been removed to show the integration of the system.

13.5 Conclusions

The processing facility identified potential high impact environmental risks rising from a change in formulation for their munitions to use new IHE-X fills including the materials DNAN and NTO which had the potential to result in negative environmental impacts. A decision framework tool was used to frame these risks in terms of the compliance, reputational, financial and safety risks to the business. Sequential application of a decision framework led to mitigation options being assessed in a thorough and transparent way using empirical data from existing scientific literature through a critical review and by gathering new scientific data through adsorption experiments. Recommendations to the processing facility were to continue with offsite disposal of the wastewater as an immediate option and investment in an activated carbon adsorptive treatment system providing as a longer-term solution.

The use of the decision frameworks has been demonstrated to provide a powerful tool for evaluating activities that could pose an environmental risk. The resulting information forms a defensible and auditable basis for decision making that becomes an integral part of the wider environmental management case. Although applied to a munitions manufacture facility for this example, the process is not specific to explosives and can be used in the case of any business activity that could pose an environmental impact.

References

Ruff v. Ensign-Bickford Industries, INC., 171 F. Supp. 2d 1226, District Court, D. Utah, 1:99-cv-00120, 2001 https://www.courtlistener.com/opinion/2488411/ruff-v-ensign-bickford-industries-inc/

Balzarolo D, Lazzara P, Colonna P, Becciu G and Rana G 2011 The implementation of the Water Framework Directive in Italy *Dialogues on Mediterranean Water Challenges: Rational water use, water price versus 167 value and lessons learned from the European Water Framework Directive* (Bari: CIHEAM) pp 155–67

Council of the European Union 1995 Directive 95/50/EC. Off. J. Eur. Communities 249, 35–40

Davies P J and Provatas A 2006 Characterisation of 2,4-dinitroanisole: an ingredient for use in low sensitivity melt cast formulations (Edinburgh (Australia): Defence Science and Technology Organisation, Weapons Systems DIV).

Fawcett-Hirst W, Temple T J, Ladyman M K and Coulon F 2021 A review of treatment methods for insensitive high explosive contaminated wastewater *Heliyon* **7** e07438

Fawcett-Hirst W, Temple T J, Ladyman M K and Coulon F 2020 Adsorption behaviour of 1,3,5-trinitroperhydro-1,3,5-triazine, 2,4-dinitroanisole and 3-nitro-1,2,4-triazol-5-one on commercial activated carbons *Chemosphere* **255** 126848

Felt D, Johnson J L, Larson S, Hubbard B, Henry K, Nestler C and Ballard J H 2013 Evaluation of treatment technologies for wastewater from insensitive munitions production ERDC/EL TR-13-20, US Army Engineer Research and Development Center, Environmental Laboratory, ERDC/EL TR-13-20, Vicksburg

Fleming E C, Cerar R and Christenson K 1996 Removal of RDX, TNB, TNT and HMX from Cornhusker AAP waters using adsorption technologies Vicksburg

Kottek M, Grieser J, Beck C, Rudolf B and Rubel F 2006 World map of the Köppen–Geiger climate classification updated *Meteorol. Zeitschrift* **15** 259–63

Ladyman M K K, Temple T J J, Piperakis M M M, Fawcett-Hirst W, Gutierrez-Carazo E and Coulon F 2019 Decision framework for the environmental management of explosive contaminated land *Sci. Total Environ.* **690** 730–8

London J O and Smith D M 1985 *A Toxicological Study of NTO* (Los Alamos, NM: Los Alamos National Laboratories)

Okuda I and Thomson V E 2007 Regionalization of municipal solid waste management in Japan: balancing the proximity principle with economic efficiency *Environ. Manage.* **40** 12–9

Ramachandran G, Behnassi M and Tripathi G C 2021 The anatomy of uncertainty and implications for preventive environmental management *Social-Ecological Systems (SES)* (Cham: Springer International Publishing) pp 239–53 https://doi.org/10.1007/978-3-030–76247-6_13

Reddy G, Erexson G L, Cifone M A, Major M A and Leach G J 2005 Genotoxicity assessment of hexahydro-1,3,5-trinitro-1,3,5-triazine (RDX) *Int. J. Toxicol.* **24** 427–34

Reddy G, Williams M and Quinn M J 2020 *Wildlife Toxicity Assessment for 2,4-Dinitroanisole* (DNAN) (Houston, TX: US Army Environmental Command)

Taylor S, Dontsova K, Walsh M E M R and Walsh M E M R 2015 Outdoor dissolution of detonation residues of three insensitive munitions (IM) formulations *Chemosphere* **134** 250–6

Taylor S, Walsh M E, Becher J B, Ringelberg D B, Mannes P Z and Gribble G W 2017 Photodegradation of 2,4-dinitroanisole (DNAN): an emerging munitions compound *Chemosphere* **167** 193–203

Tennant M, Chew S C, Krämer T, Mai N, McAteer D and Pons J F 2019 Practical colorimetry of 3-nitro-1,2,4-triazol-5-one *Propellants Explos. Pyrotech.* **44** 198–202

United States Environmental Protection Agency 2018 *2018 Edition of the Drinking Water Standards and Health Advisories* (Washington, DC: Office of Water, U.S. Environmental Protection Agency)

US Department for Health and Human Services 2012 Toxicological Profile for RDX, ATSDR's Toxicological Profiles (Atlanta, GA: US Department of Health: Agency for Toxic Substances and Disease Registry Division of Toxicology and Human Health Sciences) https://doi.org/10.1201/9781420061888_ch137

Walsh M E 2016 Analytical methods for detonation residues of insensitive munitions *J. Energ. Mater.* **34** 76–91

Walsh M R, Walsh M E, Ramsey C A, Thiboutot S, Ampleman G, Diaz E and Zufelt J E 2014 Energetic residues from the detonation of IMX-104 insensitive munitions *Propellants Explos. Pyrotech.* **39** 243–50

Chapter 14

Risk recognition and continuous Risk Management

Z Rustom

Having a mature safety culture is a good thing, but only to a point. Regular safety conversations and posters serves a purpose until they become so familiar we fail to recognise the risk and hazard they are alerting us to. The system does not alter risk tolerance or aversion, it tends to make an organisational statement whereas the protection of individuals is contingent on the individual(s) recognising the hazard and acknowledging the consequence.

This chapter builds an assessment and continuous review model that recognises the experiential needs of andragogy and combats the safety fatigue that can impact an over delivered safety culture. The model builds on the experience of a Commercial Response Organisation (CRO) established to respond to incidents and events in hostile working environments. It accepts event similarity can lead to hazard complacency and a workforce's tendency to see the hazard and not the consequence. The model recognises common failings in workgroup Risk Management and avoids miscommunication by recognising stakeholders and concurrent events/risks. The model recognises that Risk Assessment significantly loses value if it is without Risk Management.

Stakeholder recognition and engagement allows the model to provide an understanding of the risk consequences and the mechanics that might allow occurrence; in this manner, the model enables proactive Risk Management.

14.1 Introduction

This chapter sets out to describe a Risk Assessment process for commercial application when responding to significant or complex incidents. It is built on a need to provide an evidenced and consistent approach to risk recognition and mitigation.

The perspective for this discussion is that of a commercial emergency response organisation and for undertaking Risk Assessment following a mobilisation to an

doi:10.1088/978-0-7503-3643-7ch14 © IOP Publishing Ltd 2022

event where something has already failed and normal systems may no longer be in place.

The evolution of the process is built on experience in responding to diverse major incidents across all risks and all hazards. It is a process established by professional responders to meet their operational needs and satisfy commercial needs by keeping everyone safe, avoiding litigation and ensuring invoices are not rejected. It was also important as the process evolved that it could be undertaken without being an unworkable distraction or burden for the people responding.

There has to be recognition that the Risk Assessment process is just one of many tools and systems used when responding to incidents, any system needs to be compatible with the overall management system either worldwide (IPIECA 2014), or specific to the United States of America (FEMA 2004). This ensures the Risk Assessment process has value and adoption without resistance. Being able to integrate with recognised management systems makes it easier to defend the process from legal and financial scrutiny.

The system comes together as a two part, multiple element, model to give a Risk Assessment process and an ongoing management process allowing the assessment and mitigation measures to be confidently updated. The elements it uses are all standard industry practices and it leans on international standards (ISO 2018), to allow it to easily be adopted as a standard operating procedure.

The origins of the process are in significant incident events, this is because major crisis and disaster events give an ability to experiment and develop systems. The incident command applied to significant incidents brings together a large number of people and allows for an exchange of ideas and the development of efficient and effective systems. The model and process discussed remains applicable for Risk Assessment and continuous Risk Management for reactive missions in response to events of all magnitudes.

14.2 Discussion

Preparation and delivery of Risk Assessment satisfy a variety of needs. In the simplest outlook, the completion of assessment forms discharge obligations under health and safety legislation and governance for employee's rights. However, an assessment without interpretation, communication and regular review has a very limited value. A Risk Assessment needs to be made to work and not just exist as a tick-box activity.

To give a Risk Assessment a beneficial purpose it has to be appropriate to and useable by all people impacted by the events associated with the assessment. The UK Health and Safety Executive (HSE) states clearly 'a Risk Assessment is not about creating huge amounts of paperwork, but rather about identifying sensible measures to control the risks in your workplace' (HSE 2014).

The context of this assessment and discussion is of a commercial emergency response organisation. It is delivered through the window of a commercial entity that regularly sends response personnel to unknown and known hazardous situations. It is the ambition of every member of the leadership to ensure every employee returns home in the same or better condition than they came to work. This can only be achieved where Risk Management is a natural element of operational management.

Industry has not always had this outlook, to get to where we are now there has been a dark past. Workplace harm and mortality was common across all industries as recent as the Victorian era; the Lancet medical journal suggested 145 000 people died in Victorian workhouses each year (Wakley 1841). Thomas Wakely as editor of the Lancet wrote that workhouses were 'antechambers of the grave'. While industry has come a long way in risk and hazard management, it does no harm to look back and learn from pioneering changes in behaviour and standards.

Despite the great engineering advances of the Victorian era, occupational health was completely disconnected from occupational safety. David Eves (2014), indicates knowledge of work-related diseases was incredibly limited and the first record of a medical practitioner becoming involved in work-related disease was in 1830. This started an understanding of 'industrial diseases' such as lead poisoning in the decorating industry and respiratory failure through dust inhalation in factories (Thackrah 1831).

We now find safety culture and risk aversion is a core topic at Leadership levels. While Wakley evidences this was not always the case; we can, in the United Kingdom, see legislative triggers that have taken Risk Management to the Board Room. Preceding Wakley's statement was the ratification of the Factories Act in 1833, an Act designed to prevent worker injury. This was then enhanced by the Mines Act in 1842 and the Quarries Act in 1894, workplace Risk Management started to emerge.

The Acts of the 1800s triggered a change in the conscience of industry, but it was a slow change and it was not until 1974 when the Health and Safety at Work Act came into being. The 1974 Act was progressive in that it sought consultation and involvement rather than being prescriptive (HSE 2021). Writing with a commercial perspective this is a good situation because it allows challenge and justification for actions, but it also allowed an acceptable degree of slow or partial adoption.

For risk recognition and Risk Management to be a mainstream element of operations took time and the application of many lessons learned from incidents and events. The journey from the 1800s to present time has seen numerous events that have served to force industry to change its ways. When researching aviation safety, Andrew Weir shares the words of an anonymous US Aviation Official as saying, 'we regulate by counting tombstones', indicating safety changes don't get made until enough people have been killed to warrant them (Weir 1999).

After the Cross-Channel Ferry 'Herald of Free Enterprise' capsized off Zeebrugge in 1987 with the loss of 193 lives, the operating company was prosecuted for corporate manslaughter. The prosecution failed and a long campaign to strengthen the law began, resulting in the passing of the Corporate Manslaughter and Corporate Homicide Act in 2007. Before this a company's liability for manslaughter depended on the conduct of a senior person, which was usually difficult to prove. For example, between 1992 and 2004 there were only 34 prosecutions for work-related manslaughter and these resulted in just six convictions.

The new Act came into force in 2008, enabling companies to be convicted for corporate manslaughter if found guilty of serious management failures resulting in a gross breach of a duty of care and, if so, imposition of unlimited fines by the courts. Government bodies could also face an unlimited fine for the first time if found to have caused death due to corporate health and safety failures. With so much now at stake, the documented Risk Assessment and evidence for risk reduction gained prominence at all levels of management.

To demonstrate a safe workplace and an employment environment where the management demonstrates it is protecting the workforce an archived trail of completed Risk Assessment is a benefit. To achieve this, the assessment process has to be a comfortable procedure for the user.

During the early years of the century the Labour Government considered making the provision of information about health and safety performance a statutory requirement under the Companies Act but the Chancellor Gordon Brown dropped the proposal in the face of business resistance. However, some larger companies wishing to protect their brands and reputations began publishing health and safety information in their annual reports voluntarily, adopting the principles of 'Corporate Social Responsibility'. This was essentially a 'third way' between the free market and state regulation, enabling companies to manage risk and reputation and demonstrate their sound stewardship of matters important to the public such as protection of the environment and health and safety (Eves 2014).

There are many industries that believe they have learned from past events to build an inclusive and strong safety culture. Cataclysmic events such as the 1986 Chernobyl reactor meltdown, the 1988 Piper Alpha explosions or the 2017 Grenfell Tower fire give rise to global reactions across associated industry groups. This builds on the corporate social responsibility proposed by Eves, and invites public displays of compliance and excellence. The mantra from BP is 'no accidents, no harm to people and no damage to the environment' (Coleman 2002).

Employees of organisations that strive to demonstrate a mature safety culture can become subject to information overload and the most important messages can be lost in the surrounding cacophony. This concept is not new, over 100 years ago the German sociologist Georg Simmel suggested the city dwellers of the time were unable to react to new situations as the City was host to an overload of sensations and information (Simmel 1950). In this manner where industry has a powerful safety culture it is too easy to be overloaded with risk, hazard and safety messaging.

When preparing and delivering a Risk Assessment it is vital that the key information and mitigations are communicated to the people who need it, in a useable and understandable manner.

A well-known Risk Assessment and mitigation programme instigated for children in the UK was the Green Cross Code (NRSC 1970), established to promote safe road crossing. The code has been promoted in different ways since its launch, but the basic principles remain the same Stop, Look, Listen, Think, Live (RoSPA 2019). The Green Cross Code serves as a simple reminder that any Risk Assessment and its reaction needs to communicate a clear understanding of the nature of the risk and the potential impacts.

In its current form, the Green Cross Code is a six stage code, as follows (RoSPA 2019):

1. THINK! First find the safest place to cross
2. STOP! Stand on the pavement near the kerb
3. USE YOUR EYES AND EARS! Look all around for traffic and listen
4. WAIT UNTIL IT IS SAFE TO CROSS! If traffic is coming, let it pass
5. LOOK AND LISTEN! When it is safe, go straight across the road—do not run
6. ARRIVE ALIVE! Keep looking and listening while you cross.

While the code is a dynamic operation and it does not provide any retained evidence as with an occupational Risk Assessment, it is a useful reminder of the clarity and procedures that make for a good Risk Assessment.

The code allows for context and the variables of time, place and circumstances. It also makes it implicit as to when the operative should proceed rather than interpretation of a scored value of Risk = Severity × Likelihood (CHAS 2022).

Having highlighted the above evolution in the risk awareness landscape and further recognised the context of this discussion is from the point of a commercial emergency response organisation, it is vital that we recognise the start point for Risk Assessment is somewhat uncomfortable. The reason most people will call a commercial emergency response organisation is because something went wrong, and they were not prepared for it. It becomes essential the Risk Assessment and management process is robust and reliable.

To ensure simplicity and easy export, the model starts by considering ISO3100:2018 Risk Management Guidelines as the primary platform (ISO 2018). The Risk Management Guidelines are then supported by ISO22320:2011 Emergency Management (ISO 2011). These give us a recognised foundation for the assessment model and the use of a common vocabulary (figure 14.1).

With the high consequences a commercial emergency response organisation is exposed to, there has to be a standard operating process in place. To ensure ample transparency for third party scrutiny the assessment model and proforma documents

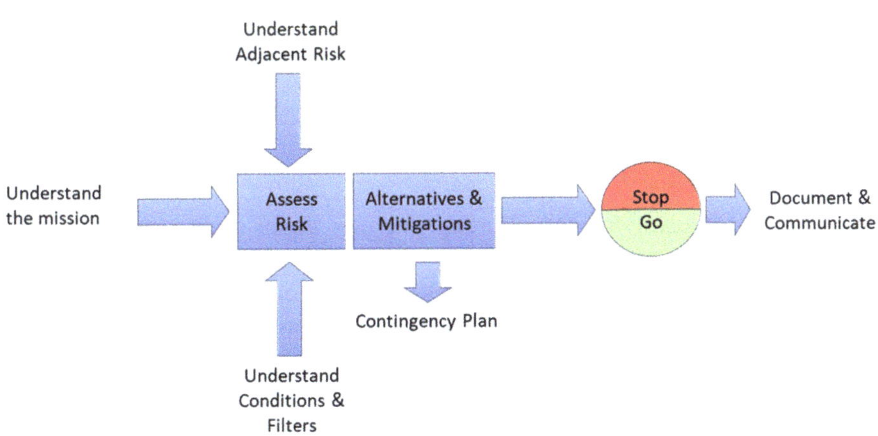

Figure 14.1. Risk Assessment process.

ensure the same process is used every time whilst being mindful of interpretation. At the heart is the proven and defendable Risk = Severity × Likelihood equation, despite comments to the contrary expressed by Gherardi and Turner (1987) as they suggest few people know how to use numbers correctly, also Functowicz and Ravetz (1990) who suggest numerical information can seriously mislead. To strengthen the quantitative assessment, this is augmented by addressing causal factors and the activities of other stakeholders. Looking at adjacent risk and changing local conditions allows us to learn from NASA's Risk-Informed Decision Making (RIDM) process, (NASA 2011).

Understand the mission

It is essential to challenge the first impression and ask why we are here. Understanding or recognising causal factors is a major first step. Commercial emergency response organisations are normally called when things go wrong, or protective measures fail. While this could be an easy fix to removing risk the process asks to check again. Recognising the Swiss Cheese model identifies multiple layers of failure (Reason 1990), including latent conditions, active failures and limited windows of opportunity.

Understand adjacent risk

Recognition of adjacent risk is not enough, there needs to be an understanding of the near (or near in time) activities, substances and decisions that may have an impact on the delivery of the Risk Assessment.

Adjacent risk or activity can also be a draw on the resources originally believed to be part of a successful response.

It becomes important to recognise all stakeholders, their levels of involvement and their expectations.

Understand conditions and filters

NASA's RIDM process recognises the importance of conditions and filters as part of the Risk Assessment and Risk Management process. Situations can change or fail due to uninfluenceable conditions, these include but are not limited to:

- Funding Environment
- Operating Environment
- Limited Data
- Technology developments
- Design Capacity
- Test & Production Processes.

An additional consideration while leading into the initial assessment of risk is the workgroup. Hubris can create failures from within the workgroup, even after recognising and working on mitigations. In his 2006 presentation to the Royal Aeronautical Society, James Reason suggested human error may be propagated by 'forgetting to be afraid' (Reason 2006). This was echoed in the independent safety review following the 2005 Texas City Refinery explosion, saying 'Workers and supervisors can increasingly rely on how things were done before, rather than rely on sound engineering principles and other controls. People can forget to be afraid!' (Baker *et al* 2007).

Once the risk environment is [fully] understood the risks to the mission can be assessed.

Our first process to understand the mission and recognise filters will identify or expose many of the risks, nonetheless, knowledge of the mission and techniques is important to give the assessor a credible eye and competency. Over familiarity is a problem that should be avoided as it leads to risk normalisation, making us partially blind to some elements of risk.

> Risk normalisation is what we do when we ride an elevator or fly in an airplane without worrying about the risk of falling. Over time, we become so accustomed and comfortable with such activities that we end up with the wrong feeling of being safe, without taking into account the risks we are taking (Ciola 2018).

Going back to where this discussion started, much of what is done is for legislative compliance and to demonstrate good corporate governance, the Risk Assessment needs to form part of a chain of evidence and demonstrate scoring parity when compared with other missions.

Use of a matrix allows consistency when making an assessment and it quantifies the information which makes it easier to assess change over time. The X-Y values on the matrix are open to variation although consistency is always recommended. The energy industry will tend towards a 5 × 5 matrix, as below (figure 14.2).

The basic 5 × 5 matrix can be adapted to meet specific needs or provide better detail to describe the decision-making process. This can be done by applying detailed knowledge of the operational landscape and the mission activities.

The example below (figure 14.3) was prepared for a specific energy exploration programme and utilises priorities of People, Environment, Assets and Reputation (PEAR) to allow the Risk Assessment to reflect on corporate values. The use of terms such as 'Extreme, High, Medium and Low' are presented to be consistent with

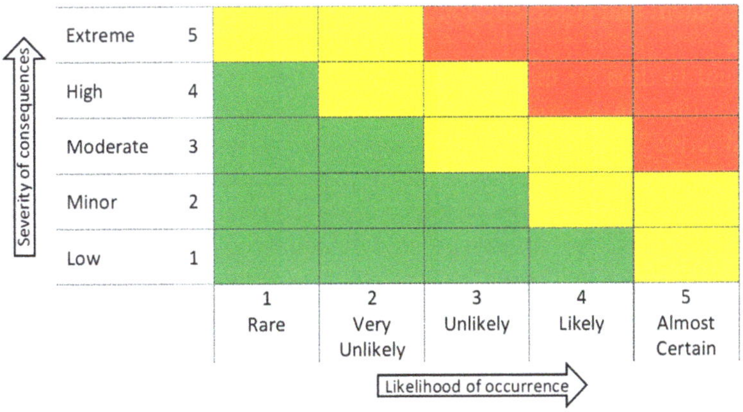

Figure 14.2. 5 × 5 risk matrix

	Potential Severity				Increasing Probability				
	People	Environment	Assets	Reputation	A Happened more than 3 times in the Organisation	B Happened 2 or 3 times in Organisation	C Happened in the Organisation	D Known in Industry	E Unknown but possible in Industry
5	Multiple Fatality	Massive Effect	Extensive Damage >$50M	International Impact	Extreme	Extreme	High	Medium	Medium
4	Single Fatality	Major Effect	Major Damage $10M - $50M	National Impact	Extreme	High	Medium	Medium	Low
3	Serious Injury	Localised Effect	Local Damage $1M - $10M	Industry Impact	High	Medium	Medium	Low	Low
2	Minor Injury	Minor Effect	Minor Damage <$1M	Local Impact	Medium	Medium	Low	Low	Low
1	Slight Injury	Slight Effect	Slight Damage	Slight Impact	Medium	Low	Low	Low	Low
0	No Injury	Zero Effect	Zero Damage	No Impact	Low	Low	Low	Low	Low

Figure 14.3. Example PEAR risk matrix.

the organisation's contingency plans, there is no specific lexicon required to populate the grid.

Generally, we do not do Risk Assessments to assess what the risks are, we do them to find and communicate mitigations to risks. This is what gives the process a purpose, it is to make the mission safer and share with the operative the control measures required to maintain a safer operation.

If all we communicate is the mitigation, we desensitise ourselves from the risk, by taking away the understanding of the risk we can actually put ourselves at greater danger. Looking back at the Green Cross Code (RoSPA 2019), we have a sequence of risk mitigations and barriers to harm but without the road safety campaigns we can lose knowledge of what and how severe the danger is.

We have to be smart in delivery of our assessment and avoid letting the process of Risk Assessment and risk reduction prevent important information being lost to the workforce.

The point in time we make a Risk Assessment could be considered as pre-crisis and our mitigation is crisis prevention. The road with moving vehicles would become a crisis situation if the person crossing was to walk into the traffic. Application of the Green Cross Code creates process and barriers to prevent the crisis occurring and allow safe passage across the road.

Barry Turner (1976) defines an understanding of the pre-crisis period (and disaster incubation) in his six stage 'systems failure and cultural readjustment model'. Turner (1976) identifies in Stage 2 of his model that events and occurrences need to be identified, understood and communicated in a meaningful manner. It is this notion of identification, understanding and communication that allows crisis events to be prevented. Fink (1986, p 22) comments, 'If you look out into the distance, and you see a storm coming, you can batten down the hatches'.

From Fink's statement we can overlay a set of conditions that are required to echo the needs of Turner's model and adequately prevent a crisis event:

- We have to be able to identify that we are looking at the correct storm, and that the incoming storm is the potential disaster element, host or trigger. In the case of the Green Cross Code, is the moving traffic the risk that can create a crisis event or is it something else, such as a missing drain cover, or a combination of several?
- We have to know in advance of the storm that the appropriate action to mitigate its effect is to batten down the hatches.
- We have to know that our reaction to the storm (i.e. the battening down of hatches) is not going to propagate an alternative crisis event. In the case of the Green Cross Code we have a single sequence, in professional Risk Management we should be able to reconsider the location and cross elsewhere if our first location is not as safe as originally thought. For this purpose we introduced 'contingency plan' to our Risk Assessment process (figure 14.1).
- We must ensure the rest of our crew know the situation and the remedial actions taken.

Even with professional training and prior experience proposed mitigations should be reviewed regularly to ensure they remain effective. It must be recognised that many factors could change, for example lighting, fatigue, weather.

NASA (2011) defines proactive Risk Management as a combination of Risk-Informed Decision Making and Continuous Risk Management. Although not like for like, our Risk Assessment process, including the processes for understanding the mission and allowances for third party activities and filters, covers the needs of NASA's Risk-Informed Decision Making process. The Continuous Risk Management process adopted by NASA fits nicely with the standards and guides of the International Organization for Standards (ISO 2011, 2015), starting with five cyclical functions of Identify, Analyse, Plan, Track, and Control. NASA makes strong mention of two central elements to the Continuous Risk Management process as document and communicate, these hold great weight for commercial organisations as compliance with legislation and good corporate governance should be evidenced.

Having communication embedded in the process ensures value to the mission and its participants. It ensures the content of the assessment and mitigation plan is shared with all who need to know it.

There is no benefit in communicating if the information is not being received and processed. The goal is to keep people safe and prevent consequential damage, delivery of the assessment information should be part of the process design (figure 14.4).

Malcolm Knowles (1984) presented four principles that are key to andragogy. These are involvement of adult learners, adult learners' experience, relevance/impact to the learner and problem centric learning (Knowles 1984). Using this strategy to deliver the Risk Assessment and mitigations to an adult workforce greatly assists in the information being well received and used.

Figure 14.4. Risk Management cycle

Involvement

The adult recipient will need to be involved in the learning process, including the planning and evaluation of the event. Delivery of the Risk Assessment and mitigations should be done so the recipient is part of the briefing process and not just an observer, the recipient needs to be able to recognise and understand their part in the process.

Learner's experience

The adult recipient will take in fresh information easier if it relates to previous life experiences. The briefing should refer to or reflect previous events and activities that allow the recipients to use their own experiences.

Relevance/impact

The adult recipient will need to know there is a purpose and some self-development in the information being passed. This should be easy for the Risk Assessment briefing, but it is easy to overlook that the information is for the benefit of all recipients. Encouragement to challenge and apply the briefing information should be built into the process, it is not acceptable to simply make the information available.

Problem centric

The adult recipient will want to problem solve and explore options rather than receive raw data that does not have direct relevance to the activity in hand.

Communication needs to go beyond the responding people, it is important that all stakeholders are suitably engaged. In our initial process of understanding adjacent risk we identified and recognised all stakeholders; throughout the mission it is key to success to ensure the stakeholder list is continually reviewed and updated (figure 14.5).

The communications model hosts its own internal communications and then upward, downward and outward communications (IMO 1989). When reviewing stakeholders it should be noted that different groups will need different levels of

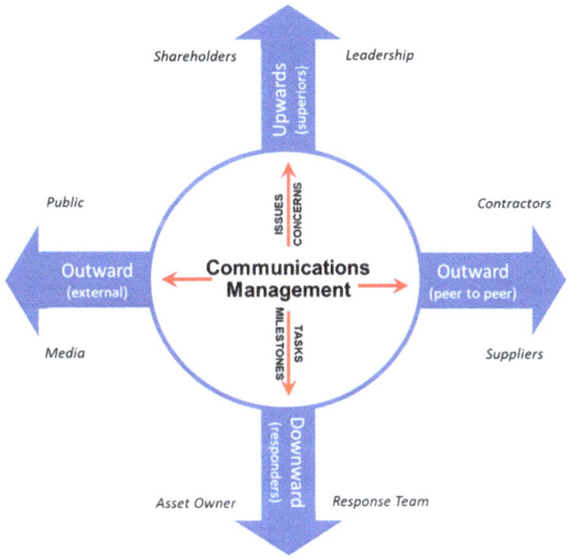

Figure 14.5. Communications model (source: reproduced with the permission of the International Maritime Organization (IMO)).

information, often passed in a multitude of different ways (Coombs 1995). Nonetheless, the information issued must be consistent. For significant events, this element of Risk Management may require a dedicated team to maintain communications and relationships with stakeholders as part of the process for maintaining risk mitigation processes.

Outward, external communications of risk and mitigations becomes more important when a newsworthy event takes place. Keeping people away from the location or stimulating sympathetic support from the public can be done by communicating the Risk Assessment and mitigations in preference to simply reporting a news story. Proactive Risk Communications following a windfarm incident generated a very favourable response from the public and demonstrated the impeccable social governance of the host company (Vattenfall 2021). The safety and environmental messages were passed clearly without naming the responsible contractor.

Information passed through the media and to the public must be robust and open to challenge, reputational damage can be done if the messaging and support data is incorrect or misleading, (Deacon 1999). The communications management element of the process must have an approvals sequence to ensure the right information is shared and inconclusive information as part of the decision-making process should not be released (figures 14.6 and 14.7).

Bringing the two systems together, Risk Assessment and mitigation with NASA's RIDM Risk Management cycle, gives a complete and resilient continuous Risk Management programme.

Each of the elements on either system should be regarded as an evidence gathering and decision-making gateway.

Figure 14.6. Public statement (Credit: Vattenfall 2021).

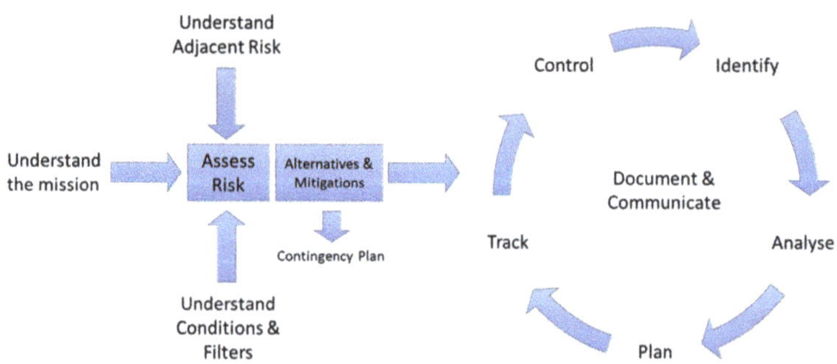

Figure 14.7. Full continuous Risk Management process.

14.3 Conclusion

Risk assessments work and they have the potential to save lives, but to do this they need to be used correctly, their findings and deliberations need to be recorded and communicated clearly.

When a workforce deploys to undertake a mission, it is an easy route to use a Risk Assessment form as an aide memoire for a 'toolbox talk' type of briefing. This gives an absence of evidence and a lack of consistency.

In most cases Risk Assessments are simple and can be completed by applying elementary observations and some on-the-job experience, but that could be a massive trap if things go wrong. Robust and defendable evidence is necessary. We have seen through the evolution of legislation how the Board Room is becoming fully accountable for the occupational health, safety and wellbeing of people involved in company activities.

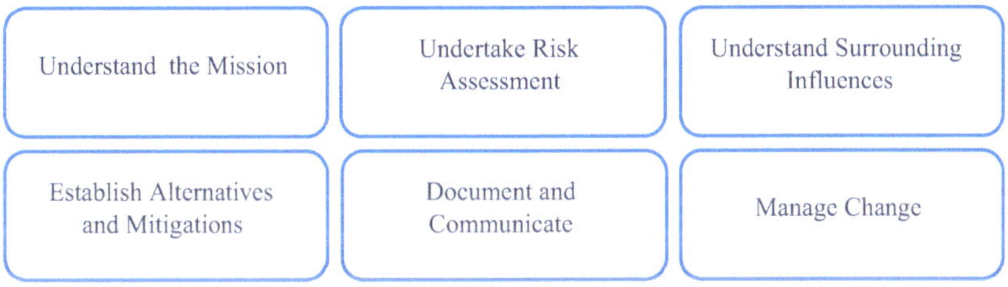

Figure 14.8. System elements.

Workplace harm to people was steadily brought under control by establishment of the UK Health and Safety Executive. To meet this reform, processes had to emerge that allow for sound governance and workplace protection.

To make the process work for all, satisfying the needs of all stakeholders, a modular working system emerged. Most importantly, the system works in assessing risk, agreeing mitigations, maintaining evidence and monitoring change. The six elements of the system can then be identified and revisited regularly throughout the mission (figure 14.8).

At the heart of the system remains the equation Risk = Severity x Likelihood. Where resilience is added to the system is that an intimate understanding of the mission and an engagement with surrounding filters forms the start point.

Built into the system is a perpetual cycle to assess success and change giving a process of continuous Risk Management.

Documentation and communication ensures the system builds a collection of evidence throughout the mission and at closure ensuring openness for scrutiny at all stages and ensuring the information gets to where it is needed.

References

Baker *et al* 2007 *The Report of the BP U.S. Refineries Independent Safety Review Panel* (Houston, TX: BP U.S. Refineries Independent Safety Review Panel)

CHAS 2022 *Contractors Health and Safety Assessment Scheme* London https://chas.co.uk/help-advice/risk-management-compliance/risk-safety-statement-types/

Ciola G 2018 Shell Research Centre for Greenhouse Gas Innovation (RCGI) Safety Day Presentation (10th May 2018) at University of São Paulo (Poli-USP), São Paulo

Coleman G 2002 *Getting HSE Right, A Guide for BP Managers* (London: BP plc)

Coombs W T 1995 The development of guidelines for the selection of the 'appropriate' crisis response strategies *Manag. Commun. Quart.* **4** 447–76

Deacon J and Deacon B 1999 *Perception Is Reality, Managing the Media when Emergencies Strike* (Bedford: JB Media Management)

Eves D 2014 Two steps forward, one step back *A Brief History of the Origins, Development and Implementation of Health and Safety Law in the United Kingdom, 1802–2014* (London: Institute of Risk and Safety Management)

FEMA 2004 *National Incident Management System and the Incident Command System*

Fink S 1986 *Crisis Management: Planning for the Inevitable* (New York: AMACOM)

Functowicz S O and Ravetz J R 1990 *Uncertainty and Quality in Science for Policy* (Dordrecht: Kluwer)

Gherardi S and Turner B A 1987 Real men don't collect soft data. Quaderno 13, D. di Politica Sociale, Universita di Trento

HSE 2014 *Risk Assessment a Brief Guide to Controlling Risks in the Workplace* www.hse.gov.uk/pubns/indg163.htm

HSE 2021 The History of HSE https://hse.gov.uk/aboutus/timeline/index.htm

IMO 1989 *Response to Marine Oil Spills, IMO Level 2 Model Course* (London: Int. Maritime Organisation)

IPIECA 2014 *Incident Management System for the Oil and Gas Industry* (London: Int. Petroleum Industry Environmental Conservation Association)

ISO 2011 *ISO 22320:2011 Emergency Management* (Geneva: Int. Organization for Standards)

ISO 2015 *ISO 9001:2015 Quality Management Systems* (Geneva: Int. Organization for Standards)

ISO 2018 *ISO 31000:2018 Risk Management—Guidelines* (Geneva: Int. Organization for Standards)

Knowles M 1984 *The Adult Learner: A Neglected Species* 3rd edn (Houston, TX: Gulf Publishing)

NASA 2011 *NASA Risk Management Handbook* (Washington, DC: US National Aeronautics and Space Administration, NASA Headquarters)

NRSC (National Road Safety Committee) 1970 *The Green Cross Code* (London: HM Government, Central Office of Information)

Reason J T 1990 *Human Error* (Cambridge: Cambridge University Press)

Reason J 2006 Human factors risk culture *A Presentation at Management of Human Factors Risk in Safety-Critical Industries at the Royal Aeronautical Society*, 11th May 2006

RoSPA 2019 *The Royal Society for the Prevention of Accidents* Birmingham https://rospa.com/rospaweb/docs/campaigns-fundraising/family-safety-week/2019/pedestrian-training.pdf

Simmel G 1950 *The Sociology of Georg Simmel* (Glencoe, IL: Free Press)

Thackrah D C 1831 *The Effects of the Principal Arts, Trades and Professions* (Leeds: Porter)

Turner B A 1976 The organisational and interorganisational development of disasters *Adm. Sci. Quart.* **21** 378–97

Vattenfall 2021 Ormonde Windfarm, Environmental Notice. Posted on Facebook, 24th October 2021

Wakley T 1841 *Lancet* 193–6

Weir A 1999 *The Tombstone Imperative: The Truth about Air Safety* (London: Simon and Schuster UK Ltd)

Chapter 15

Conclusion: challenges in Risk Analysis

M Ladyman and T Temple

15.1 Introduction

From the chapters presented in this book, it is evident that risk practitioners across multiple industries and varying risk themes agree there are significant outstanding challenges in Risk Management. This is not surprising, nor the first time that this has been highlighted [1]. In fact, since the formalisation of Risk Analysis as a discipline, practitioners and researchers have recognised the need for a holistic framework and attempted to construct glossaries of terminology [2]. As recently as 2021 the British Standards Institution (BSI) published a list of terminology, although the list included notes where terminology differed or had different uses [3]. Therefore, the aim of this book was not so grand as to assume that it would be possible to produce a single holistic solution, or a definitive list of terminology. Rather, the aim was to gather diverse and multi-disciplinary examples of Risk Analysis and highlight common challenges and identify diversity in language that may serve as the first step in developing a holistic approach.

15.1.1 Challenge: a holistic approach

Holistic is a term that appears in several chapters (chapters 4, 8 and 10), and is implied in several others (chapters 3, 6 and 12) and tends to arise from the need for interconnectedness between different types of risk i.e. safety, environmental, legislative, social, financial etc… The need for interconnectedness arises as these compartmentalised risks may contribute to the overall operational or business risk of the organisation. Indeed, it is rare that a risk in one area does not influence risks in other areas of a business. For example, realisation of an environmental risk (such as a pollution incident) may result in fines that present an economic or reputational risk to the business or organisation as was discussed in chapter 1 through the example of the Deepwater Horizon oil spill [4]. In recent years, standardised Risk Management frameworks such as ISO14001: 2015 (Standard for Environmental Management Systems) have been modified to include evaluation of the consequences of

doi:10.1088/978-0-7503-3643-7ch15

© IOP Publishing Ltd 2022

environmental risk on organisational reputation, compliance and finances etc… demonstrating a growing understanding of the need for more holistic approaches [5]. The **systems** approach presented in chapter 4 takes this a step further by suggesting the need for a holistic systems-based approach to safety risk in the design phase to minimise risks through life. However, this definition of holistic is from the point of view of safety risk, much like the ISO14001 definition is from the environmental point of view. As such, these holistic approaches suggest a 'bottom-up' approach i.e. escalating domain specific risks to business level risks.

An alternative way of looking at the concept of 'all-inclusivity' in Risk Assessment is through sustainability. Chapter 10 explored the need for a sustainable approach to Risk Assessment which takes equal consideration of the three pillars: economic, social and environmental, and can be extended throughout a product lifecycle e.g. planting to waste management in the Acai tree lifecyle. However, since its first modern definition in the Brundtland Commission report in 1987 [6], the meaning of sustainability has been adapted for use in different industries and more recently has been conflated with the concept of resilience. In addition, in recent years sustainability has increasingly been interpreted within the confines of specific groups or organisations that counteracts the principles underpinning the concept of sustainability [7]. Therefore, while a search for sustainability 'Risk Assessment' returns over 600 000 results, suggesting a global desire for methods to assess sustainability, it is difficult to imagine how this can be achieved without a harmonised definition of what 'sustainability' means. It is also increasingly common to see isolation of the 'environmental' or 'social' pillar for assessment as it is difficult for practitioners to become suitably qualified in all themes of sustainability and there is increasing reliance on internal and external collaboration and communication.

15.1.2 Challenge: subjectivity of language

With the recognition that an organisation needs a holistic overview or a sustainable approach to Risk Assessment, efficient communication between colleagues, organisations and stakeholders becomes essential. However, it is evident from a review of the chapters included in this book, that different disciplines use different terms for similar concepts. This means that while strict definitions exist within the standards and regulation for specific types of risk (safety, environmental, financial etc…), the same words may have alternative meanings, or nuances in literature for another type of risk. For example, while the concept of 'as low as reasonably practicable' (ALARP) in safety risk and 'best practicable environmental option' (BPEO) in environmental risk may seem comparable, ALARP is a legal mandate, whereas BPEO was developed as a tool for best practice in waste management [8, 9]. Table 15.1 groups terms from the different chapters in this book with similar conceptual meanings. Though these terms have strict definitions within the different types of Risk Assessment as agreed through standardisation and professional societies, their conceptual meanings are similar. However, when communicating between disciplines the similarity in their conceptual meanings is irrelevant if the words have different definitions in the two communities. Even more challenging is

Table 15.1. Terms collated from chapters within this book that are used to refer to similar concepts within different types of Risk Assessment i.e. safety, environmental, sustainability, operational, etc.

Receptor	Impact	Aspect	Mitigation	Prioritisation	Frequency
Exposure	Consequence	Vulnerability	Control	Analysis	Probability
	Severity	Hazard	Protection	Assessment	Likelihood
	Effect	Cause	Risk response	Evaluation	
		Failure mechanism			
		Peril			

the communication of risk beyond the organisation and to the public where the technical meaning of these words can superseded by standard usage e.g. the word hazard may be more familiar to people as a term used to mean 'danger' or 'warning' rather than 'environmental stimuli that causes an explosive event' (chapter 3), 'potential occurrence of a natural or human-induced physical event …' (chapter 10) or 'negative action/ event that may cause potential harm' (chapter 12).

In addition, to similar concepts being referred to by different terminology, the chapters have also highlighted the subjectivity of language used in Risk Assessment. Risk Analysis whether it encompasses purely the assessment phase or the entire process of Risk Management is rarely an entirely objective process. Even when data is employed to support calculation of risk it is rarely entirely objective due to the decisions and assumptions made during the investigative or experimental process. Chapters 3, 6, 7, 8 and 11 discuss the need for accurate and representative data and outline some of the necessary assumptions made during the investigative process. The subjectivity of data should not be seen as 'negative' or 'positive' as it is impossible to imagine a Risk Assessment process entirely freed of subjectivity as most require human judgement at some point. However, these chapters highlight the importance of acknowledging the subjectivity of data and language used in the Risk Assessment process and to not be lured in by the objective appearance of structured Risk Assessment methodologies and frameworks. Chapter 2 highlights how even the idea of 'tolerability', which is fundamental to the ALARP concept, is subjectively determined and differs depending on whether the harm effects an employee or the public.

Other subjectively defined terms are used throughout the chapters such as the concepts of 'appropriate', 'suitable', 'efficient', 'sufficient', 'confidence' and 'justifiable'. These terms have dictionary definitions, but mean something very different in the minds of individuals. As a recent example, during the COVID-19 pandemic the UK government justified the implementation of strict regulatory restriction on freedoms to limit the spread of disease as they deemed predicted deaths of over 500 000 people to be unacceptable [10]. The resulting demonstrations suggest that some people felt these measures were not justified [11], highlighting how different people respond to the same terminology. Of course, many more factors influenced public opinion and response during the pandemic than simply their approach to risk. In some risk approaches, such as in environmental risk, words such as 'sufficient' have been defined by setting regulatory limits such as for the concentration of pollutants in water. However, chapter 7 demonstrates that even when such data is available, the limits are themselves on risk-based estimates and are subject to the same issues of subjectivity as the experimental data discussed in chapter 6.

It is not only the terminology that is important. Chapters 1 and 14 highlighted that the form and style of communication is also important. Whether the message is to internal employees or the public, important messages can quickly become lost in a crowd of competing information. This is only becoming more of a challenge in a 'post-truth' world wherein fake news spreads as fast as, if not faster, than genuine reports [12, 13]. Though, even 'genuine reports' can unintentionally promote misinformation. It is now more important than ever to communicate clearly, with a unified approach to minimise the barrage of information and ensure essential information is not lost in the noise.

15.1.3 Challenge: uncertainty in risk

Risk Analysis as a concept contains inherent uncertainty, whether this is through the interpretation of data or the existence of unknown, unknowns such as 'wicked problems' and 'black swan' events (chapters 5 and 7) [14]. While this is objectively true, the possibility of these types of events is often obscured by the confidence instilled by having completed thorough and otherwise effective Risk Assessments. It is inevitable that uncertainty (or error) will arise in data whenever assumptions and estimations are made, and it is important to keep in mind that the existence of uncertainty is neither 'good' nor 'bad' but simply an artefact of the complex process required to reach a risk-based decision. In addition, by definition unknown, unknowns are often impossible to predict and therefore can only be evaluated in retrospect. However, it is important to be cognisant of the inherent uncertainties to enable a swift and flexible response when unexpected risks arise.

15.2 Future considerations

The case studies and examples collated in this book are a small sub-set of a broad topic, focussing on Risk Analysis in science and engineering but identifying challenges that are likely to be applicable across multiple disciplines. The collected chapters formalise anecdotal and discursive evidence into concrete case studies highlighting the challenges that Risk Analysis practitioners currently face: the need for a 'joined-up' approach across sectors, the need for efficient communication and the inherent existence of uncertainty. It is only through continued cross-disciplinary collaboration and networking that finding appropriate and unified solutions can be achieved.

References

[1] Aven T and Kristensen V 2005 Perspectives on risk: review and discussion of the basis for establishing a unified and holistic approach *Reliab. Eng. Syst. Saf.* **90** 1–14

[2] Dionne G 2013 Risk management: History, definition, and critique *Risk Manag. Insur. Rev.* **16** 147–66

[3] 21/30435255 DC BS ISO 31073. Risk Management. Vocabulary, London, 2021 https://bsol.bsigroup.com/PdfViewer/Viewer?pid=000000000030435255
[4] Kanso A M, Nelson R A and Kitchen P J 2019 BP and the deepwater horizon oil spill: a case study of how company management employed public relations to restore a damaged brand *J. Marketing Commun.* **26** 703–31
[5] BS EN ISO 14001:2015 Environmental management systems—requirements with guidance for use, 2020 https://bsol.bsigroup.com/PdfViewer/Viewer?pid=000000000030408743
[6] Our common future 1987 https://are.admin.ch/are/en/home/media/publications/sustainable-development/brundtland-report.html
[7] Mebratu D 1998 Sustainability and sustainable development: historical and conceptual review *Environ. Impact Assess. Rev.* **18** 493–520
[8] *Health and Safety at Work etc Act*. UK Government, 1974 https://legislation.gov.uk/ukpga/1974/37/contents
[9] Bond A J and Brooks D J 1997 A strategic framework to determine the best practicable environmental option (BPEO) for proposed transport schemes *J. Environ. Manage.* **51** 305–21
[10] Biggs A T and Littlejohn L F 2021 Revisiting the initial COVID-19 pandemic projections *Lancet Microbe* **2** e91–92
[11] Covid: huge protests across Europe over new restrictions *BBC News* https://bbc.co.uk/news/world-europe-59363256
[12] Lazer D M J *et al* 2018 The science of fake news: addressing fake news requires a multidisciplinary effort *Science* **359** 1094–6
[13] Rousseau J 2021 Challenges to science communication in a post-truth world *South African J. Commun. Theor. Res.* **47** 122–40
[14] Elahi S 2011 Here be dragons… exploring the 'unknown unknowns' *Futures* **43** 196–201

www.ingramcontent.com/pod-product-compliance
Ingram Content Group UK Ltd.
Pitfield, Milton Keynes, MK11 3LW, UK
UKHW051901100926
8556IPUK00005B/722